genetics

genetics

VOLUME **1**
A–D

Richard Robinson

**MACMILLAN
REFERENCE
USA™**

THOMSON

GALE

New York • Detroit • San Diego • San Francisco • Cleveland • New Haven, Conn. • Waterville, Maine • London • Munich

Genetics
Richard Robinson

Volume ISBN Numbers
0-02-865607-5 (Volume 1)
0-02-865608-3 (Volume 2)
0-02-865609-1 (Volume 3)
0-02-865610-5 (Volume 4)

LIBRARY OF CONGRESS CATALOGING- IN-PUBLICATION DATA

Genetics / Richard Robinson, editor in chief.
    p.   ; cm.
Includes bibliographical references and index.
    ISBN 0-02-865606-7 (set : hd.)
    1. Genetics—Encyclopedias.
    [DNLM: 1. Genetics—Encyclopedias—English. 2. Genetic Diseases, Inborn—Encyclopedias—English. 3. Genetic Techniques—Encyclopedias—English. 4. Molecular Biology—Encyclopedias—English. QH 427 G328 2003]   I. Robinson, Richard,   1956–
    QH427 .G46 2003
    576'.03—dc21

                                    2002003560

Printed in Canada
10 9 8 7 6 5 4 3 2 1

Preface

The twentieth century has been called "the genetic century," and rightly so: The genetic revolution began with the rediscovery of Gregor Mendel's work in 1900, Watson and Crick elucidated the structure of DNA in 1953, and the first draft of the human genome sequence was announced in February 2001. As dramatic and important as these advances are, however, they will almost certainly pale when compared to those still awaiting us. Building on foundations laid over the last one hundred years, the twenty-first century will likely see discoveries that profoundly affect our understanding of our genetic nature, and greatly increase our ability to manipulate genes to shape ourselves and our environment. As more is learned, the pace of discovery will only increase, revealing not only the identities of increasing numbers of genes, but more importantly, how they function, interact, and, in some cases, cause disease.

As the importance of genetics in our daily lives has grown, so too has the importance of its place in the modern science classroom: In the study of biology, genetics has become the central science. Our purpose in creating this encyclopedia is to provide students and teachers the most comprehensive and accessible reference available for understanding this rapidly changing field.

A Comprehensive Reference

In the four volumes of *Genetics*, students will find detailed coverage of every topic included in standard and advanced biology courses, from fundamental concepts to cutting-edge applications, as well as topics so new that they have not yet become a part of the regular curriculum. The set explores the history, theory, technology, and uses (and misuses) of genetic knowledge. Topics span the field from "classical" genetics to molecular genetics to population genetics. Students and teachers can use the set to reinforce classroom lessons about basic genetic processes, to expand on a discussion of a special topic, or to learn about an entirely new idea.

✴Explore further in Gene, Polymerase Chain Reaction, and Eugenics

Genetic Disorders and Social Issues

Many advances in genetics have had their greatest impact on our understanding of human health and disease. One of the most important areas of research is in the understanding of complex diseases, such as cancer and Alzheimer's disease, in which genes and environment interact to produce or prevent disease. *Genetics* devotes more than two dozen entries to both single-gene and complex genetic disorders, offering the latest understanding of

their causes, diagnoses, and treatments. Many more entries illustrate basic genetic processes with discussion of the diseases in which these processes go wrong. In addition, students will find in-depth explanations of how genetic diseases arise, how disease genes are discovered, and how gene therapy hopes to treat them.

Advances in our understanding of genetics and improvements in techniques of genetic manipulation have brought great benefits, but have also raised troubling ethical and legal issues, most prominently in the areas of reproductive technology, cloning, and biotechnology. In *Genetics*, students will find discussions of both the science behind these advances and the ethical issues each has engendered. As with nearly every entry in *Genetics*, these articles are accompanied by suggestions for further reading to allow the student to seek more depth and pursue other points of view.

The Tools of the Trade

The explosion of genetic knowledge in the last several decades can be attributed in large part to the discovery and development of a set of precise and powerful tools for analyzing and manipulating DNA. In these volumes, students will find clear explanations of how each of these tools work, as well as how they are used by scientists to conduct molecular genetic research. We also discuss how the computer and the Internet have radically expanded the ability of scientists to process large amounts of data. These technologies have made it possible to analyze whole genomes, leading not just to the discovery of new genes, but to a greater understanding of how entire genomes function and evolve.

The Past and the Future

The short history of genetics is marked by brilliant insights and major theoretical advances, as well as misunderstandings and missed opportunities. *Genetics* examines these events in both historical essays and biographies of major figures, from Mendel to McClintock. The future of genetics will be created by today's students, and in these volumes we present information on almost two dozen careers in this field, ranging from attorney to clinical geneticist to computational biologist.

Contributors and Arrangement of the Material

The goal of each of the 253 entries in *Genetics* is to give the interested student access to a depth of discussion not easily available elsewhere. Entries have been written by professionals in the field of genetics, including experts whose work has helped define the current state of knowledge. All of the entries have been written with the needs of students in mind, and they all provide the background and context necessary to help students make connections with classroom lessons.

To aid understanding and increase interest, most entries are illustrated with clear diagrams and dramatic photographs. Each entry is followed by cross-references to related entries, and most have a list of suggested readings and/or Internet resources for further exploration or elaboration. Specialized or unfamiliar terms are defined in the margin and collected in a glossary at the end of each volume. Each volume also contains an index, and

*Explore further in Alzheimer Disease, Genetic Testing, and Gene Therapy

*Explore further in Biotechnology and Genetic Engineering, Cloning Organisms, and Cloning: Ethical Issues

*Explore further in Sequencing DNA, DNA Microarrays, and Internet

*Explore further in Morgan, Thomas Hunt, and Computational Biologist

a cumulative index is found at the end of volume four. A topical index is also included, allowing students and teachers to see at a glance the range of entries available on a particular topic.

Acknowledgments and Thanks

Genetics represents the collective inspiration and hard work of many people. Hélène Potter at Macmillan Reference USA knew how important a reference this encyclopedia would be, and her commitment and enthusiasm brought it into being. Kate Millson has provided simply outstanding editorial management throughout this long process, and I am deeply in her debt. Our three editorial board members—Ralph R. Meyer, David A. Micklos, and Margaret A. Pericak-Vance—gave the encyclopedia its broad scope and currency, and were vital in ensuring accuracy in this rapidly changing field. Finally, the entries in *Genetics* are the product of well over one hundred scientists, doctors, and other professionals. Their willingness to contribute their time and expertise made this work possible, and it is to them that the greatest thanks are due.

Richard Robinson
Tucson, Arizona
rrobinson@nasw.org

For Your Reference

The following section provides a group of diagrams and illustrations applicable to many entries in this encyclopedia. The molecular structures of DNA and RNA are provided in detail in several different formats, to help the student understand the structures and visualize how these molecules combine and interact. The full set of human chromosomes are presented diagrammatically, each of which is shown with a representative few of the hundreds or thousands of genes it carries.

NUCLEOTIDE STRUCTURE

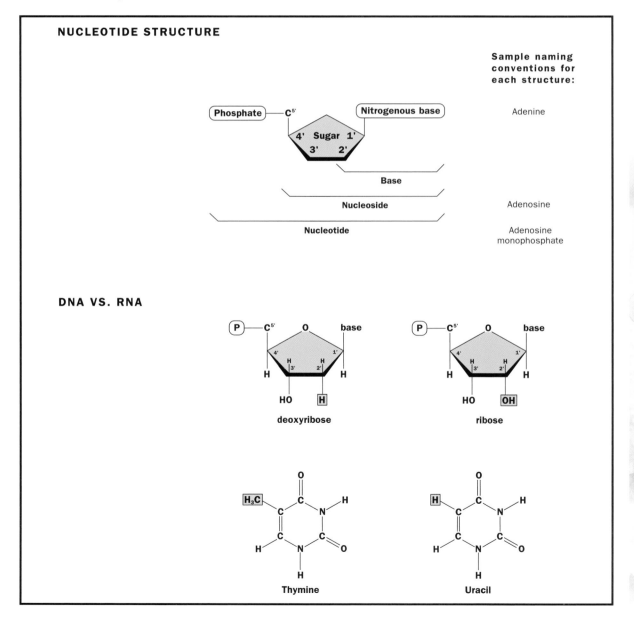

Sample naming conventions for each structure:

Adenine

Adenosine

Adenosine monophosphate

DNA VS. RNA

deoxyribose

ribose

Thymine

Uracil

NUCLEOTIDE STRUCTURES

Purine-containing DNA nucleotides

Adenine

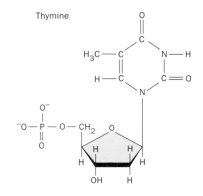

Guanine

Pyrimidine-containing DNA nucleotides

Thymine

Cytosine

CANONICAL B-DNA DOUBLE HELIX

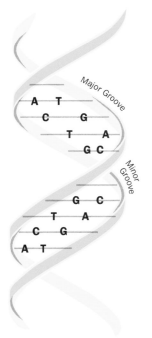

Major Groove

A T
C G
T A
G C

Minor Groove

G C
T A
C G
A T

Ribbon model

Ball-and-stick model

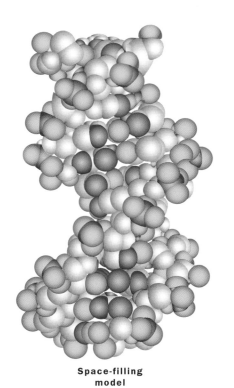

Space-filling model

DNA NUCLEOTIDES PAIR UP ACROSS THE DOUBLE HELIX; THE TWO STRANDS RUN ANTI-PARALLEL

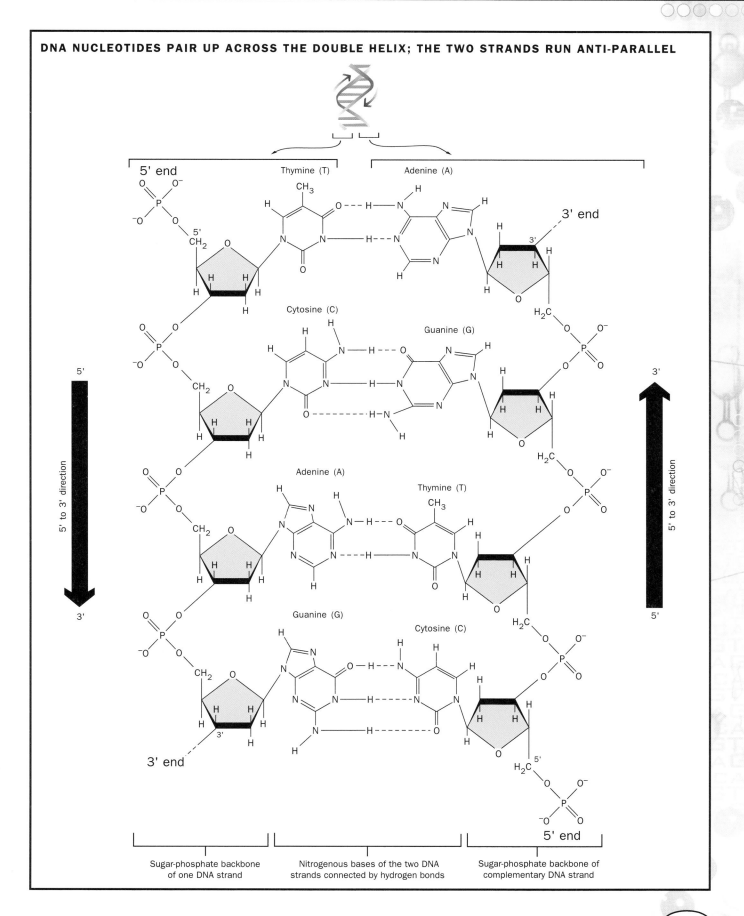

Sugar-phosphate backbone of one DNA strand

Nitrogenous bases of the two DNA strands connected by hydrogen bonds

Sugar-phosphate backbone of complementary DNA strand

SELECTED LANDMARKS OF THE HUMAN GENOME

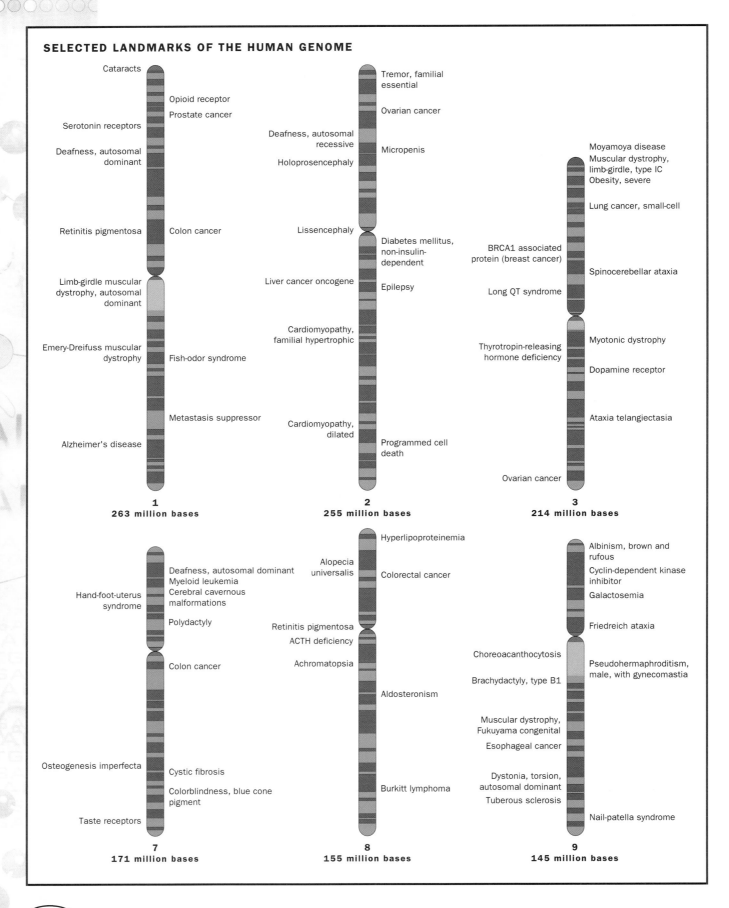

Cataracts

Opioid receptor

Prostate cancer

Serotonin receptors

Deafness, autosomal dominant

Retinitis pigmentosa

Colon cancer

Limb-girdle muscular dystrophy, autosomal dominant

Emery-Dreifuss muscular dystrophy

Fish-odor syndrome

Metastasis suppressor

Alzheimer's disease

1
263 million bases

Tremor, familial essential

Ovarian cancer

Deafness, autosomal recessive

Micropenis

Holoprosencephaly

Lissencephaly

Diabetes mellitus, non-insulin-dependent

Liver cancer oncogene

Epilepsy

Cardiomyopathy, familial hypertrophic

Cardiomyopathy, dilated

Programmed cell death

2
255 million bases

Moyamoya disease
Muscular dystrophy, limb-girdle, type IC
Obesity, severe

Lung cancer, small-cell

BRCA1 associated protein (breast cancer)

Spinocerebellar ataxia

Long QT syndrome

Thyrotropin-releasing hormone deficiency

Myotonic dystrophy

Dopamine receptor

Ataxia telangiectasia

Ovarian cancer

3
214 million bases

Hand-foot-uterus syndrome

Deafness, autosomal dominant
Myeloid leukemia
Cerebral cavernous malformations

Polydactyly

Colon cancer

Osteogenesis imperfecta

Cystic fibrosis

Colorblindness, blue cone pigment

Taste receptors

7
171 million bases

Hyperlipoproteinemia

Alopecia universalis

Colorectal cancer

Retinitis pigmentosa
ACTH deficiency

Achromatopsia

Aldosteronism

Burkitt lymphoma

8
155 million bases

Albinism, brown and rufous

Cyclin-dependent kinase inhibitor

Galactosemia

Friedreich ataxia

Choreoacanthocytosis

Pseudohermaphroditism, male, with gynecomastia

Brachydactyly, type B1

Muscular dystrophy, Fukuyama congenital

Esophageal cancer

Dystonia, torsion, autosomal dominant

Tuberous sclerosis

Nail-patella syndrome

9
145 million bases

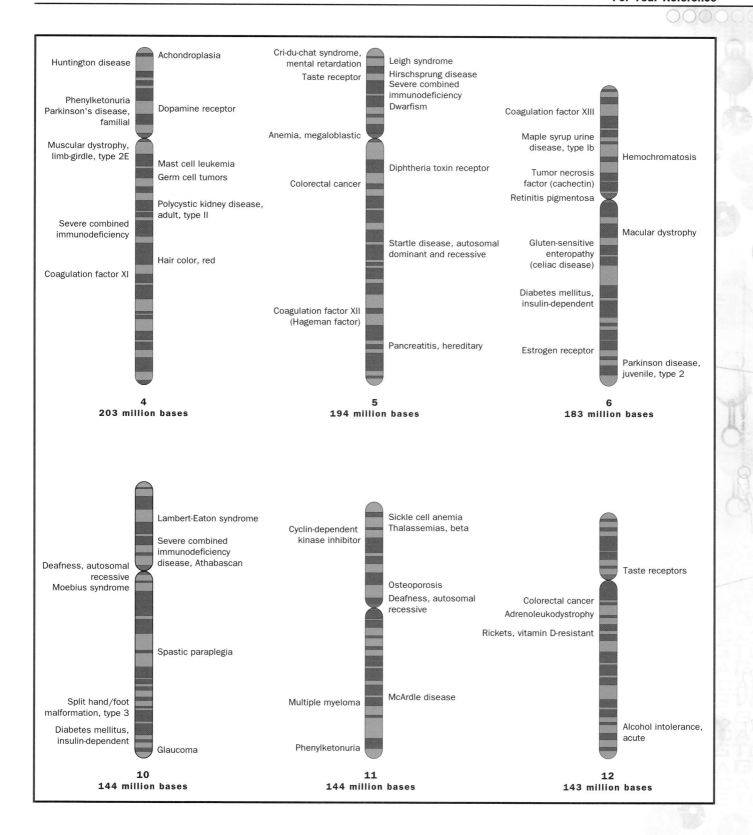

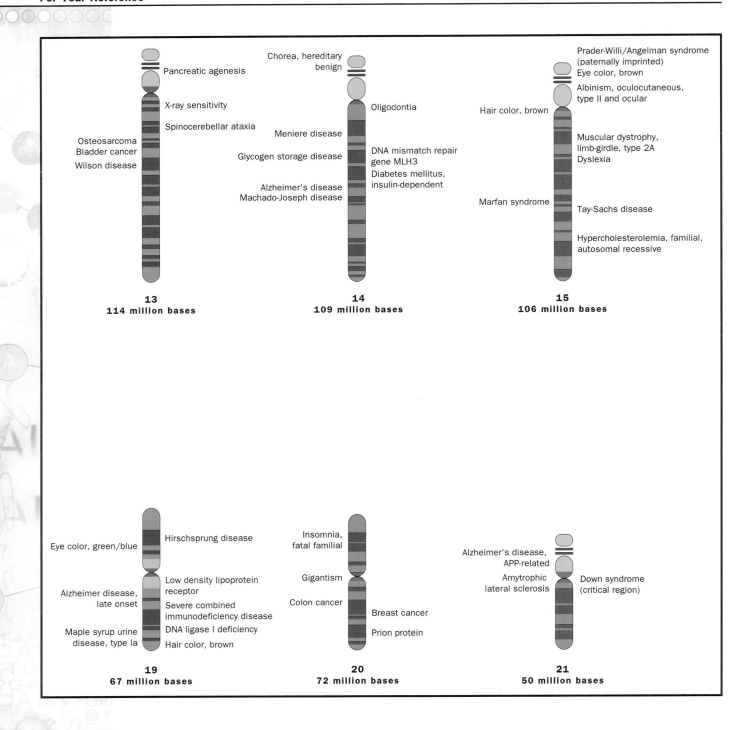

Pancreatic agenesis

X-ray sensitivity

Spinocerebellar ataxia

Osteosarcoma
Bladder cancer
Wilson disease

13
114 million bases

Chorea, hereditary
benign

Oligodontia

Meniere disease

Glycogen storage disease

Alzheimer's disease
Machado-Joseph disease

DNA mismatch repair
gene MLH3
Diabetes mellitus,
insulin-dependent

14
109 million bases

Prader-Willi/Angelman syndrome
(paternally imprinted)
Eye color, brown
Albinism, oculocutaneous,
type II and ocular

Hair color, brown

Muscular dystrophy,
limb-girdle, type 2A
Dyslexia

Marfan syndrome

Tay-Sachs disease

Hypercholesterolemia, familial,
autosomal recessive

15
106 million bases

Eye color, green/blue

Hirschsprung disease

Alzheimer disease,
late onset

Low density lipoprotein
receptor
Severe combined
immunodeficiency disease
DNA ligase I deficiency

Maple syrup urine
disease, type Ia

Hair color, brown

19
67 million bases

Insomnia,
fatal familial

Gigantism

Colon cancer

Breast cancer

Prion protein

20
72 million bases

Alzheimer's disease,
APP-related
Amytrophic
lateral sclerosis

Down syndrome
(critical region)

21
50 million bases

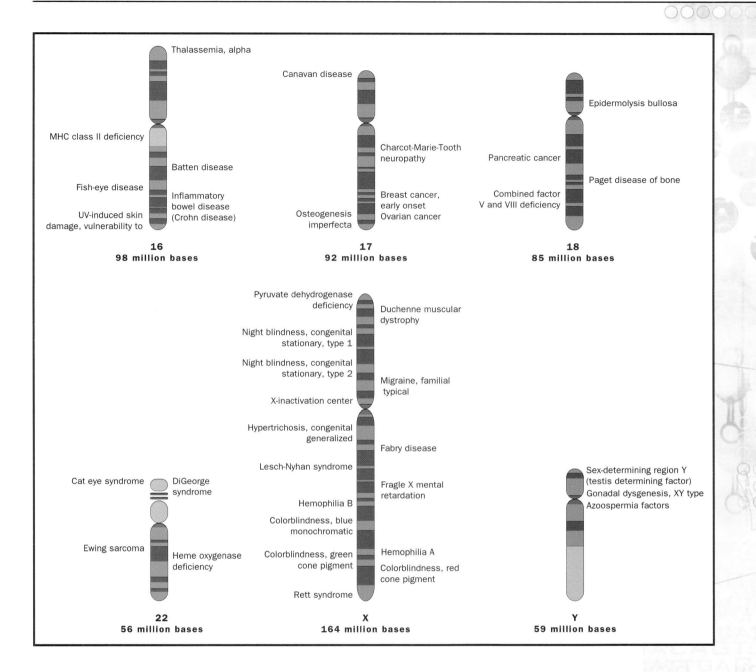

Contributors

Eric Aamodt
Louisiana State University Health Sciences Center, Shreveport
Gene Expression: Overview of Control

Maria Cristina Abilock
Applied Biosystems
Automated Sequencer
Cycle Sequencing
Protein Sequencing
Sequencing DNA

Ruth Abramson
University of South Carolina School of Medicine
Intelligence
Psychiatric Disorders
Sexual Orientation

Stanley Ambrose
University of Illinois
Population Bottleneck

Allison Ashley-Koch
Duke Center for Human Genetics
Disease, Genetics of
Fragile X Syndrome
Geneticist

David T. Auble
University of Virginia Health System
Transcription

Bruce Barshop
University of California, San Diego
Metabolic Disease

Mark A. Batzer
Louisiana State University
Pseudogenes
Repetitive DNA Elements
Transposable Genetic Elements

Robert C. Baumiller
Xavier University
Reproductive Technology
Reproductive Technology: Ethical Issues

Mary Beckman
Idaho Falls, Idaho
DNA Profiling
HIV

Samuel E. Bennett
Oregon State University Department of Genetics
DNA Repair
Laboratory Technician
Molecular Biologist

Andrea Bernasconi
Cambridge University, U.K.
Multiple Alleles
Nondisjunction

C. William Birky, Jr.
University of Arizona
Inheritance, Extranuclear

Joanna Bloom
New York University Medical Center
Cell Cycle

Deborah Blum
University of Wisconsin, Madison
Science Writer

Bruce Blumberg
University of California, Irvine
Hormonal Regulation

Suzanne Bradshaw
University of Cincinnati
Transgenic Animals
Yeast

Carolyn J. Brown
University of British Columbia
Mosaicism

Michael J. Bumbulis
Baldwin-Wallace College
Blotting

Michael Buratovich
Spring Arbor College
Operon

Elof Carlson
The State Universtiy of New York, Stony Brook
Chromosomal Theory of Inheritance, History
Gene
Muller, Hermann
Polyploidy
Selection

Regina Carney
Duke University
College Professor

Shu G. Chen
Case Western Reserve University
Prion

Gwen V. Childs
University of Arkansas for Medical Sciences
In situ Hybridization

Cindy T. Christen
Iowa State University
Technical Writer

Patricia L. Clark
University of Notre Dame
Chaperones

Steven S. Clark
University of Wisconsin
Oncogenes

Nathaniel Comfort
George Washington University
McClintock, Barbara

P. Michael Conneally
Indiana University School of Medicine
Blood Type
Epistasis
Heterozygote Advantage

Howard Cooke
Western General Hospital: MRC Human Genetics Unit
Chromosomes, Artificial

Denise E. Costich
Boyce Thompson Institute
Maize

Terri Creeden
March of Dimes
Birth Defects

Kenneth W. Culver
Novartis Pharmaceuticals Corporation
Genomics
Genomics Industry
Pharmaceutical Scientist

Mary B. Daly
Fox Chase Cancer Center
Breast Cancer

Pieter de Haseth
Case Western Reserve University
Transcription

Rob DeSalle
American Museum of Natural History
Conservation Geneticist
Conservation Biology: Genetic Approaches

Elizabeth A. De Stasio
Lawerence University
Cloning Organisms

Danielle M. Dick
Indiana University
Behavior

Michael Dietrich
Dartmouth College
Nature of the Gene, History

Christine M. Disteche
University of Washington
X Chromosome

Gregory Michael Dorr
University of Alabama
Eugenics

Jennie Dusheck
Santa Cruz, California
Population Genetics

Susanne D. Dyby
U.S. Department of Agriculture: Center for Medical, Agricultural, and Veterinary Entomology
Classical Hybrid Genetics
Mendelian Genetics
Pleiotropy

Barbara Emberson Soots
Folsom, California
Agricultural Biotechnology

Susan E. Estabrooks
Duke Center for Human Genetics
Fertilization
Genetic Counselor
Genetic Testing

Stephen V. Faraone
Harvard Medical School
Attention Deficit Hyperactivity Disorder

Gerald L. Feldman
Wayne State University Center for Molecular Medicine and Genetics
Down Syndrome

Linnea Fletcher
Bio-Link South Central Regional Coordinater, Austin Community College
Educator
Gel Electrophoresis
Marker Systems
Plasmid

Michael Fossel
Executive Director, American Aging Association
Accelerated Aging: Progeria

Carol L. Freund
National Institute of Health: Warren G. Magnuson Clinical Center
Genetic Testing: Ethical Issues

Joseph G. Gall
Carnegie Institution
Centromere

Darrell R. Galloway
The Ohio State University
DNA Vaccines

Pierluigi Gambetti
Case Western Reserve University
Prion

Robert F. Garry
Tulane University School of Medicine
Retrovirus
Virus

Perry Craig Gaskell, Jr.
Duke Center for Human Genetics
Alzheimer's Disease

Theresa Geiman
National Institute of Health: Laboratory of Receptor Biology and Gene Expression
Methylation

Seth G. N. Grant
University of Edinburgh
Embryonic Stem Cells
Gene Targeting
Rodent Models

Roy A. Gravel
University of Calgary
Tay-Sachs Disease

Nancy S. Green
March of Dimes
Birth Defects

Wayne W. Grody
UCLA School of Medicine
Cystic Fibrosis

Charles J. Grossman
Xavier University
Reproductive Technology
Reproductive Technology: Ethical Issues

Cynthia Guidi
University of Massachusetts Medical School
Chromosome, Eukaryotic

Patrick G. Guilfoile
Bemidji State University
DNA Footprinting
Microbiologist
Recombinant DNA
Restriction Enzymes

Richard Haas
University of California Medical Center
Mitochondrial Diseases

William J. Hagan
College of St. Rose
Evolution, Molecular

Jonathan L. Haines
Vanderbilt University Medical Center
Complex Traits
Human Disease Genes, Identification of

Mapping
McKusick, Victor

Michael A. Hauser
Duke Center for Human Genetics
DNA Microarrays
Gene Therapy

Leonard Hayflick
University of California
Telomere

Shaun Heaphy
University of Leicester, U.K.
Viroids and Virusoids

John Heddle
York University
Mutagenesis
Mutation
Mutation Rate

William Horton
Shriners Hospital for Children
Growth Disorders

Brian Hoyle
Square Rainbow Limited
Overlapping Genes

Anthony N. Imbalzano
University of Massachusetts Medical School
Chromosome, Eukaryotic

Nandita Jha
University of California, Los Angeles
Triplet Repeat Disease

John R. Jungck
Beloit College
Gene Families

Richard Karp
Department of Biological Sciences, University of Cincinnati
Transplantation

David H. Kass
Eastern Michigan University
Pseudogenes
Transposable Genetic Elements

Michael L. Kochman
University of Pennsylvania Cancer Center
Colon Cancer

Bill Kraus
Duke University Medical Center
Cardiovascular Disease

Steven Krawiec
Lehigh University
Genome

Mark A. Labow
Novartis Pharmaceuticals Corporation
Genomics
Genomics Industry
Pharmaceutical Scientist

Ricki Lewis
McGraw-Hill Higher Education; The Scientist
Bioremediation
Biotechnology: Ethical Issues
Cloning: Ethical Issues

Genetically Modified Foods
Plant Genetic Engineer
Prenatal Diagnosis
Transgenic Organisms: Ethical
  Issues
Lasse Lindahl
  University of Maryland, Baltimore
  Ribozyme
  RNA
David E. Loren
  *University of Pennsylvania School of
  Medicine*
  Colon Cancer
Dennis N. Luck
  Oberlin College
  Biotechnology
Jeanne M. Lusher
  *Wayne State University School of
  Medicine; Children's Hospital of
  Michigan*
  Hemophilia
Kamrin T. MacKnight
  *Medlen, Carroll, LLP: Patent,
  Trademark and Copyright Attorneys*
  Attorney
  Legal Issues
  Patenting Genes
  Privacy
Jarema Malicki
  Harvard Medical School
  Zebrafish
Eden R. Martin
  Duke Center for Human Genetics
  Founder Effect
  Inbreeding
William Mattox
  *University of Texas/Anderson
  Cancer Center*
  Sex Determination
Brent McCown
  University of Wisconsin
  Transgenic Plants
Elizabeth C. Melvin
  Duke Center for Human Genetics
  Gene Therapy: Ethical Issues
  Pedigree
Ralph R. Meyer
  University of Cincinnati
  Biotechnology and Genetic Engi-
    neering, History of
  Chromosome, Eukaryotic
  Genetic Code
  Human Genome Project
Kenneth V. Mills
  College of the Holy Cross
  Post-translational Control
Jason H. Moore
  Vanderbilt University Medical School
  Quantitative Traits
  Statistical Geneticist
  Statistics
Dale Mosbaugh
  *Oregon State University: Center for
  Gene Research and Biotechnology*

DNA Repair
Laboratory Technician
Molecular Biologist
Paul J. Muhlrad
  University of Arizona
  Alternative Splicing
  Apoptosis
  Arabidopsis thaliana
  Cloning Genes
  Combinatorial Chemistry
  Fruit Fly: *Drosophila*
  Internet
  Model Organisms
  Pharmacogenetics and Pharma-
    cogenomics
  Polymerase Chain Reaction
Cynthia A. Needham
  *Boston University School of
  Medicine*
  Archaea
  Conjugation
  Transgenic Microorganisms
R. John Nelson
  University of Victoria
  Balanced Polymorphism
  Gene Flow
  Genetic Drift
  Polymorphisms
  Speciation
Carol S. Newlon
  *University of Medicine and
  Dentistry of New Jersey*
  Replication
Sophia A. Oliveria
  *Duke University Center for Human
  Genetics*
  Gene Discovery
Richard A. Padgett
  Lerner Research Institute
  RNA Processing
Michele Pagano
  *New York University Medical
  Center*
  Cell Cycle
Rebecca Pearlman
  Johns Hopkins University
  Probability
Fred W. Perrino
  *Wake Forest University School of
  Medicine*
  DNA Polymerases
  Nucleases
  Nucleotide
David Pimentel
  *Cornell University: College of
  Agriculture and Life Sciences*
  Biopesticides
Toni I. Pollin
  *University of Maryland School of
  Medicine*
  Diabetes
Sandra G. Porter
  Geospiza, Inc.
  Homology

Eric A. Postel
  Duke University Medical Center
  Color Vision
  Eye Color
Prema Rapuri
  Creighton University
  HPLC: High-Performance Liq-
    uid Chromatography
Anthony J. Recupero
  Gene Logic
  Bioinformatics
  Biotechnology Entrepreneur
  Proteomics
Diane C. Rein
  BioComm Consultants
  Clinical Geneticist
  Nucleus
  Roundworm: *Caenorhabditis ele-
    gans*
  Severe Combined Immune Defi-
    ciency
Jacqueline Bebout Rimmler
  Duke Center for Human Genetics
  Chromosomal Aberrations
Keith Robertson
  *Epigenetic Gene Regulation and
  Cancer Institute*
  Methylation
Richard Robinson
  Tucson, Arizona
  Androgen Insensitivity Syndrome
  Antisense Nucleotides
  Cell, Eukaryotic
  Crick, Francis
  Delbrück, Max
  Development, Genetic Control of
  DNA Structure and Function,
    History
  Eubacteria
  Evolution of Genes
  Hardy-Weinberg Equilibrium
  High-Throughput Screening
  Immune System Genetics
  Imprinting
  Inheritance Patterns
  Mass Spectrometry
  Mendel, Gregor
  Molecular Anthropology
  Morgan, Thomas Hunt
  Mutagen
  Purification of DNA
  RNA Interferance
  RNA Polymerases
  Transcription Factors
  Twins
  Watson, James
Richard J. Rose
  Indiana University
  Behavior
Howard C. Rosenbaum
  *Science Resource Center, Wildlife
  Conservation Society*
  Conservation Geneticist
  Conservation Biology: Genetic
    Approaches

Astrid M. Roy-Engel
Tulane University Health Sciences Center
Repetitive DNA Elements

Joellen M. Schildkraut
Duke University Medical Center
Public Health, Genetic Techniques in

Silke Schmidt
Duke Center for Human Genetics
Meiosis
Mitosis

David A. Scicchitano
New York University
Ames Test
Carcinogens

William K. Scott
Duke Center for Human Genetics
Aging and Life Span
Epidemiologist
Gene and Environment

Gerry Shaw
MacKnight Brain Institute of the University of Flordia
Signal Transduction

Alan R. Shuldiner
University of Maryland School of Medicine
Diabetes

Richard R. Sinden
Institute for Biosciences and Technology: Center for Genome Research
DNA

Paul K. Small
Eureka College
Antibiotic Resistance
Proteins
Reading Frame

Marcy C. Speer
Duke Center for Human Genetics
Crossing Over
Founder Effect
Inbreeding
Individual Genetic Variation
Linkage and Recombination

Jeffrey M. Stajich
Duke Center for Human Genetics
Muscular Dystrophy

Judith E. Stenger
Duke Center for Human Genetics
Computational Biologist
Information Systems Manager

Frank H. Stephenson
Applied Biosystems
Automated Sequencer
Cycle Sequencing
Protein Sequencing
Sequencing DNA

Gregory Stewart
State University of West Georgia
Transduction
Transformation

Douglas J. C. Strathdee
University of Edinburgh
Embryonic Stem Cells
Gene Targeting
Rodent Models

Jeremy Sugarman
Duke University Department of Medicine
Genetic Testing: Ethical Issues

Caroline M. Tanner
Parkinson's Institute
Twins

Alice Telesnitsky
University of Michigan
Reverse Transcriptase

Daniel J. Tomso
National Institute of Environmental Health Sciences
DNA Libraries
Escherichia coli
Genetics

Angela Trepanier
Wayne State University Genetic Counseling Graduate Program
Down Syndrome

Peter A. Underhill
Stanford University
Y Chromosome

Joelle van der Walt
Duke University Center for Human Genetics
Genotype and Phenotype

Jeffery M. Vance
Duke University Center for Human Genetics

Gene Discovery
Genomic Medicine
Genotype and Phenotype
Sanger, Fred

Gail Vance
Indiana University
Chromosomal Banding

Jeffrey T. Villinski
University of Texas/MD Anderson Cancer Center
Sex Determination

Sue Wallace
Santa Rosa, California
Hemoglobinopathies

Giles Watts
Children's Hospital Boston
Cancer
Tumor Suppressor Genes

Kirk Wilhelmsen
Ernest Gallo Clinic & Research Center
Addiction

Michelle P. Winn
Duke University Medical Center
Physician Scientist

Chantelle Wolpert
Duke University Center for Human Genetics
Genetic Counseling
Genetic Discrimination
Nomenclature
Population Screening

Harry H. Wright
University of South Carolina School of Medicine
Intelligence
Psychiatric Disorders
Sexual Orientation

Janice Zengel
University of Maryland, Baltimore
Ribosome
Translation

Stephan Zweifel
Carleton College
Mitochondrial Genome

Table of Contents

genetics

Accelerated Aging: Progeria

Human progeria comes in two major forms, Werner's syndrome (adult-onset progeria) and Hutchinson-Gilford syndrome (juvenile-onset progeria). Werner's patients are usually diagnosed in early maturity and have an average life span of forty-seven years. Hutchinson-Gilford patients are usually diagnosed within the first two years of life and have an average life span of thirteen years. The latter syndrome is often simply termed "progeria" and both are sometimes lumped together as progeroid syndromes.

Progeria's Effects

There is considerable controversy as to whether or not progeria is a form of aging at all. Most clinicians believe that progeria is truly a form of early aging, although only a segmental form in which only certain specific tissues and cell types of the body age early. Hutchinson-Gilford children show what appears to be early aging of their skin, bones, joints, and cardiovascular system, but not of their immune or central nervous systems.

Clinical problems parallel this observation: They suffer from thin skin and poor skin healing, **osteoporosis**, arthritis, and heart disease, but do not have more infections than normal children and they do not have early **dementia**. Death is usually due to cardiovascular disease, especially heart attacks and strokes, yet Hutchinson-Gilford children lack normal risk factors associated with these diseases, such as smoking, high cholesterol, hypertension, or diabetes.

Clinically, the children appear old, with thin skin, baldness, swollen joints, and short stature. They do not go through puberty. The face is strikingly old in appearance. The typical Hutchinson-Gilford child looks more like a **centenarian** than like other children, and may look more like other progeric children than like members of their own families. There is no effective clinical intervention.

Inheritance of Progeria

The segmental nature of progeria is perhaps its most fascinating feature. If progeria is actually a form of aging gone awry, then this implies that aging is more than merely wear and tear on the organism. If progeria is a genetically mediated, segmental form of aging, this may imply that aging itself is

osteoporosis thinning of the bone structure

dementia neurological illness characterized by impaired thought or awareness

centenarian person who lives to age 100

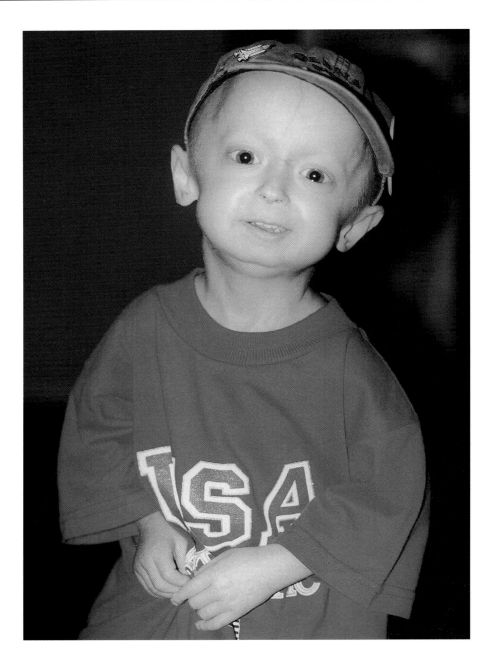

This five year old boy has Hutchinson-Gilford progeria, a fatal, "premature aging" disease in which children die of heart failure or stroke at an average age of thirteen. Photo courtesy of The Progeria Research Foundation, Inc. and the Barnett Family.

genetically mediated and, like other genetic disease, is not only the outcome of genetic error but might be open to clinical intervention.

Supporting this observation, there are a number of other less well-known forms of progeria, including acrogeria, metageria, and acrometageria, as well as several dozen human clinical syndromes and diseases with features that have been considered to have progeroid aspects. The latter category includes Wiedemann-Rautenstrauch, Donohue's, Cockayne's, Klinefelter's, Seip's, Rothmund's, Bloom's, and Turner's syndromes, ataxia telangiectasia, cervical lipodysplasia, myotonic dystrophy, dyskeratosis congenita, and trisomy 21 (Down syndrome). In each of these cases, there are features that are genetic and that have been considered segmental forms of aging.

In the most well-known of these, trisomy 21, the immune and central nervous systems both appear to **senesce** early, in contrast with Hutchinson-Gilford progeria, in which the opposite occurs. Bolstering the suggestion that this is a form of segmental progeria, trisomy 21 patients are prone to both infections and early onset of a form of Alzheimer's dementia.

The gene that is mutated in Werner's syndrome is known to code for a DNA helicase. This enzyme unwinds DNA for replication, transcription, recombination, and repair. The inability to repair DNA may explain the features of premature aging, as well as the increased rate of cancer in Werner's syndrome patients. Another mutated helicase is responsible for Bloom's syndrome. Both conditions are inherited as **autosomal** recessive disorders.

Data suggesting that Hutchinson-Gilford progeria is genetic is circumstantial. The disease is presumptively caused by a sporadic (one in eight million live births), autosomal dominant mutation, although a rare autosomal recessive mutation is not impossible. The helicase abnormality that causes Werner's syndrome is not present in Hutchinson-Gilford cells. There is a slight correlation with the paternal age at conception. Whatever the mechanism, it appears to operate prior to birth; several **neonatal** cases have been reported.

Germinal Mosaicism

Cellular data, particularly regarding structures called telomeres, suggests that some of the cells from Hutchinson-Gilford patients are prone to early cell senescence. Telomeres are special DNA structures at the tips of the chromosomes. These telomeres gradually shorten over time, and this shortening is associated with some aspects of cellular aging. Skin fibroblasts from Hutchinson-Gilford patients have shorter than normal telomeres and consequently undergo early cell senescence. At birth, the mean telomere length of these children is equivalent to that of a normal eighty-five-year-old.

Introduction of human telomerase into such cells leads to reextension of the telomeres and results in normal immortalization of these progeric cell cultures. Clinical interventional studies using this strategy in humans are pending. Predictably, circulating **lymphocytes** of Hutchinson-Gilford children have normal telomere lengths, in keeping with their normal immune function. Research thus far suggests that progeria may not be so much a genetic disease as it is an "epigenetic mosaic disease." In progeria, this means that the genes are normal, but the abnormally short telomere length in only certain cells lines causes an abnormal pattern of gene expression. The senescent pattern of gene expression in specific tissues results in the observed clinical disease of progeria.

Although consistent with all known laboratory and clinical data, the actual genetic mechanisms that underlie Hutchinson-Gilford progeria are still uncertain and arguable (the gene for Werner's syndrome, however, has been cloned). The question of what causes progeria holds a fascination largely for what it may tell us about the course of aging itself. SEE ALSO AGING AND LIFE SPAN; ALZHEIMER'S DISEASE; DISEASE, GENETICS OF; DNA REPAIR; DOWN SYNDROME; INHERITANCE PATTERNS; MOSAICISM; TELOMERE.

Michael Fossel

senesce a state in a cell in which it will not divide again, even in the presence of growth factors

autosomal describes a chromosome other than the X and Y sex-determining chromosomes

neonatal newborn

lymphocytes white blood cells

Bibliography

Fossel, Michael. *Reversing Human Aging.* New York: William Morrow & Company, 1996.

———. "Telomeres and the Aging Cell: Implications for Human Health." *Journal of the American Medical Association* 279 (1998): 1732–1735.

Hayflick, Leonard. *How and Why We Age.* New York: Ballantine Books, 1994.

Addiction

Addiction in its broadest sense can be defined as the habituation to a practice considered harmful. A more narrow definition of the term refers to chronic use of a chemical substance in spite of severe psychosocial consequences. Terms such as "workaholic," "sex addict," and "computer junkie" arose to describe behaviors that have features in common with alcoholism and other substance addictions. The most convincing data supporting a role of genetics in addiction has been collected for alcoholism, although genetics most likely has a role in other forms of addiction.

Definitions

In order to assess alcoholism, or any form of addiction, a clear definition of the condition is necessary. The American Psychiatric Association and the World Health Organization have developed clinical criteria (DSM-IV and ICD10, respectively) that are widely used for the diagnosis of substance-use related disorders. DSM-IV criteria recognizes ten classes of substances (alcohol, amphetamines, cannabis, hallucinogens, inhalants, nicotine, opioids, phencyclidine, and sedatives) that lead to substance dependence, another term for addiction.

The precise diagnostic criteria for dependence vary among substances. DSM-IV defines dependence as manifesting, within a twelve-month period, at least three of the following criteria:

- Tolerance (increased dose needed to achieve the same affect, or reduced response to the same dose)
- Withdrawal symptoms
- Progressive increase in dose or time used
- Persistent desire for, or failure to reduce substance use
- Increasing efforts made to obtain substance
- Social, occupational, or recreational activity is replaced by activity associated with substance use
- Continued substance use despite recognized physical and psychological consequences

Heritability in Humans

Most family, twin, and adoption studies have shown that addiction to alcohol has significant heritability. For example, there is an increased risk for alcoholism in the relatives of alcoholics. Depending on the study, the risk of alcoholism in siblings of alcoholics is between 1.5 and 4 times the risk for the general population. The identical twins of alcoholics (who share 100

percent of their genes) are more likely to be alcoholics than the fraternal twins of alcoholics (who share only about 50 percent). Adoption study data suggest that the risk for developing alcoholism for adopted children is influenced more by whether their biological parents were alcoholics than whether their adopted parents are alcoholics, suggesting that genes contribute to alcoholism more than environment. Similar but less extensive data has been collected for nicotine addiction. Very little genetic epidemiological data has been collected for illegal drugs.

The only genes that have been conclusively shown to affect susceptibility to addiction in humans are genes that encode proteins responsible for the metabolism of alcohol. In the body, ethanol ("drinking" alchohol) is oxidized by enzymes to acetaldehyde and then to acetate. Certain **alleles** of aldehyde dehydrogenase genes that are common in some populations, such as Asians, lead to increased levels of acetaldehyde when alcohol is consumed. Acetaldehyde causes an unpleasant flushing reaction that leads to a voluntary reduction of alcohol consumption. The systematic search for other genes that affect susceptibility to alcohol and nicotine addiction in humans has lead to the identification of chromosome **loci** that may contain genes that affect susceptibility to addiction, but has not lead to the identification of any specific genes.

alleles particular forms of genes

loci sites on a chromosome (singular, locus)

Models of Addiction

Progress in genetic analysis of addiction in animal models has been more successful. The pharmacologic effects of abused substances can readily be demonstrated in many model systems, from worms to rodents. Rodents can be trained to voluntarily consume alcohol and other abused substances. Once trained, these rodents will expend energy to continue to receive drugs and will display withdrawal symptoms when denied drugs. Chromosomal regions with naturally occurring variants that affect voluntary consumption, intoxication, and withdrawal have been mapped in mice. The specific genes responsible for these effects have not yet been identified.

Cell biology and neurochemistry studies in humans and model systems have identified many molecules that have altered abundance and distribution, enzymes with altered activity, and genes with altered expression resulting from substance abuse. In particular, the dopamine and serotonin **neurotransmitter** systems have been the focus of intense studies. These are brain systems directly involved in many basic responses, including pleasure and reward systems.

neurotransmitter molecule released by one neuron to stimulate or inhibit a neuron or other cell

To directly test the role of specific genes and pathways, mice have been engineered to delete or over-express genes. Mice lacking any of these genes (called PKC_ϵ, DRD2, and DBH) are more sensitive to the effects of alcohol and consume less alcohol. In contrast, mice lacking any one of four other genes (PKA regulatory II_β, NPY, or 5-HT_{1b}) are less sensitive to the effects of alcohol and consume more alcohol. Mice cannot be trained to self-administer alcohol if they lack the Mu opioid receptor, which is involved in transmitting signals to the body's own internal opiate system.

Mutant fruit flies with altered responses to alcohol intoxication have also been created. Two mutants, called "cheapdate" and "amnesiac," arise from different mutations in the same gene. These mutations affect the cellular level of the signal transduction molecule cyclic-AMP. As the names imply,

flies with cheapdate mutations are very sensitive to the affects of alcohol, and flies with amnesiac mutations are unable to learn.

The major conclusion from work in model systems is that the pathways and systems involved in addiction are central to normal behaviors with instinctive reward processes, such as feeding and procreation. Addiction is a process that involves learning and the subversion of these basic reward pathways. SEE ALSO COMPLEX TRAITS; DISEASE, GENETICS OF; GENE AND ENVIRONMENT; SIGNAL TRANSDUCTION; TWINS.

Kirk C. Wilhelmsen

Bibliography

American Psychiatric Association Task Force on DSM-IV. *Diagnostic and Statistical Manual of Mental Disorders*, 4th ed. Washington, DC: American Psychiatric Association, 1994.

Begleiter, Henri, and Benjamin Kissan, eds. *The Genetics of Alcoholism.* New York: Oxford University Press, 1995.

Tamara J. Phillips, et al. "Alcohol Preference and Sensitivity Are Markedly Reduced in Mice Lacking Dopamine D2 Receptors." *Nature Neuroscience* 1 (1998): 610–615.

Theile, Todd, et al. "Ethanol Consumption and Resistance Are Inversely Related to Neuropeptide Y Levels." *Nature* 396 (1998): 366–369.

Aging and Life Span

Aging is, simply put, the act of getting older. Aging is part of the natural life cycle of an organism. From birth, through maturation, and eventually to death, aging is the element that ties all segments of life together.

Life Span and the Aging Process

How long an organism lives is called its life span. In 1998, the average life span for a human, worldwide, was sixty-six years. However, life span is a complex trait, meaning that many factors, including family history, lifestyle, disease, and residence in a developed nation, determine how long an individual's life will be. The average life span in a particular population changes as these factors change. For example, the average life span in the United States in 1900 was forty-nine; in 1998 it was seventy-seven.

This increase was likely due to several factors, but perhaps the most important was the improvement of sanitation, hygiene, and public health from 1900 to 1998. These improvements included purification of drinking water, treatment of wastewater, widespread vaccination, and improved access to health care. However, even as these sanitary measures were adopted, other elements of modern life emerged as strong influences on life span, such as diet, exercise, and socioeconomic status. Studies have shown that individuals who exercise regularly, eat a diet lower in saturated fats, and avoid unnecessary risk-taking live longer. This may be because such a lifestyle reduces the risk of developing cardiovascular disease and cancer, the top causes of death in developed countries.

Finally, life span is in part genetically determined. Studies of life span in large families have shown that longevity is, to some degree, inherited. This may be due to shared genetic risks of diseases or behaviors that shorten the life span, or it may reflect direct genetic influences in longevity separate from risk of disease.

The aging process causes many changes, both visible and invisible. In humans, these changes take several forms. In the first two decades of life, from birth to adulthood, aging involves physical growth and maturation and intellectual development. These changes are fairly noticeable and relatively swift compared to the rest of the life span. After reaching physical maturity, humans begin to show subtle signs of physical aging that grow more pronounced over time. Long-term exposure to sunlight and the outdoors may begin to toughen the skin and produce wrinkles on the face and body. The senses change: Sight, hearing, taste, and smell become less acute. Gradual changes in the eye cause many older adults to need glasses to read. Hair begins to thin and turn gray. Individuals with less active lifestyles often begin to gain weight, particularly around the waist and hips. Beginning in their 40s (or, rarely, in their late 30s), many women experience menopause, which marks the end of childbearing years. Less visible or noticeable changes associated with aging are the loss of bone density over time (particularly in women), slower reflexes, less acute mental agility, and declining memory.

Diseases Associated with Aging

Many of the diseases common in older adults, such as cardiovascular disease, cancer, **dementia**, arthritis, blindness, and deafness, are consequences of the acceleration or distortion of these "natural" changes associated with aging. These complex diseases associated with aging are caused by the interaction of genetic and environmental factors. For example, Alzheimer's disease is an illness caused by changes in the brain that impair thinking and memory much more severely than the natural decline that all humans experience during aging. It is the most common form of dementia in adults over age sixty. Certain very rare **alleles** are associated with the development of Alzheimer's disease, and other much more common alleles (of different genes) increase the risk of a variety of other diseases. Environmental factors such as exposure to toxins have also been implicated. A more rare example is progeria, a disease in which the tissues of the body age about seven times more rapidly than normal. In this case, a person who is chronologically only a teenager looks much older.

dementia neurological illness characterized by impaired thought or awareness

alleles particular forms of genes

Genetics and Aging

Many scientists have hypothesized that some genes may control aspects of aging separate from the development of disease. These hypotheses are based on experimental studies of non-human organisms and the observation that longevity in humans appears to run in families. Studies of yeast and roundworm have identified over ten genes in each that are associated with longevity and aging, and more recent studies have suggested similar genes exist in the fruit fly. The exact function of these genes is unknown, but one or more may help slow down the metabolic rate. Studies in mice have shown that reducing metabolism by reducing food intake can increase life span. Finally, shortening of the **telomeres** decreases longevity in some model organisms.

telomeres chromosome tips

Finding similar genes in humans is more complicated, since scientists cannot experimentally control genes to test their effects on longevity in

Life span is determined partly by genetics: These grandchildren are more likely to live longer based on the long life span of their grandparents.

centenarians people who live to age 100

humans. Therefore, genetic studies of human longevity require a more observational approach. One study design is to examine large numbers of long-lived individuals such as **centenarians** and see what factors they have in common, such as lifestyle, medical history, and genetics.

Studies of centenarians have suggested that variants in multiple genes, including the human leukocyte antigen (HLA) genes of the immune system, apolipoprotein E (APOE), angiotensin-converting enzyme (ACE), plasminogen activating inhibitor 1 (PAI-1), and p53, are associated with living past age ninety. Forms of several of these genes, such as APOE, ACE, and p53, are associated with increased risk of developing Alzheimer's disease, cardiovascular disease, and cancer, respectively. The association of these genes with longevity may be due to these disease associations, or it may be due to their direct influence on extending the human life span. Regardless, genes clearly influence aging and longevity, whether it is by influencing the development of life span-shortening diseases, or by positively influencing longevity independently of causing disease. SEE ALSO ACCELERATED AGING: PROGERIA; ALZHEIMER'S DISEASE; CANCER; CARDIOVASCULAR DISEASE; COMPLEX TRAITS; TELOMERE.

William K. Scott

Bibliography

Anderson, Robert N. "United States Life Tables, 1998." *National Vital Statistics Reports* 48, no. 18. Hyattsville, MD: National Center for Health Statistics, 2001.

Finch, Caleb E., and Rudolph E Tanzi. "Genetics of Aging." *Science* 278 (1997): 407–411.

Schächter, François. "Causes, Effects, and Constraints in the Genetics of Human Longevity." *American Journal of Human Genetics* 62 (1998): 1008–1014.

Agricultural Biotechnology

Biotechnology is the use of living organisms—microbes, plants, or animals—to provide useful new products or processes. In a broad sense, biotechnology continues a process that is thousands of years old. Using traditional plant breeding techniques, humans have altered the genetic composition of almost every crop by only planting seeds from plants with desired traits, or by controlling pollination. As a result, most commercial crops bear little resemblance to their early relatives. Current maize varieties are so changed from their wild progenitors that they cannot survive without continual human intervention.

The 1970s heralded recombinant DNA technology, which gave researchers the ability to cut and recombine DNA fragments from different sources to express new traits. Genes and traits previously unavailable through traditional breeding became available through DNA recombination.

Techniques

Modern plant genetic engineering involves transferring desired genes into the DNA of some plant cells and regenerating a whole plant from the transformed tissue. New DNA may be introduced into the cell via biological or physical means.

The most widely used biological method for transferring genes into plants capitalizes on a trait of a naturally occurring soil bacterium, *Agrobacterium tumefaciens*, which causes crown gall disease. This bacterium, in the course of its natural interaction with plants, has the ability to infect a plant cell and transfer a portion of its DNA into a plant's **genome**. This leads to an abnormal growth on the plant called a gall. Scientists take advantage of this natural transfer mechanism by first removing the disease-causing genes and then inserting a new beneficial gene into *A. tumefaciens*. The bacteria then transfer the new gene into the plant.

genome the total genetic material in a cell or organism

Another gene transfer technique involves using a "gene gun" to literally shoot DNA through plant cell walls and membranes to the cell nucleus, where the DNA can combine with the plant's own genome. In this technique, the DNA is made to adhere to microscopic gold or tungsten particles and is then propelled by a blast of pressurized helium.

Advantages

Depending on which genes are transferred, agricultural biotechnology can protect crops from disease, increase their yield, improve their nutritional content, or reduce pesticide use. In 2000, more than half of American soybeans and cotton and one-fourth of American corn crops were genetically

A golden rice field in Kaili, China. Golden rice is part of a "second generation" of genetically modified foods.

modified by modern biotechnology techniques. Genetically modified foods may also help people in developing countries. One in five people in the developing world do not have access to enough food to meet their basic nutritional needs. By enhancing the nutritional value of foods, biotechnology can help improve the quality of basic diets.

"Golden rice" is a form of rice engineered to contain increased amounts of vitamin A. Researchers are also developing rice and corn varieties with enriched protein contents, as well as soybean and canola oils with reduced saturated fat. Other potential benefits include crops that can withstand drought conditions or high salinity, allowing populations living in harsh regions to farm their land.

Agricultural biotechnology also provides benefits for the manufacture of pharmaceutical products. Because plants do not carry human diseases, plant-made vaccines and antibodies require less screening for bacterial toxins and viruses. In addition to plants, animals may also be engineered to produce beneficial genes. In order to produce large quantities of **monoclonal antibodies** for research on new therapeutic drugs, several compa-

monoclonal antibodies immune system proteins derived from a single B cell

nies have genetically engineered cows and goats to secrete antibodies into their milk. One company has inserted a spider gene into dairy goats. The spider silk extracted from the goat's milk is expected to produce fibers for bulletproof vests and medical supplies, such as stitch thread, and other applications where flexible and extremely strong fibers are required.

Concerns

Despite the benefits of genetic engineering, there are concerns about whether recombinant DNA techniques carry greater risks than traditional breeding methods. Consumer acceptance of food derived from genetically engineered crops has been variable. Many individuals express concerns regarding the environmental impact and ethics of the new technology, and about food safety. One of the major food safety concerns is that there is a risk that crops expressing newly inserted genes may also contain new **allergens**.

allergens substances that trigger an allergic reaction

Some groups have expressed concern that widespread use of plants engineered for specific types of pest resistance could accelerate the development of pesticide-resistant insects or have negative effects on organisms that are not crop pests. Another environmental concern is that transgenic, pest-protected plants could **hybridize** with neighboring wild relatives, creating "superweeds" or reducing genetic **biodiversity**.

hybridize to combine two different species

biodiversity degree of variety of life

Regulations

To address these concerns, agricultural biotechnology products are regulated by a combination of three federal agencies: the U.S. Department of Agriculture (USDA), the Environmental Protection Agency (EPA), and the Food and Drug Administration (FDA). Together, these agencies assess genetically modified crops, as well as products that use those crops. They test the crops and products for safety to humans and to the environment, and for their efficacy and quality. SEE ALSO BIOPESTICIDES; GENETICALLY MODIFIED FOODS; PLANT GENETIC ENGINEER; TRANSGENIC ANIMALS; TRANSGENIC MICROORGANISMS; TRANSGENIC PLANTS.

Barbara Emberson Soots

Bibliography

Ferber, Dan. "Risks and Benefits: GM Crops in the Cross Hairs." *Science* 286 (1999): 1662–1666.

Internet Resources

Agricultural Biotechnology. U.S. Department of Agriculture. <http://www.usda.gov/agencies/biotech>.

Transgenic Plants and World Agriculture. Royal Society of London, U.S. National Academy of Sciences, Brazilian Academy of Sciences, Chinese Academy of Sciences, Indian National Science Academy, Mexican Academy of Sciences, and Third World Academy of Sciences. <http://stills.nap.edu/html/transgenic>.

Alternative Splicing

When molecular biologists began analyzing the complete sequence of the human genome in mid-2001, one surprising observation was that humans have relatively few genes. We may have as few as 30,000 genes, only about two

Introns are removed by the spliceosomes.

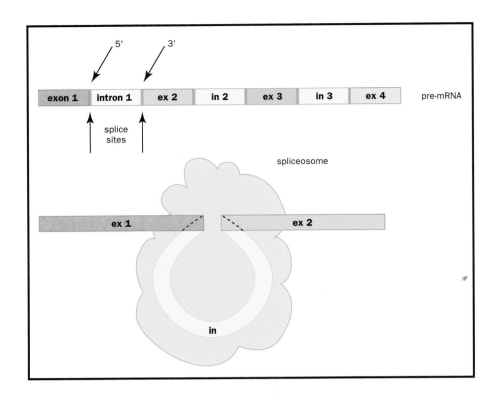

The sensitivity of the human ear to a wide range of sound frequencies is due to alternative splicing of a potassium channel gene, giving rise to a set of related proteins whose exact form varies with the position in the cochlea.

mRNA messanger RNA

spliceosomes RNA-protein complexes that remove introns from RNA transcripts.

times as many as the much simpler fruit fly, *Drosophila melanogaster.* How can the much greater size and complexity of humans be encoded in only twice the number of genes required by a fly? The answer to this paradox is not fully understood, but it appears that humans and other mammals may be more adept than other organisms at encoding many different proteins from each gene. One way they do this is through alternative splicing, the processing of a single RNA transcript to generate more than one type of protein.

In most eukaryotic genes, the protein-coding sequences, termed exons, are interrupted by stretches of sequence, termed introns, that have no protein-coding information. After the gene is copied, or transcribed, to RNA, the introns are removed from this "pre-mRNA," and the exons are spliced together to form a mature **mRNA**, consisting of one contiguous protein-coding sequence. In addition, the complete mRNA contains upstream and downstream sequences flanking the coding sequences. These sequences do not encode protein, but help to regulate translation of the mRNA into protein. Variations in the splice pattern lead to alternative transcripts and alternative proteins.

Splicing is accomplished in the cell's nucleus by **spliceosomes**, which are molecular machines composed of proteins and small RNA molecules. The boundaries between exons and introns in a pre-mRNA are marked very subtly. Certain segments of the pre-mRNA, termed splice sites, direct the spliceosomes to the precise positions in the transcript where they can excise introns and splice together exons. Splice sites are short sequences, typically less than ten bases long. 5′ splice sites mark the 5′ end of introns; 3′ splice sites define the 3′ end of introns. ("Five prime" and "three prime" refer to the upstream and downstream ends of the RNA.)

Although splice sites often can be recognized as such by common patterns in their base sequence, there are many variations on the basic splice site consensus sequence. These differences affect how readily a particular splice site is recognized and processed by the splicing machinery. Many other molecules within the cell, called splicing factors, also participate in the splicing reaction. The combination of all of these determines the pattern of splicing for a particular pre-mRNA molecule.

For many genes the pattern of splicing is always the same. These genes encode many copies of their corresponding pre-mRNA molecules. The introns are removed in a consistent pattern, producing mature mRNA molecules of identical sequence, all of which encode identical proteins.

For other genes the splice pattern varies depending on the tissue in which the gene is expressed, or the stage of development the organism is in. Because the choice of splice sites depends on so many different factors, the same pre-mRNAs from these genes may become spliced into several, or even many, different mature mRNA variants. 5′ splice sites may be ignored, converting intron sequences into exons; 3′ splice sequences can be ignored, converting exon sequences into introns; or different sequences, ordinarily not recognized as splice sites, can function as new splice sites. (To understand why ignoring a 5′ splice site would convert an intron to an exon, recall that transcription of RNA proceeds from 5′ to 3′.) The production of such mRNA variations through the use of different sets of splice sites is known as alternative splicing. It has been estimated that at least one-third of all human genes are alternatively spliced.

Alternative splicing can have profound effects on the structure and function of the protein encoded by a gene. Many proteins are comprised of several domains, or modules, that serve a particular function. For example, one domain may help the protein bind to another protein, while another domain gives the protein enzymatic activity. By alternative splicing, exons, and, therefore, protein domains, can be mixed and matched, altering the nature of the protein. By regulating which splice patterns occur in which tissue types, an organism can fine-tune the action of a single gene so it can perform many different roles.

The various forms of a protein are known as isoforms. Isoforms are often tissue-specific. The dystrophin gene, for example, has one form in muscle and another in brain tissue. Defects in alternative splicing are associated with several important human diseases, including amyotrophic lateral sclerosis, dementia, and certain cancers.

Alternative splicing can also act to turn genes off or on. In mRNA, codons, consisting of three adjacent **nucleotides**, either encode an amino acid or signal the ribosome to stop synthesizing a **polypeptide**. Normally, exon sequences must not encode stop codons (AUG, UAG, or UAA) until after the final amino-acid-coding codon. Alternative splicing can introduce a stop codon in the beginning or middle of a protein-coding sequence, resulting in an mRNA that encodes a prematurely truncated polypeptide.

Human hearing offers a dramatic illustration of how important alternative splicing is in everyday life. Microscopic hair cells lining the inner ear vibrate when stimulated by sound. One of the proteins in the hair cells that plays a role in the hearing sensation is a calcium-activated potassium channel. The gene for this protein can generate more than five hundred different mRNA variants through alternative splicing. The resulting potassium-

The protozoan *Trypanosome brucei,* which causes African sleeping sickness, edits some of its messenger RNA molecules after they are transcribed. Uracil nucleotides are added in some locations in the mature RNA and deleted from others. Similar cases of RNA editing occur in other organisms, and even in humans. The human apolipoprotein B gene is edited in the intestine but not in the liver, leading to two distinct forms of the protein, serving different functions in the two organs.

nucleotides the building blocks of RNA or DNA

polypeptide chain of amino acids

Alternative splicing creates protein isoforms or may lead to no protein production.

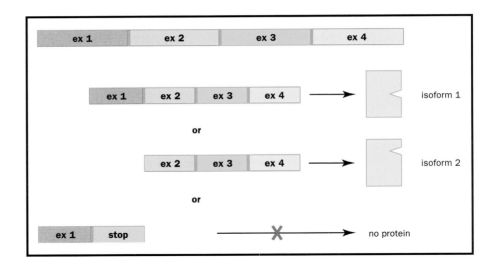

channel proteins have slightly differing physiological properties. This is in part what tunes hair cells to different frequencies. SEE ALSO GENE; PROTEINS; RNA PROCESSING; TRANSCRIPTION.

Paul J. Muhlrad

Bibliography

Alberts, Bruce, et al. *Molecular Biology of the Cell*, 4th ed. New York: Garland Science, 2002.

Griffiths, Anthony J. F., et al. *An Introduction to Genetic Analysis*, 7th ed. New York: W. H. Freeman, 2000.

Lodish, Harvey, et al. *Molecular Cell Biology*, 4th ed. New York: W. H. Freeman, 2000.

Alzheimer's Disease

Alzheimer's disease (AD) is a diagnosis applied to a group of degenerative brain disorders with similar clinical and pathological characteristics. It is the most common cause of dementia, with onset of symptoms after the age of fifty-five years. It is recognized as a major public health concern in societies with an aging population. AD affects four million people in the United States. At least 90 percent of those affected are over sixty-five years of age. In 1998 direct health care costs were estimated to be $50 billion. Indirect costs, such as lost productivity and absences from work, were estimated to be $33 billion.

First Description of AD

In 1907, Alois Alzheimer, a German physician from Bavaria, published the case of one of his patients. The patient, Mrs. Auguste D., at the age of fifty-one years developed an unfounded jealousy regarding her husband. This behavioral change was followed closely by a subtle and slow decline in other cognitive abilities, including memory, orientation to time and to physical location, language, and the ability to perform learned behaviors. All of her difficulties gradually progressed in severity. Within three years, the patient did not recognize her family or herself, could not maintain her self-care,

Eugenie Bonenfant, left, is a resident in Rhode Island's first assisted living community designed exclusively for people with Alzheimer's disease. The unit supervisor, Margaret Knight, visits, and she is surrounded by her own, familiar furniture.

and was institutionalized. She died a short four and a half years after her illness began. Her brain was removed at autopsy. Using a novel (at the time) silver stain to highlight changes in brain sections, Dr. Alzheimer viewed the tissue under his microscope. He described what are now the pathologic lesions of the disease that bears his name: loss of **neurons**, senile plaques found in the brain substance but outside of the neurons, and neurofibrillary tangles found inside neurons.

neurons brain cells

Dr. Alzheimer's patient had developed dementia. Dementia is an acquired and continuing loss of thinking abilities in three or more areas of cognition (which include memory, language, orientation, calculation, judgment, personality, and other functions) severe enough that the individual can no longer function independently at work or in society. There is no decrease in level of consciousness. Early in the illness, physical strength is maintained, though later the individual may "forget" how to perform certain physical functions, such as using tools or utensils, dressing, or performing personal hygiene activities. Onset of dementia may occur over days, months, or years. Its course may be static or progressive. Causes of dementia, other than AD, include other neurodegenerative disease, central nervous system infection, brain tumor, metabolic disease, vitamin deficiency, and **cerebrovascular disease**.

cerebrovascular disease stroke, aneurysm, or other circulatory disorder affecting the brain

An Evolving Understanding of Dementia

Within three years of the publication of Dr. Alzheimer's first case, the term "Alzheimer's disease" was applied to patients who developed significant difficulty in memory and other areas of cognition at an age less than sixty-five years. Individuals who developed such symptoms later in life, generally after the age of sixty-five, were said to be suffering from senility, a process

considered a normal part of aging. The phrase "hardening of the arteries," implying narrowing of arterial size with a reduction in blood flow to the brain, was used by physicians and by laypersons to designate the reason for senility. However, a causal relationship between arterial narrowing and senility had not been established scientifically.

Critical research reports were published in 1968 and 1970 providing evidence that senility and the disease Alzheimer described were similar both clinically and pathologically. Patients in each category developed similar and multiple cognitive deficits. Patients in each category developed plaques and tangles, and the majority of those diagnosed with senility did not have evidence of "hardening of the arteries." Over the next decade senile dementia, Alzheimer's type, would replace senility as the accepted common cause of late-life dementia.

In 1984, consensus criteria for a clinical diagnosis of AD were established. Cardinal features include the insidious onset of decline in at least two areas of cognition, gradual progression of severity in these spheres resulting in dementia, onset of symptoms between the ages of forty and ninety years (most often after age sixty-five), and absence of another medical condition that by itself could cause dementia. **Pathological** study of tissue after death should reveal the characteristic findings of senile plaques in age-associated numbers (numbers larger than expected for the individual's age) and of neurofibrillary tangles. Using these criteria, both Alzheimer's disease as a presenile disorder and senile dementia, Alzheimer type, are subsumed into the broader diagnosis, Alzheimer's disease.

pathological altered or changed by disease

Genetics of Alzheimer's Disease

There are three areas of evidence that indicate a genetic basis for AD. First, it occurs as a Mendelian, autosomal dominant disease of early onset (occurring before the age of sixty) in multiple families. However, the number of such families with autosomal dominant inheritance is small. Second, it is generally the case that if an individual has a first-degree relative (parent or sibling) with AD, he or she has a greater risk of developing the disease than a person with no affected first-degree relative. Finally, AD is more likely to occur in each of a pair of **identical twins** than it is to occur in a pair of **fraternal twins**.

identical twins monozygotic twins who share 100 percent of their genetic material

fraternal twins dizygotic twins who share 50 percent of their genetic material

Recognizing these observations, in the mid-1980s researchers initiated scientific efforts to identify genes of importance in the disease, using the then-emerging recombinant DNA technology. By 1995, three causative genes and one susceptibility gene had been identified: *APP*, *PS1/2*, and *APOE*.

APP. In 1991, a British research group identified mutations in the APP gene that occurred only in patients with AD in very rare families. (Less than twenty such families have been reported in the medical literature.) The mutations were not found in family members who did not have AD. The APP gene codes for amyloid precursor protein, one of whose degradation products is a main constituent of the senile plaques of AD.

linkage analysis examination of co-inheritance of disease and DNA markers, used to locate disease genes

PS1 and *PS2.* In 1992, using **linkage analysis** of data from early-onset, autosomal-dominant families, researchers in Seattle, Washington; Jacksonville, Florida; and Antwerp, Belgium, almost simultaneously determined

GENES FOR ALZHEIMER'S DISEASE

Age at Onset	Inheritance	Chromosome	Gene	Protein	% AD
Early Onset	AD	14	PS1	presenilin 1	< 2
Early Onset	AD	21	APP	amyloid precursor protein	< 20 families*
Early Onset	AD	1	PS2	presenilin 2	3 families*
Early Onset	AD	?	?	?	?
Late Onset	Familial/ Sporadic	19	APOE	apolipoprotein E	~50
Late Onset	Familial	12p11-q13	?	?	?
Late Onset	Familial	9p22.1	?	?	?
Late Onset	Familial	10q24	?	?	?
Late Onset	?	?	?	?	?

Age of Onset: Early Onset: < 60 years, late onset: > 60 years; **Inheritance:** AD: autosomal dominant, familial: disease in at least one first-degree relative, sporadic: disease in no other family member; **Chromosome:** number, arm, and region; **Gene:** designation of identified gene; **Protein:** name of protein coded for by the gene; **% AD:** percent of AD caused by or * number of families identified with AD for each gene.

that a then-unknown gene for early-onset AD was located on chromosome 14. In 1995, a research scientist in Toronto, Canada, identified this gene as *PS1*, which codes for the protein called presenilin1. Individuals who have mutations in the gene consistently develop AD. Also in 1995, using comparative genomic techniques, the Seattle research group cited above identified the *PS2* gene, which codes for the protein termed presenilin 2. Using data from a few large, genetically isolated families with early- and late-onset disease, they determined that mutations in the gene consistently occur only in patients with AD.

APP, *PS1*, and *PS2* are causative genes: When mutated, each causes AD. If a person has a mutated gene, he or she will develop the disease at about the same age as others who have the same mutation. The risk of developing the disease approaches 100 percent.

APOE. In 1993 researchers in Durham, North Carolina, reported that one form (**allele**) of the *APOE* gene occurred more commonly in patients with late onset AD than was expected given its occurrence in the population as a whole. Numerous additional research groups corroborated the finding. The *APOE* gene occurs in three forms (alleles), determined by the DNA sequence. The three forms are termed *APOEε2*, *APOEε3*, and *APOEε4*, and they code for apolipoprotein E molecules differing from one another by only one or two amino acids. APOE is a susceptibility gene; it imparts an increased risk of disease occurrence but by itself does not cause the disease. The presence of the ε4 form (*APOEε4*) in either one or two copies in an individual increases the likelihood that the individual will develop AD. Occurrence may depend on other genetic factors or environmental factors or some combination from each category.

Additional families exist with early-onset, autosomal-dominant AD with no *APP*, *PS1*, or *PS2* mutations. Such families provide evidence that there may be additional causative genes. Whole-genome-scan analyses reported in the late 1990s provide evidence of additional susceptibility genes on chromosomes 9, 10, and 12. The genes located on these chromosomes have yet to be identified.

allele particular form of genes

Rationale for a Genetic Approach to Alzheimer's Disease

Alzheimer's disease, broadly defined, is a complex genetic disorder: Multiple causative and susceptibility genes acting singly or in concert produce similar symptoms and pathologic changes in patients. In each of its forms, it manifests age-dependent penetrance, meaning that the older an individual becomes, the more likely it is that he or she will develop the disease. Disease manifestations (such as age of onset or rate of progression) may be influenced by environmental exposures (alcohol use, head injury) or other health conditions (such as cerebrovascular disease). Identification of AD genes will lead to a better understanding of the cellular processes that cause dementia.

Currently, amyloid production from amyloid precursor protein is the focus of much research, although debate continues about its role. Amyloid production and deposition in the brain are affected by each of the four known AD genes. Decrease in amyloid production or increase in amyloid metabolism with a resulting decrease in deposition may result in delayed age of onset or slower progression of disease. Thus, alteration of amyloid processing of sufficient magnitude might result in disease prevention. Once process-altering treatments become available, knowing who is at risk for the disease will be important.

Genetic Testing and Alzheimer's Disease

DNA testing can be performed to determine whether an individual has a mutation in one of the causative genes and/or whether he or she carries one or two copies of the *APOEε4* susceptibility gene. Whether to test and which test to perform will depend on three conditions: family history of dementia, age of onset of disease, and clinical status of the individual. If a person has dementia, the test result could be useful in determining that the cause of the dementia is a form of AD. If a person has no symptoms of dementia, an estimate of the individual's risk could be developed, using the test. In the case of such estimates, both the actual accuracy of the test and the tested individual's understanding of its accuracy are of concern. While the consensus is that presymptomatic testing for causative mutations may be performed with appropriate counseling, debate over the safety and utility of *APOE* testing for individuals who do not show symptoms of Alzheimer's is ongoing.

In 2001, there was no treatment that prevented, much less cured, AD. Information regarding the risk of developing AD is useful only in life planning activities (such as purchasing or offering health insurance coverage or long-term care insurance coverage, or choosing retirement age) or in family planning. An individual's ability to cope with either an increased or a decreased risk may vary. Misuse of the information resulting in insurance or employment discrimination is possible. Absence of a causative gene mutation or of an *APOEε4* susceptibility gene in either symptomatic or presymptomatic disease does not preclude AD as the cause of dementia or mean that the individual has no risk of developing AD in later years. SEE ALSO COMPLEX TRAITS; DISEASE, GENETICS OF; GENE DISCOVERY; GENETIC TESTING; INHERITANCE PATTERNS; PSYCHIATRIC DISORDERS.

P. C. Gaskell Jr.

The first preimplantation testing for the APP mutation was announced in February 2002. Four gene-negative embryos from a gene-positive woman were selected and implanted, and she gave birth to one child who was free of the gene mutation, which causes early-onset Alzheimer's disease.

Bibliography

Mace, Nancy L., and Peter V. Rabins, eds. *The 36-Hour Day*, 3rd ed. Baltimore: The Johns Hopkins University Press, 1999.

St. George-Hyslop, Peter H. "Piecing Together Alzheimer's." *Scientific American* (Dec. 2000): 76–83.

Terry, Robert D., et al., eds. *Alzheimer Disease*, 3rd ed. Philadelphia, PA: Lippincott, Williams & Wilkins, 1999.

Internet Resources

"Ethical, Legal, and Social Issues." Human Genome Project, U.S. Department of Energy Office of Science. <http://www.ornl.gov/TechResources/Human_Genome/home.html>.

"Progress Report on Alzheimer's Disease, 1999." National Institute on Aging. Bethesda: National Institutes of Health, 1999. <http://www.nih.gov/nia/>.

Ames Test

The Ames test is a **protocol** for identifying mutagenic chemical and physical agents. Mutagens generate changes in DNA. Many mutagenic agents modify the chemical structure of adenine, thymine, guanine, and cytosine, the bases in DNA, changing their base-pairing properties and causing mutations to accumulate during DNA synthesis.

Ethyl methanesulfonate (EMS), for example, is a very potent mutagen. The ethyl group of EMS reacts with guanine in DNA, forming the abnormal base O^6-ethylguanine. During DNA **replication**, DNA **polymerases** that **catalyze** the process frequently place thymine, instead of cytosine, opposite O^6-ethylguanine. Following subsequent rounds of replication, the original G:C base pair can become an A:T pair. This changes the genetic information, is often harmful to cells, and can result in disease. Many mutagens cause a wide variety of cancers in humans.

During the 1960s the biologist Bruce Ames developed a test that still carries his name and that is still used as a relatively inexpensive way to assess the mutagenic potential of many chemical compounds. The procedure uses the bacteria *Salmonella typhimurium*. Wild-type *S. typhimurium* grows well on agar that contains only minimal nutrients. It can thrive on agar that contains only sugar, ammonium salts, phosphate, sulfate, and some trace metal ions. Amino acids are not needed because the bacteria have genes that encode enzymes that can make all twenty amino acids.

Ames developed strains of *S. typhimurium* that contain mutations in genes that the bacteria use to make the amino acid histidine. Such *his⁻* strains cannot survive unless histidine is added to their agar. Ames reasoned that mutagenic agents could cause changes in the aberrant gene that encodes the defective *his⁻* enzyme, causing it to revert back to the normal form, encoding the active protein. (The mutagen would likely also cause many other, undetected mutations.) A mutation that returns a function to a mutant is called a reverse mutation. The Ames test measures the ability of *his⁻ S. typhimurium* to grow on agar that does not contain histidine. Growth indicates that a reverse mutation has reverted the *his⁻* gene back to an active form.

A typical Ames test involves exposing *his⁻ S. typhimurium* to a test agent and then placing the exposed bacteria in petri dishes that contain agar with

protocol laboratory procedure

replication duplication of DNA

polymerases enzyme complexes that synthesize DNA or RNA from individual nucleotides

catalyze aid in the reaction of

O^6-ethylguanine.

AMES II ASSAY

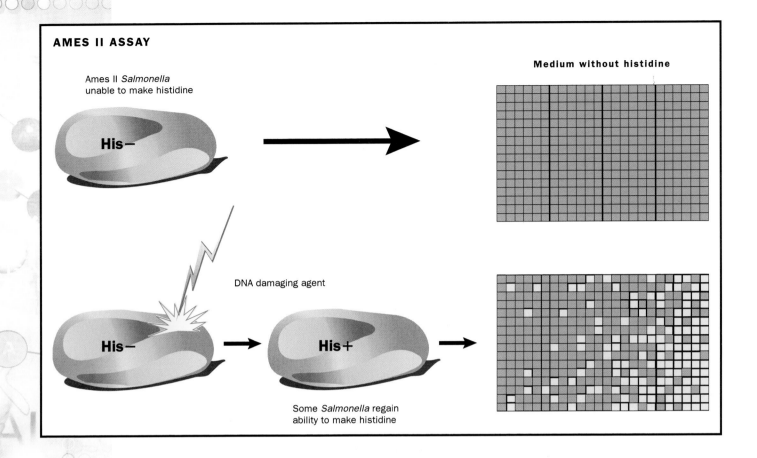

Ames II *Salmonella* unable to make histidine

His−

DNA damaging agent

His−

His+

Some *Salmonella* regain ability to make histidine

Medium without histidine

Modern robotic technology is used to perform the Ames test today, but the fundamental procedure remains the same. Each square represents a single well in a plastic growth plate. Mutated *Salmonella* (yellow) are recorded by computer-driven detectors. Courtesy of Xenometrix, Inc., a subsidiary of Discovery Partners International, Inc.

incubating heating to optimal temperature for growth

no histidine. After **incubating** the dishes, the bacteria that have grown are counted. This number, which reflects the bacteria that undergo a reverse mutation from *his⁻* to *his S. typhimurium*, is compared to the number of bacteria that undergo reverse mutations when they are not exposed to the agent. If the agent causes too many reverse mutations above those measured as spontaneous, it is considered to be mutagenic.

The Ames test can detect mutagens that work directly to alter DNA. In humans, however, many chemicals are *promutagens*, agents that must be activated to become true mutagens. Activation, involving a chemical modification, often occurs in the liver as a consequence of normal liver activity on unusual substances. Bacteria such as *S. typhimurium* do not produce the enzymes required to activate promutagens, so promutagens would not be detected by the Ames test unless they were first activated. An important part of the Ames test also involves mixing the test compound with enzymes from rodent liver that convert promutagens into active mutagens. These potentially activated promutagens are then used in the Ames test. If the liver enzymes convert the agent to a mutagen, the Ames test will detect it, and it will be labeled as a promutagenic agent.

The Ames test is widely used by the pharmaceutical industry to test drugs prior to using them in clinical trials. When a drug is mutagenic in the Ames test, it is usually rejected for further development and will probably not be tested in animals or used therapeutically in humans. The cosmetic industry also uses the Ames test to assess the mutagenic potential of makeup and other hygienic products. The Food and Drug Administration requires companies to perform the Ames test before marketing most drugs or cosmetics. SEE ALSO CANCER; CARCINOGENS; MUTAGEN; MUTAGENESIS; MUTATION; NUCLEOTIDE.

David A. Scicchitano

Bibliography

Ames, Bruce N., and Lois S. Gold. "The Causes and Prevention of Cancer: The Role of Environment." *Biotherapy* 11 (1998): 205–220.

Mortelmans, Kristien, and Errol Zeiger. "The Ames *Salmonella* Microsome Mutagenicity Assay." *Mutation Research: Fundamental and Molecular Mechanisms of Mutagenesis* 455 (2000): 29–60.

Androgen Insensitivity Syndrome

Androgen insensitivity syndrome (AIS) is a disorder caused by mutation of the gene for the **androgen** receptor. This protein binds testosterone and regulates the expression of other genes that stimulate male sexual development. Testosterone is the principal male androgen. AIS is an X-linked recessive disorder that completely or partially prevents development of male sexual characteristics despite the presence of the Y chromosome. Thus, the **phenotype** of a person with AIS, typified by female or ambiguous sexual characteristics, is at odds with the **genotype**, which includes the presence of both the X and Y, or male-determining, chromosomes.

The extent of the syndrome ranges from complete androgen insensitivity and development of normal external (but not internal) female sexual anatomy, to partial insensitivity, with altered or ambiguous male or female genitals, to mild insensitivity, with normal male genitals, enlarged breasts, and possibly impotence. Treatments depend on the extent of the syndrome, and may include hormone therapy, surgery, and psychological counseling. Gene testing and genetic counseling are available for families with affected members.

Sexual Development

AIS can best be understood against the background of normal human sexual development, which begins in the womb. The **gonads** arise from the same embryonic tissue, which is differentiated into one or the other by the actions of several genes not involved in AIS. In males, the most important gene is *SRY*, located on the Y chromosome. When present, this causes testis development. The genes responsible for ovary development are not as well characterized.

Once differentiated, the ovaries produce estrogen, and testes produce testosterone. These two hormones provide crucial signals for the differentiation of other sex-related characteristics, including an important set of primitive ducts.

The tubes and cavities that will house the adult's eggs or sperm after they leave the gonads develop from two different sets of ducts. Early in development, every fetus has both sets of ducts. One set, called the Wolffian ducts, has the capacity to develop into the male vas deferens and accessory

androgen testosterone or other masculinizing hormone

phenotype observable characteristics of an organism

genotype set of genes present

gonads testes or ovaries

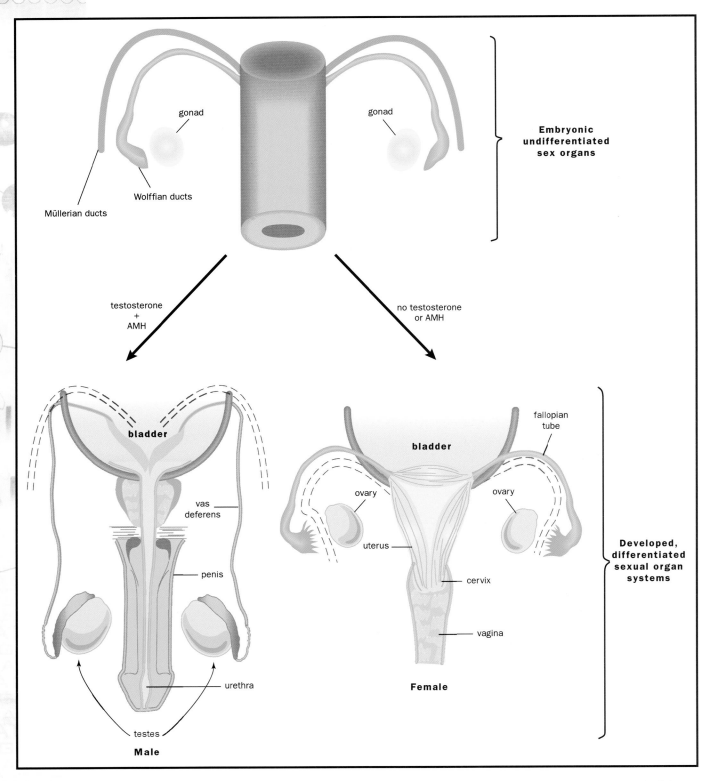

Normal sexual development is governed by presence or absense of testosterone and AMH. In males, degeneration of the Mullerian ducts is triggered by AMH. Development of the Wolffian ducts is triggered by testosterone. In individuals with androgen insensitivity syndrome, AMH acts but testosterone does not.

structures, which store, nourish, and ejaculate sperm. The other set, called the Müllerian ducts, has the capacity to become the female fallopian tubes, uterus, cervix, and upper vagina.

In males, testosterone from the testes stimulates the development of the Wolffian ducts. Testosterone also stimulates nearby tissue to swell and form the penis and scrotum. A second hormone made by the testes, called anti-Müllerian hormone (AMH), induces the Müllerian ducts to undergo **apoptosis**, causing them to degenerate. During puberty, testosterone stimulates the development of other male secondary sex characteristics, including facial hair and a deepening of the voice.

apoptosis programmed cell death

In females, the absence of testosterone and AMH causes the Müllerian ducts to develop and the Wolffian ducts to degenerate. The same tissue that forms the penis and scrotum in males forms the clitoris, labia, and lower vagina in females. At puberty, estrogen stimulates development of female secondary sex characteristics, including enlargement of the breasts and onset of menstruation.

Testosterone and Its Receptor

Testosterone is a hormone, a molecule released in one set of cells that regulates the action of other cells. Testosterone exerts its action on these target cells by first binding with a receptor, called the androgen receptor (AR). The receptor is a protein that resides within the target cell. The testosterone-receptor complex moves to the nucleus of the target cell, where it acts as a transcription factor. **Transcription factors** bind to DNA to control the rate of gene expression. In the case of testosterone, the genes affected are those that "masculinize" the fetus, triggering the transformation of the Wolffian ducts into the mature male sexual anatomy and causing other, more subtle changes, including in the brain. Thus, the interaction of testosterone with its receptor is the principal means by which the male genotype (presence of a Y chromosome) leads to the development of the male phenotype (presence of a vas deferens, penis, and accessory structures).

transcription factors proteins that increase the rate of gene transcription

The Consequences of Androgen Insensitivity

With an understanding of normal sexual development, consider the consequences of complete androgen insensitivity on the events of development in a person with the XY genotype. Since the Y chromosome is present, there will be testes. The testes will produce testosterone and AMH. AMH will cause degeneration of the Müllerian ducts, and so there will be no fallopian tubes, uterus, cervix, or upper vagina. But the defective receptor means that testosterone cannot exert its effects, so the Wolffian ducts also degenerate, and there will be no vas deferens. There will also be no penis or scrotum. Instead, the testes remain in the abdomen (where they originate), and the exterior tissue develops a short vagina that ends in a blind pocket. In milder forms of the syndrome, with only partial insensitivity to androgens, the genital structures may be ambiguous, with varying degrees of male versus female predominance.

The Androgen Receptor Gene and Protein

The AR gene is located on the long arm of the X chromosome, at a location that is designated as Xq11-q12. The gene is about 90,000 nucleotides

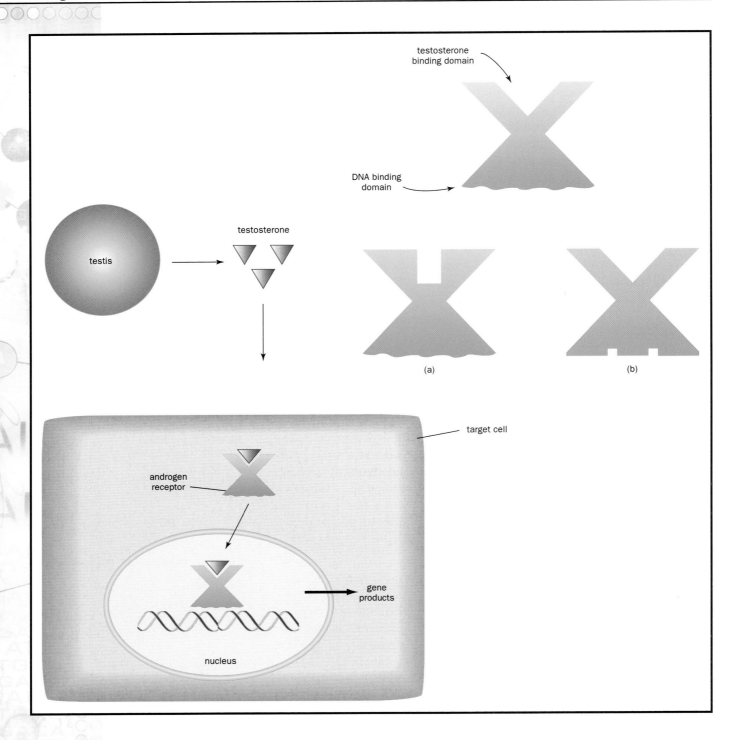

testosterone
binding domain

DNA binding
domain

testosterone

testis

(a)

(b)

target cell

androgen
receptor

gene
products

nucleus

Testosterone from the testis binds to the androgen receptor. This complex then triggers gene expression. Mutation in the receptor prevents either testosterone binding (a), DNA binding (b), or some other activity.

long, though fewer than 3,000 of these actually code for amino acids in the protein. The protein formed from the gene has different **domains** that perform different functions. One region binds testosterone, another regulates the movement of the complex to the nucleus, and a third binds the complex to the DNA. Other regions, some overlapping, control other functions.

Mutations to the coding portion for any one of these domains can prevent the receptor complex from functioning properly. All known mutations

domains regions

in the AR gene cause a loss of function and exhibit the recessive inheritance pattern. A male carries only one X chromosome, and receives only one copy of the AR gene. If this gene is mutated, the male will have androgen insensitivity syndrome.

Women with one mutated AR gene will not exhibit the syndrome but instead will be carriers, whose male children have a 50 percent risk of inheriting the mutant gene. Since affected individuals are sterile, they cannot pass it on to offspring. It is believed that about one-third of all cases are due to new mutations, which are not present in the mother's genes but which arise in the development of the early embryo. Genetic testing is available for women who desire children but who have a family history of androgen insensitivity. For women who are carriers, prenatal testing is available to determine if a fetus has inherited the mutant gene.

The Range of Androgen Insensitivity Syndromes

AIS occurs in a range of forms, from complete to mild. Most mutations to the testosterone-binding region, and some other types of mutations, cause complete androgen insensitivity syndrome (CAIS). In this form of the syndrome, the XY person is born phenotypically female, and from birth is raised as a girl. Gender identification (the internal sense of being male or female) is female. **Sexual orientation** is typically heterosexual, and so most CAIS individuals are attracted to males. At puberty, estrogen production by the adrenal glands causes breasts to develop. However, no pubic or armpit hair develops, since in males and females this is controlled by testosterone, and no menstruation occurs. It is at this point that the condition is usually diagnosed. Once discovered, the testes are usually surgically removed to prevent the possibility of testicular cancer, which is more common in people with CAIS. The woman is infertile, but may be able to enjoy sexual relations if the vagina is long enough to prevent pain during intercourse; or, if the vagina is not long enough, it can be surgically lengthened. CAIS is thought to occur in 2 to 5 births per 100,000.

sexual orientation attraction to one sex or the other

Those with partial androgen insensitivity syndrome (PAIS) have androgen receptors that are partially responsive to testosterone, and a range of outcomes may result. A person with PAIS may be born with external genital structures that are not typically male or typically female, a condition called intersexuality. The appearance of the genitals may range from predominantly male to predominantly female. There may be a very small penis or enlarged clitoris, abnormalities in the location of the urethra, and partial fusion of the labia. Breasts may develop in males at puberty. Internal gender identification may be with either sex. PAIS is thought to be as common as CAIS.

Ambiguous genitals are often surgically altered at birth. Problems arise when the surgically assigned sex conflicts with the internal gender identification, which develops early and becomes even more pronounced through late childhood and puberty. Increasing understanding of PAIS and sensitivity to the issues of gender identification have brought new awareness about the potential for these problems, but the practice of surgical sex assignment is still common. In 2001 the British Association of Paediatric Surgeons recommended that surgery "only be undertaken with considerable caution and following full multidisciplinary investigation and counseling of the parents."

Other therapies include hormone treatments and psychological counseling, including family counseling, and these are often part of the treatment, with or without surgery.

Kennedy Disease

Kennedy disease is a neurological condition that is also due to a mutation of the androgen receptor gene. Affected individuals are phenotypically normal males who are fertile, although after puberty they may develop enlarged breasts, consistent with very mild androgen insensitivity. The disorder causes progressive weakness over several decades, along with tremor, difficulty swallowing, and some sensory problems. The mutation that causes Kennedy disease is an expanded "triplet repeat" of CAG nucleotides, making this condition one of the family of triplet repeat diseases that includes Huntington's disease. SEE ALSO APOPTOSIS; HORMONAL REGULATION; INHERITANCE PATTERNS; SEX DETERMINATION; SEXUAL ORIENTATION; TRANSCRIPTION FACTORS; TRIPLET REPEAT DISEASE; X CHROMOSOME; Y CHROMOSOME.

Richard Robinson

Bibliography

Gilbert, Scott. *Developmental Biology*, 5th ed. Sunderland, MA: Sinauer Associates, 1997.

Internet Resources

Androgen Insensitivity Syndrome. <http://www.emedicine.com/PED/topic2222.htm>.

Statement of the British Association of Paediatric Surgeons Working Party on the Surgical Management of Children Born with Ambiguous Genitalia. <http://www.baps.org.uk/documents/Intersex%20statement.htm>.

Warne, Garry. *Complete Androgen Insensitivity Syndrome.* Victoria, Australia: Department of Endocrinology and Diabetes, Royal Children's Hospital. <http://www.rch.unimelb.edu.au/publications/CAIS.pdf>.

Aneuploidy   *See Chromosomal Aberrations*

Antibiotic Resistance

Antibiotic resistance is the ability of a bacterium or other microorganism to survive and reproduce in the presence of antibiotic doses that were previously thought effective against them. Examples of microbe resistance to antibiotics dot the countryside, plaguing humankind. For instance, in February 1994 dozens of students at La Quinta High School in southern California were exposed to the pathogenic (disease-causing) agent, *Mycobacterium tuberculosis*, and eleven were diagnosed with active tuberculosis. Many strains of this bacterium are multi-drug resistant (MDR). As for the sexually transmitted pathogen *Neisseria gonorrhea*, which causes gonorrhea, the antibiotics penicillin and tetracycline that were used against it in the 1980s can no longer be the first lines of defense, again because of antibiotic resistance. If only 2 percent of a *N. gonorrhea* population is antibiotic resistant, a community-wide infection of this persistent strain can occur.

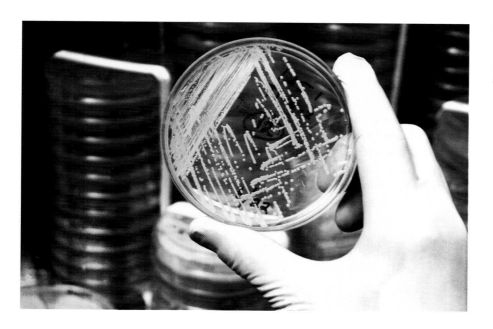

Antibiotic resistant *Staphylococcus aureus* bacteria grow on this medium filled with antibiotics.

Mechanisms of Resistance

Antibiotics, whether made in the laboratory or in nature by other microbes, are designed to hinder metabolic processes such as cell wall synthesis, protein synthesis, or transcription, among others. If humans are to prosper against microbial disease, it is necessary to understand how and why bacteria are able to mount their clever defenses. Aided with the knowledge of the genetics and mechanisms of resistance, scientists can discover new ways to combat the resistant bacteria.

The phenomenon of antibiotic resistance in some cases is innate to the microbe. For instance, penicillin directly interferes with the synthesis of bacterial cell walls. Therefore, it is useless against many other microbes such as fungi, viruses, wall-less bacteria like *Mycoplasma* (which causes "walking pneumonia"), and even many **Gram negative** bacteria whose outer membrane prevents penicillin from penetrating them. Other bacteria change their "genetic programs," which allows them to circumvent the antibiotic effect. These changes in the genetic programs can be in the form of chromosomal mutations, acquisition of R (resistance) **plasmids**, or through transposition of "**pathogenicity** islands."

An example of a chromosomal mutation is the increasing number of cases of penicillin-resistant *Neisseria gonorrhae*. This bacterium mutated the gene coding for a porin protein in its outer membrane, thereby halting the transport of penicillin into the cell. This is also termed "vertical evolution," meaning that the spread occurs through bacterial population growth. The most common method by which antibiotic resistance is acquired is through the **conjugation** transfer of R plasmids, also termed "horizontal evolution." In this method the bacteria need not multiply to spread their plasmid. Instead the plasmid is moved during conjugation. These plasmids often code for resistance to several antibiotics at once.

The third method is transfer due to transposable elements on either side of a "pathogenicity island," which is group of genes that appear on the DNA

Gram negative bacteria that do not take up Gram stain, due to membrane structure

plasmids small rings of DNA found in many bacteria

pathogenicity ability to cause disease

conjugation a type of DNA exchange between bacteria

and carry the codes for several factors which make the infection more successful. These transposable elements allow the genes to jump from bacteria to bacteria or simply from chromosome to plasmid within the organism.

The "road blocks" that bacteria have evolved which result in antibiotic resistance employ several mechanisms. One strategy is simply to destroy or limit the activity of the antibiotic. The beta-lactamases are enzymes which render the penicillin-like antibiotics dysfunctional by cleaving a vital part of the molecule. Some bacteria can deactivate antibiotics by adding chemical groups to them; for instance, by changing the electrical charge of the antibiotic through the addition of a phosphate group. Other bacteria accomplish a similar effect by bulking themselves up with the addition of an acetyl group.

Still other bacteria acquire resistance by simply not allowing the antibiotic to enter the cell. The bacterium mentioned above, *Neisseria gonorrhea*, has altered porin proteins, thereby stopping uptake of the antibiotic. Some bacteria acquire intricate pumping mechanisms to expel the drug when it gains entry to their cell.

macromolecule large molecule such as a protein, a carbohydrate, and a nucleic acid

Finally, bacteria may mutate the gene for the target **macromolecule** with which the antibiotic is supposed to bind. For example, tetracycline binds to and inhibits ribosomes, so a mutation in the ribosomal genes may cause conformational alterations in the ribosomal proteins that prevent tetracycline from binding but still allow the ribosome to function.

Resistance and Public Health

The effects of antibiotic resistance are reflected in the agriculture, food, medical, and pharmaceutical industries. Livestock are fed about half of the antibiotics manufactured in the United States as a preventative measure, rather than in the treatment of specific diseases. Such usage has resulted in hamburger meat that contains drug-resistant and difficult-to-treat *Salmonella Newport*, which has led to seventeen cases of gastroenteritis including one death. Some MDR-tuberculoid strains arise because patients are reluctant to follow the six-months or more of treatment needed to effectively cure tuberculosis. If the drug regimen is not followed, less sensitive bacteria have the chance to multiply and gradually emerge into resistant strains. In other cases the "shotgun" method of indiscriminately prescribing/taking several antibiotics runs the risk of creating "super MDR-germs." Moreover, millions of antibiotic prescriptions are written by physicians each year for viral infections, against which antibiotics are useless. The patient insists on a prescription, and many doctors willingly go along with the request.

pandemics diseases spread throughout entire populations

Because global travel is common, the potential of creating **pandemics** is looming. In many Third World countries, diluted antibiotics are sold on the black market. The dosage taken is often too low to be effective, or the patient takes the drug for a very short time. All these behaviors contribute to the development of resistant strains of infectious organisms. If humans are to gain the upper hand against MDR bacteria, it is the responsibility of these industries and the public to educate themselves and to engage in careful practices and therapy. SEE ALSO CONJUGATION; EUBACTERIA; MUTAGEN; PLASMID; TRANSPOSABLE GENETIC ELEMENTS.

Paul K. Small

Bibliography

Garrett, Laurie. *The Coming Plague.* New York: Farrar, Strauss, Giroux, 1996.

Ingraham, John, and Caroline Ingraham. *Introduction to Microbiology*, 2nd ed. Pacific Grove, CA: Brooks/Cole, 2000.

Nester, Eugene W., et al. *Microbiology: A Human Perspective*, 3rd ed. Boston: McGraw Hill, 2001.

Schaechter, Moselio, et al. *Mechanisms of Microbial Disease*, 3rd ed. Baltimore: Williams and Wilkins, 1998.

Antisense Nucleotides

Antisense nucleotides are strings of RNA or DNA that are **complementary** to "sense" strands of nucleotides. They bind to and inactivate these sense strands. They have been used in research, and may become useful for therapy of certain diseases.

complementary matching opposite, like hand and glove

Antisense RNA

Messenger RNA (mRNA) is a single-stranded molecule used for protein production at the **ribosome**. Because its sequence is used for **translation**, mRNA is called a "sense" strand or sense sequence. A complementary sequence to that mRNA is an "antisense" sequence. For instance, if the mRNA sequence was AUGAAACCCGUG, the antisense strand would be UACUUUGGGCAC. Complementary sequences will pair up in RNA just as they do in DNA. When this happens to an mRNA, however, it can no longer be translated at the ribosome, no protein synthesis occurs, and the "duplex" RNA is degraded.

ribosome protein-RNA complex at which protein synthesis occurs

translation synthesis of protein using mRNA code

This phenomenon has been used experimentally and commercially to block the synthesis of specific proteins in transgenic organisms (ones to which a foreign gene has been added). The strategy is to add a synthetic gene that, when transcribed, will make the antisense RNA sequence for the target protein's mRNA.

This technique was first used commercially in 1988 for the Flavr-Savr tomato. The gene chosen for inactivation was polygalacturonase (*PG*), whose enzyme unlinks pectins in the plant cell wall, thereby softening it. The intent was to increase the time the fruit could be left to ripen without softening, thus increasing flavor of commercial tomatoes. The Calgene company created a transgenic tomato plant expressing the antisense RNA for PG mRNA, and reduced PG production by up to 90 percent. Although the tomato was not a commercial success, it demonstrated the potential for this strategy.

Antisense RNA is currently being investigated as a human therapy for certain forms of cancer. The goal is to use gene therapy techniques to insert an antisense gene into tumor cells. Many cancers are due to overexpression of the genes that promote cell proliferation, called tumor suppressor genes. Antisense RNA might be able to inhibit this overexpression. Another target is the *BCL-2* gene, whose protein prevents apoptosis, or programmed cell death. In certain cancers, the *BCL-2* gene is overactive, preventing death of cells and leading to their proliferation. Antisense therapy against *BCL-2* is currently being tested under the trade name Genazyme.

Antisense nucleotides bind to "sense" mRNA, preventing translation of the RNA message at the ribosome, thus preventing protein synthesis. Adapted from <http://www.cem.msu.edu/~cem181h/projects/97/antisense/dia1.gif>.

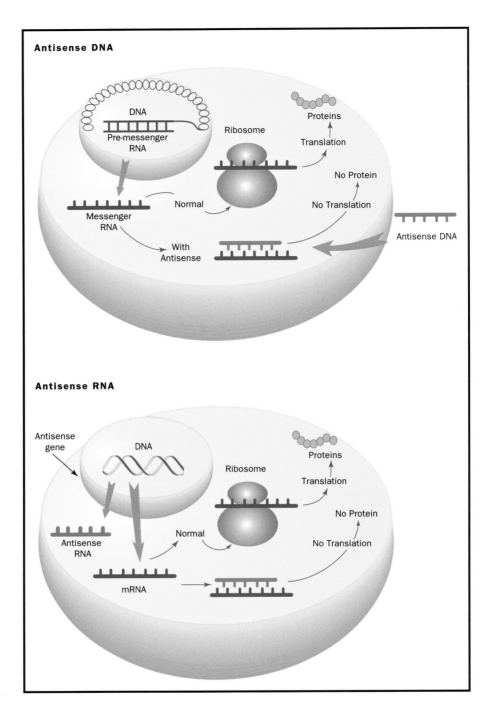

Antisense DNA

Antisense DNA strands can also be made (note that in the double helix, the side of the DNA that is transcribed is itself antisense). Short antisense strands of DNA can be introduced into cells, which then bind with target mRNA. Antisense DNA is currently an approved therapy for cytomegalovirus infections of the eye, under the trade name Vitravene. Vitravene targets two different viral proteins. Antisense DNA is also being explored for therapy of HIV, some cancers, and other diseases.

One advantage of using antisense therapy in treating infectious diseases such as virus infections is that it can be tailored to the particular strain in circulation, and then modified as the virus mutates. One difficulty in applying this therapy is successfully delivering the antisense DNA or RNA to all target tissues (for instance, making sure the antisense strands reach infected blood cells for HIV). Another problem is maintaining prolonged suppression of target protein expression, since the antisense molecule will eventually be degraded by the cell's **nuclease** enzymes. One strategy to prevent degradation is to chemically modify the DNA to interfere with nuclease action.

nuclease enzyme that cuts DNA or RNA

RNA Interference

Investigation of the mechanism of action of antisense RNA led to the surprising discovery that naturally occurring double-stranded RNA molecules (dsRNA) suppress gene expression as well as or better than antisense sequences. This suppression by dsRNA of expression of the related gene is called RNA interference. dsRNA molecules are cut into short segments by nucleases; the antisense strand of such a segment then peels off and binds with its complementary mRNA. This new, double-stranded RNA is then subject to further nuclease attack. RNA interference is believed to be an ancient means of protecting against double-stranded RNA viruses. Further understanding of RNA interference may lead to improvements in or replacement of antisense therapies. SEE ALSO GENE THERAPY; NUCLEASES; NUCLEOTIDE; RNA INTERFERENCE; TRANSGENIC PLANTS.

Richard Robinson

Bibliography

Smith C. J. S., et al. "Antisense RNA Inhibition of Polygalacturonase Gene Expression in Transgenic Tomatoes." *Nature* 334 (1988): 724–726.

Tamm I., B. Dorken, and G. Hartmann. "Antisense Therapy in Oncology: New Hope for an Old Idea?" *Lancet* 358, no. 9280 (2001): 489–497.

Internet Resource

"Antisense DNA." Michigan State University. <http://www.cem.msu.edu/~cem181h/projects/97/antisense/dia1.gif>.

Apoptosis

Death is an inevitable fact of life for organisms. Increasingly, biologists have come to realize that death is also, in many cases, an important and predestined fate of individual cells of organisms. Apoptosis is a process by which cells in a multicellular organism commit suicide. While cells can die as a result of **necrosis**, apoptosis is a form of death that the cell itself initiates, regulates, and executes using an elaborate arsenal of cellular and molecular machinery. For this reason, the term apoptosis is often used interchangeably with the term "programmed cell death," or PCD (although technically, apoptosis is but one particular form of programmed cell death). There is some disagreement on the origins of the word. The word apoptosis has ancient Greek origins, referring to the falling of leaves, or possibly "dropping of scabs" or "falling off of bones." There is even less agreement on its proper pronunciation, and even specialists in the field seem to use every possible way to say the word. "A-pop-TOE-sis" and "AP-oh-TOE-sis" are both common.

necrosis cell death from injury or disease

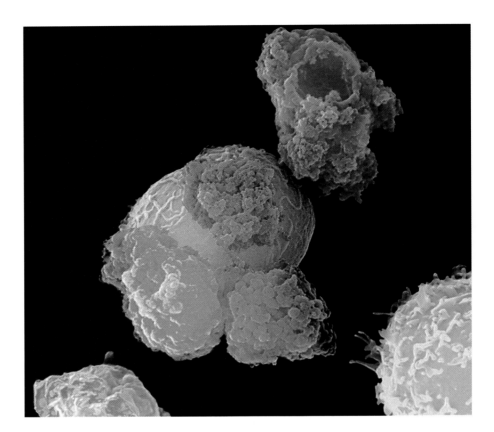

Magnification (6000×) of apoptosis, or cell death. Programmed cell death in the human body allows, among other things, for young children's brains to develop and for a female's monthly menstruation to occur.

APOPTOSIS GENES IN *C. ELEGANS*

Much of our understanding of what causes apoptosis comes from genetic studies in *Caenorhabditis elegans*. Several cell death proteins (CED) proteins were identified in *C. elegans* by studying apoptosis-defective mutants. The main executioner is CED-3, a caspase, which becomes activated by CED-4, another caspase. The central guardian protecting cells against apoptosis is CED-9, which inhibits the actions of CED-4 and CED-3. CED-9 has a mammalian homolog, called BCL-2, which serves a similar role in mammals.

Why Cells Commit Suicide

Why do cells commit apoptosis? There seem to be two major reasons. First, apoptosis is one means by which a developing organism shapes its tissues and organs. For instance, a human fetus has webbed hands and feet early on its development. Later, apoptosis removes skin cells, revealing individual fingers and toes. A fetus's eyelids form an opening by the process of apoptosis. During metamorphosis, tadpoles lose their tails through apoptosis. In young children, apoptosis is involved in the processes that literally shape the connections between brain cells, and in mature females, apoptosis of cells in the uterus causes the uterine lining to slough off at each menstrual cycle.

Cells may also commit suicide in times of distress, for the good of the organism as a whole. For example, in the case of a viral infection, certain cells of the immune system, called cytotoxic T lymphocytes, bind to infected cells and trigger them to undergo apoptosis. Also, cells that have suffered damage to their DNA, which can make them prone to becoming cancerous, are induced to commit apoptosis.

The Regulatory Mechanism

The cellular mechanisms that regulate and cause apoptosis were first elucidated by genetic studies of the roundworm, *Caenorhabditis elegans*. Normally, in the development of a *C. elegans* worm, one out of every eight body cells produced is eliminated by programmed cell death. By studying mutants in which either too many or too few cells died, worm geneticists identified many of the proteins that control apoptosis. Subsequently, the critical medical relevance of apoptosis became clear when biologists discovered that mammals contain many of the same genes that control apoptosis in worms. More strikingly, they found that many of these genes were mutated in tumors from cancer patients. Other genes often found to be mutated in cancers are those which regulate the cell cycle, which is the complex set of processes controlling how and when cells divide. These two findings led cancer researchers to recognize that cancer, a disease of uncontrolled cell proliferation, can result either from too much cell division or not enough apoptosis. Because of this important finding, apoptosis has become the subject of intense medical research, and molecules that regulate apoptosis are being studied as potential targets for anti-cancer drug therapies.

A cell can be triggered to undergo apoptosis either by external signaling molecules, such as so-called "death activator" proteins, or through molecules that reside within the cell and monitor events that might commit the cell to suicide, such as damage to DNA. There are several biochemical pathways that lead to apoptosis. One of the major pathways involves inducing **mitochondria** to leak one of their proteins, cytochrome c, into the cystosol. This in turn activates a set of related proteases (enzymes that degrade proteins) called caspases. Ultimately, the caspases degrade proteins in the cell and activate enzymes that degrade other cell constituents, such as the DNA. Cells undergoing apoptosis exhibit characteristic morphological and biochemical traits, which can be recognized by microscopic examination or biochemical assays. Apoptosis can occur in as little as twenty minutes, after which the cell "corpse" typically becomes engulfed and completely degraded by neighboring **phagocytic** cells that are present in the tissue and attracted to the apoptotic cell. SEE ALSO CANCER; CELL CYCLE; ROUNDWORM: *CAENORHABDITIS ELEGANS*; SIGNAL TRANSDUCTION.

Paul J. Muhlrad

Caspase inhibitors are being investigated as a possible means to slow the progress of Huntington's disease, a degenerative brain disease.

mitochondria energy-producing cell organelle

phagocytic cell-eating

Bibliography

Lodish, Harvey, et al. *Molecular Cell Biology*, 4th ed. New York: W. H. Freeman, 2000.

Nature 407, no. 12 (Oct., 2000). (Issue devoted to review articles on apoptosis).

Internet Resource

The WWW Virtual Library of Cell Biology. "Apoptosis." <http://vlib.org/Science/Cell_Biology/apoptosis.shtml>.

Arabidopsis thaliana

Arabidopsis thaliana, or thale cress, is a small flowering plant in the mustard family. *Arabidopsis* has no inherent agricultural value and is even considered a weed, but it is one of the favored model organisms of plant geneticists and molecular biologists, and it is the most thoroughly studied plant species at the molecular level. Model organisms have traits that make them attractive and convenient for biologists, who anticipate being able to extend their findings to other, less easily studied species. *Arabidopsis* is small and easy to grow, allowing researchers to cultivate it with minimal investments in effort and laboratory space. It has a short generation time, taking about six weeks for a seed to grow into a mature plant that produces more seeds. This rapid maturation

The small genome and short life cycle of the *Arabidopsis thaliana* help to make it a useful tool for plant geneticists. Study of this plant's genome could help to shape development of the commercial mustard plant crops, and could provide useful information for study of other, more complex, flowering plants.

pollen male plant sexual organ

enables biologists to conduct genetic cross experiments in a relatively short period of time. A single mature plant can produce over 5,000 seeds, another property that makes *Arabidopsis* convenient for use in genetic analysis.

A Small and Simple Genome

Beyond these basic traits, other attributes of *Arabidopsis* make it particularly well-suited for analysis by modern molecular genetic methods. Its genome (the amount of DNA in each set of chromosomes) is only about 125 million base pairs. This is small compared to many other plants, and makes searching for particular genes easier in *Arabidopsis* than in plants with larger genomes. For comparison, the genome sizes for rice (*Oryza sativa*), wheat (*Triticum aestivum*), and corn (*Zea mays*) are 420 million, 16 billion, and 2.5 billion base pairs, respectively. Furthermore, the *Arabidopsis* genome is contained on just five pairs of chromosomes, making it easier for geneticists to locate specific genes.

Geneticists can carry out crosses (interbreeding two different plant strains) with *Arabidopsis* by introducing the **pollen** from one plant to the

stigma on another. This mode of reproduction, called outcrossing, is useful for combining mutations from different plants. Alternatively, *Arabidopsis* can reproduce by a process called selfing, in which an individual plant uses its own pollen to fertilize its **ovules**.

Selfing, which is not possible in many plants, is very useful for geneticists who wish to study mutations. Most mutations are recessive, which means that they physically manifest themselves (display a **phenotype**) only when they are present on the chromosomes contributed by both the ovule and the fertilizing pollen. In selfing, **heterozygous** mutations (which are present on only one of the two sets of chromosomes) will become **homozygous** (present on both) in one quarter of the progeny produced in this manner.

Arabidopsis and Transformation

Another property that endears *Arabidopsis* to plant molecular biologists is that it is easily transformed. Transformation is a method for introducing foreign DNA into an organism. This technique is invaluable for studying how genes function and interact with other genes. Biologists usually transform plants by infecting them with genetically engineered varieties of a bacterium, *Agrobacterium tumefaciens*. In nature, when *Agrobacterium* infects plants, it inserts certain genes directly into the plant cells, causing a disease called crown gall. The genetically engineered *Agrobacterium* strains have had their disease-causing genes removed. They can still infect a plant and insert their DNA, but do not cause a disease. To transform plants, the molecular biologist inserts the foreign gene to be studied into *Agrobacterium*, which will then transfer the gene to a plant that it infects. This transformation technique does not work well on many other plant species, limiting the utility of those plants for molecular genetic analysis.

Arabidopsis researchers also use a variation on the *Agrobacterium*-mediated transformation technique to introduce mutations in the plant. Studying the effect of a mutation in a particular gene often yields critical information about the normal function of that gene. Because *Agrobacterium* inserts its transforming DNA randomly in the genome, in many cases the DNA gets inserted directly within a gene sequence. This usually destroys the function of the disrupted gene, resulting in a "knockout mutant." Furthermore, the piece of transformed DNA (T-DNA) that is inserted in the disrupted plant gene can serve as a flag for tracking down the gene by molecular biology methods. Large-scale projects using this T-DNA insertion technique are underway to mutate, identify, and characterize every gene in the *Arabidopsis* genome.

The First Completely Sequenced Plant Genome

At the end of 2000, an international team of researchers announced that *Arabidopsis* was the first plant to have its complete genome sequence—the exact order of essentially all 125 million DNA base pairs—determined. The project revealed that *Arabidopsis* contains over 25,500 genes. By identifying and studying these genes, biologists are learning lessons about plant biology that could provide important advances in agriculture, such as improved crop resistance to pathogens, salt, light stress, and drought, and to the production of more healthful edible oils, pharmaceuticals, and biodegradable plastics.

stigma female plant sexual organ

ovules eggs

phenotype observable characteristics of an organism

heterozygous characterized by possession of two different forms (alleles) of a particular gene

homozygous containing two identical copies of a particular gene

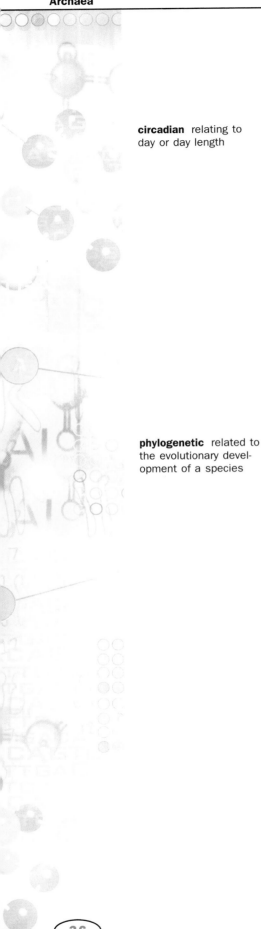

Arabidopsis research has also produced important discoveries in fundamental plant science, such as the identification of a plant hormone receptor, a clearer understanding of how plants sense and respond to light, and more about the processes that induce plants to form flowers. *Arabidopsis* research may even have direct relevance to human biology. For example, a photoreceptor protein that regulates **circadian** rhythms in *Arabidopsis* was found to share sequence similarity to a retinal photoreceptor, which may perform a similar role in mammals. SEE ALSO INHERITANCE PATTERNS; MODEL ORGANISMS; TRANSFORMATION; TRANSGENIC PLANTS.

Paul J. Muhlrad

circadian relating to day or day length

Bibliography

Meinke, David W., et al. "*Arabidopsis thaliana*: A Model Plant for Genome Analysis." *Science* 283, no. 5389 (1998): 662–681.

Internet Resources

The Arabidopsis Information Resource. <http://www.arabidopsis.org/>.

Nature, Vol. 408, December 2000. (Issue devoted to *Arabidopsis thaliana*; <http://www.nature.com/genomics/papers/a_thaliana.html>).

Archaea

Members of the Archaea comprise one of the three principal domains of living organisms in the universal **phylogenetic** tree of life. The other two principal domains are the Bacteria and the Eukarya. The phylogenetic tree is a theoretical representation of all living things, constructed on the basis of comparative ribosomal RNA sequencing and reflecting evolutionary relationships rather than structural similarities.

phylogenetic related to the evolutionary development of a species

Characteristics of Archaea

Many scientists hypothesize that the Archaea are the closest modern relatives of Earth's first living cells. Called "universal ancestors," these are the cells from which all other life is believed to have evolved. This hypothesis is based on two types of evidence. Genetic analyses indicate that the Archaea domain branches off of the phylogenetic tree at a point that is closest to the tree's root. Furthermore, it has been observed that many of the Archaea prefer to live in extremes of temperature, salt concentration, and pH—environmental conditions thought to be similar to those found on Earth over 3.5 billion years ago, when life first originated.

The Archaea share certain characteristics with Bacteria, others with Eukarya, and have some characteristics that are unique. For example, cells of the Archaea are structurally more similar to Bacteria, live predominantly as single cells, and have cell walls, although the walls do not contain the complex material called peptidoglycan that is a signature molecule of the Bacteria. While some Eukarya have cell walls, it is not a universal characteristic of that domain, and the Eukarya walls are composed of chitin or cellulose, neither of which occurs in cell walls of Archaea or Bacteria. Like the Bacteria, the Archaea lack a membrane-enclosed nucleus and their DNA exists in a circular form. On the other hand, their DNA is associated with histones, a characteristic of Eukarya, and their cell machinery (such as protein-

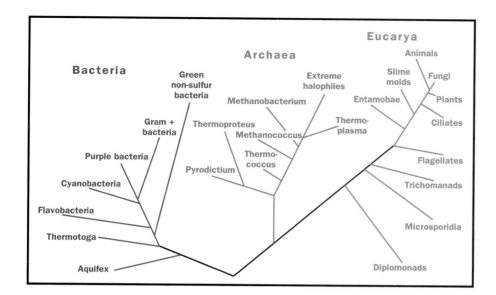

Evolutionary relatedness of the three domains of life. Distance along the tree is proportional to time since divergence from a common ancestor.

synthesizing enzymes and RNA polymerases) more closely resembles that found in the Eukarya. The lipids that comprise their membranes are unique, resembling neither the Bacteria nor the Eukarya.

Certain members of the Archaea are able to produce methane gas, another unique characteristic. Methane is one of the most important greenhouse gases. An Italian scientist named Alessandro Volta first discovered it as a type of "combustible air" over two hundred years ago. He trapped gas from marsh sediments and showed that it was flammable long before we knew that it was produced by members of the Archaea that lived in salt marsh sediment. Other important habitats for Archaea with this unique ability include the digestive tracts of animals and sewage sludge digesters.

Thriving in Environmental Extremes

The ability of many members of the Archaea to thrive in environmental conditions that we would find extreme is perhaps one of their most fascinating characteristics. There are genus like *Halobacterium*, which inhabit extremely salty environments, such as the Great Salt Lake in Utah and the Dead Sea in Israel. The salt concentration in these lakes is at least ten times that of seawater. Still other lakes, like Lake Magadi in Kenya, are not only extremely salty, but are also extremely alkaline, with pH values as high as 10 or 12. Archaea can be found even here, and their names reflect their habitat: *Natronobacterium*, *Natronosomonas*, and *Natronococcus* ("natro" means "salt"). The reddish-purple color sometimes seen in seawater-evaporating ponds, where solar salt is prepared, is the result of the growth of red-pigmented Archaea.

Extremes of temperature offer no challenge to certain members of the Archaea. A number of species, in fact, require temperatures over 80 °C in order to grow. Some live quite happily in the superheated outflow of geothermal power plants. Others thrive in the conditions of extreme acidity and temperature found in sulfur-rich, acidic hot springs like those in Yellowstone National Park, in the United States.

The high-salt density of the Dead Sea makes it difficult for humans to swim in its waters. Human bodies are much more buoyant in the Dead Sea waters than in fresh water. It is possible to lay back in the water—floating as if on an air mattress.

Archaea bacteria are also known as "extremophiles," thanks to their ability to survive in extreme environments such as very hot and very cold climates.

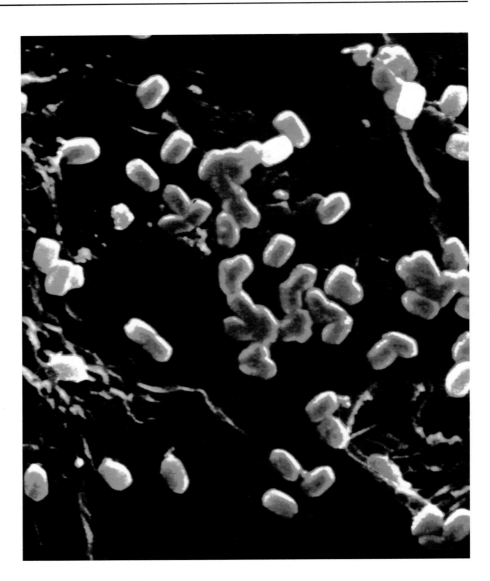

Archaea also populate the areas surrounding deep-sea vents, underwater volcanoes that form when the earth's crust opens along the ocean floor's spreading centers. The deep-sea vents have the hottest temperatures at which any living organism has been found. As of 2001, the current record for heat tolerance belonged to *Pyrolobus fumarii*, which can grow in water at a maximum temperature of 113 °C, well above boiling. At the opposite extreme, Archaea are among the few organisms found in the frigid waters of the Antarctic.

Value to Industry and Research

As a consequence of their ability to thrive in extreme conditions, the Archaea have become increasingly valuable. For example, the DNA polymerase of *Thermus aquaticus*, an Archaea found in the Yellowstone hot springs, is a heat tolerant enzyme that is crucially important in modern molecular biology laboratories, because of its use in the polymerase chain reaction. Archaea have also become important for commercial purposes. Their enzymes, sometimes called extremozymes, have made their way into laundry detergent, for example, where they digest proteins and lipids in hot water or cold, and in extremely alkaline conditions, thus helping to remove life's little messes. SEE ALSO CELL, EUKARYOTIC; EUBACTERIA; POLYMERASE CHAIN REACTION; RIBOSOME.

Cynthia A. Needham

Bibliography

Madigan, Michael T., John M. Martinko, and Jack Parker. *Brock Biology of Microorganisms*, 9th ed., Upper Saddle River, NJ: Prentice Hall, 2000.

Campbell, Neil A. *Biology*, 4th ed. Menlo Park, CA: Benjamin Cummings, 1996.

Madigan, Michael T., and Barry Marrs. "Extremophiles." *Scientific American* (April, 1997): 82–87.

Attention Deficit Hyperactivity Disorder

Attention deficit/hyperactivity disorder (ADHD) is a condition characterized by inattention and/or impulsivity and hyperactivity that begins in children prior to the age of seven. Their inattention leads to daydreaming, distractibility, and difficulties sustaining effort on a single task for a prolonged period of time. Their impulsivity disrupts classrooms and creates problems with peers, as they blurt out answers, interrupt others, or shift from schoolwork to inappropriate activities. Their hyperactivity is frustrating to those around them and poorly tolerated at school. Children with ADHD show academic underachievement and conduct problems. As they grow older, they are at risk for low self-esteem, poor peer relationships, conflict with parents, delinquency, smoking, and substance abuse.

Course, Prevalence, and Treatment

Although the longitudinal course of this condition and its prevalence in adulthood have been sources of controversy, a growing literature has documented the persistence of ADHD into adulthood, with about two-thirds of ADHD children continuing to experience impairing symptoms of the disorder though adulthood. Over time, symptoms of hyperactivity and impulsivity are more likely to diminish compared with symptoms of inattention.

Prevalence studies from North America, Europe, and Asia show that ADHD affects about 5 percent of the population. The impact of the disorder on society, in terms of financial cost, stress to families, and disruption in schools and workplaces, is enormous. Although current treatments for the disorder are not 100 percent effective, clinical trials have shown that stimulant medications, such as methylphenidate and amphetamine, relieve symptoms and lessen adverse outcomes, while showing few adverse side effects. Because these medicines increase the availability of the **neurotransmitter** dopamine in the brain, dysregulation of dopamine systems has been a primary candidate for the **pathophysiology** of ADHD. But drugs like desipramine and alomoxeline, which have their effects on other brain systems, also exert strong anti-ADHD effects. This suggests that dysregulation of dopamine systems cannot completely explain the pathophysiology of ADHD.

neurotransmitter molecule released by one neuron to stimulate or inhibit a neuron or other cell

pathophysiology disease process

neuroimaging techniques for making images of the brain

frontal lobe one part of the forward section of the brain, responsible for planning, abstraction, and aspects of personality

etiology causation of disease, or the study of causation

bipolar disorders psychiatric diseases characterized by alternating mania and depression

Neuropsychological and **neuroimaging** studies provide converging evidence for the hypothesis that brain dysfunction causes the symptoms of ADHD. Neuropsychological tests show many ADHD patients to have deficits in the executive functions needed for organizing, planning, sequencing, and inhibiting behaviors. These performance deficits are similar to, albeit milder than, the deficits seen among patients with **frontal lobe** disorders. Several structural and functional neuroimaging studies implicate networks of regions throughout the brain, not just in the frontal lobes.

The Genetic Epidemiology of ADHD

Family, twin, and adoption studies provide strong support for the idea that genes influence the **etiology** of ADHD. Family studies find the parents and siblings of ADHD children to have a five-fold increase in the risk for ADHD. Children of ADHD adults have a ten-fold increase in risk, which has led to the idea that persistent cases of ADHD may have a stronger genetic component. Consistent with a genetic theory of ADHD, second-degree relatives (such as cousins) are at increased risk for the disorder but their risk is lower than that seen in first-degree relatives.

Family studies have provided evidence for the genetic heterogeneity of ADHD. Studies that systematically assess other psychiatric disorders suggest that ADHD and major depression often occur together in families; that ADHD children with conduct and/or **bipolar disorders** might be a distinct familial subtype of ADHD; and that ADHD is familially independent from anxiety disorders and learning disabilities. It may therefore be appropriate to divide ADH children into those with and those without conduct and bipolar disorders, thus forming more familially homogeneous subgroups. In contrast, major depression may be a nonspecific manifestation of different ADHD subforms.

Several twin studies have provided evidence of genetic influence on hyperactive and inattentive symptoms. An early study found the heritability of hyperactivity to be 64 percent. A study of ADHD in twins who also had reading disabilities reported the heritability of attention-related behaviors to be 98 percent. All twin studies considered together suggest that the heritability of ADHD is about 70 percent, which makes it one of the most heritable of psychiatric disorders.

Adoption studies also implicate genes in the etiology of ADHD. Two early studies found that the adoptive relatives of hyperactive children were less likely to be hyperactive or have associated conditions than the biological relatives. Biological relatives of hyperactive children also performed more poorly on standardized measures of attention than did adoptive relatives. A study using the contemporary definition of ADHD found that biological, not adoptive, relationships account for the transmission of ADHD.

The Molecular Genetics of ADHD

Molecular genetic studies have already implicated several genes as mediating the susceptibility to ADHD. Researchers have examined candidate genes in dopamine pathways because animal models, theoretical considerations, and the effectiveness of stimulant treatment implicate dopaminergic dysfunction in the pathophysiology of this disorder. Dopamine is a neural trans-

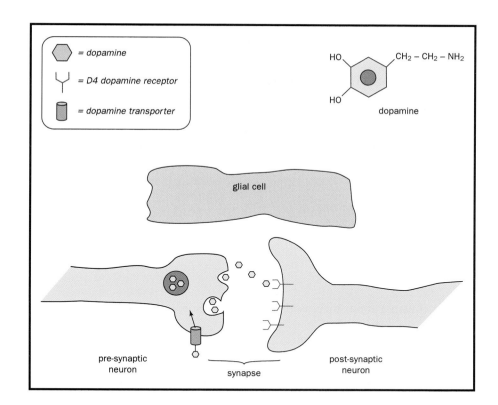

= dopamine

= D4 dopamine receptor

= dopamine transporter

dopamine

glial cell

pre-synaptic neuron

synapse

post-synaptic neuron

Dopamine is a neurotransmitter in the brain and is used in neural pathways involved in both movement control and pleasure/reward systems. Dopamine released by one neuron crosses the synapse to stimulate the adjacent neuron. It is broken down by glial cells.

mitter in the brain used in both movement control and pleasure/reward systems. In its simplest form, the dopamine hypothesis holds that excess clearance of dopamine between neurons may contribute to ADHD.

Many studies have focused on the D4 dopamine receptor gene (DRD4) which encodes a protein receptor that mediates the post-synaptic action of dopamine. A meta-analysis of these studies showed a small but statistically significant association, which could not be accounted for by any single study or by publication biases. Although the nature of the mutations in DRD4 have not been conclusively described, a version of the gene known as the 7-repeat allele has generated much interest because this allele causes a blunted response to dopamine and has been implicated in novelty seeking, a personality trait of many ADHD patients.

Several authors have reported an association between ADHD and a particular allele of the dopamine transporter (DAT) gene. This finding has been replicated by some, but not all studies. The link between the DAT gene and ADHD is further supported by a study that relates this gene to poor methylphenidate response in humans, a "knockout" mouse study showing that its elimination leads to hyperactivity in mice, and two molecular neuroimaging studies that found elevated DAT density in the **striatum** of ADHD adults.

striatum part of the midbrain

Molecular genetics studies of ADHD have also targeted other genes that are related to the dopamine system. Four studies have examined the Catechol-O-Methyltransferase (COMT) gene, whose protein product breaks down dopamine and norepinephrine. Although one study found ADHD was associated with the high-activity form of COMT, three others could not replicate the finding. Other candidate genes that show promising results for ADHD are the D5 dopamine receptor gene and the serotonin

1B receptor. This latter finding is intriguing because, although serotoner-gic medicines do not help ADHD symptoms, these systems have been implicated in animal models of the disorder. SEE ALSO BEHAVIOR; DISEASE, GENETICS OF; PSYCHIATRIC DISORDERS; TWINS.

Stephen V. Faraone

Bibliography

Faraone, S. V., D. Tsuang, and M. T. Tsuang. *Genetics of Mental Disorders: A Guide for Students, Clinicians, and Researchers.* New York: Guilford, 1999.

Faraone, S. V., and A. Doyle. "The Nature and Heritability of Attention Deficit Hyperactivity Disorder." *Child and Adolescent Psychiatric Clinics of North America* 10 (2001): 299–316.

Faraone, S. V., and J. Biederman. "Neurobiology of Attention Deficit Hyperactivity Disorder." *Biological Psychiatry* 44 (1998): 951–958.

Attorney

Attorneys involved with genetics include criminal prosecutors (district attorneys), public defenders, environmental lawyers, family lawyers, and patent attorneys. Genetics is relevant in the areas of identification of suspects and victims, identification of illegal goods (for example, items that involve the killing of endangered animals), environmental monitoring for harmful microorganisms, parentage determinations, and the patenting of genetic materials.

While all of these different types of lawyers may need to be somewhat familiar with the fundamentals of genetics, attorneys who work on gene patents must be very familiar with both genetics and biochemistry, as well as with patent law. The majority of these patent attorneys specialize in biotechnology. Most biotechnology patent attorneys have advanced degrees, with many having Ph.D.s in genetics, microbiology, molecular biology, biochemistry, or related fields. In addition to having a strong science background, patent attorneys must be licensed to practice law in at least one state, and must pass a registration examination administered by the U.S. Patent and Trademark Office (USPTO).

Patent agents, as well as patent attorneys, can represent inventors before the USPTO. Patent agents have strong science backgrounds and must pass the USPTO registration examination, but are not licensed to practice law in any state. In addition to having very strong science backgrounds and the ability to work closely with inventors, patent attorneys and agents must enjoy reading complex scientific literature and be proficient at scientific writing. A large portion of the job involves writing scientific documents in the form of patent applications. Thus, people who enjoy reading and writing about scientific topics are well suited to the profession.

Patent attorneys and agents typically work in law firms, private companies, the U.S. Patent and Trademark Office (patent examiners), or in the technology transfer offices of universities and public institutions such as hospitals and research facilities, although some work out of their homes as solo practitioners. Patent attorneys and agents often know about ground breaking developments long before the public or others in their fields. This makes the career very interesting, exciting, and enjoyable. However, because of client confidentiality, patent attorneys and agents are required to keep these developments secret until the information is made public by the client or a patent is issued.

While some patent attorneys only draft patent applications and work with patent examiners, others work as litigators, patent law experts, law school professors, or trademark lawyers. For complicated cases, patent attorneys with an education in genetics are very helpful in explaining the technology to the judge or jury. Although most patent attorneys work on patent cases in courts, their expertise may also be called upon in criminal cases, when assistance is needed to analyze and explain complex sciences such as genetics and molecular biology.

Salaries differ widely among patent attorneys and agents who work in law firms, companies, and at universities. As of 2000, new Ph.D.-level patent attorneys could expect to earn at least $100,000 per year, while new Ph.D.-level patent agents could expect to earn at least $75,000 per year. However, some patent attorneys could earn well over $500,000 per year. SEE ALSO LEGAL ISSUES; PATENTING GENES.

Kamrin T. MacKnight

Bibliography

U.S. Patent and Trademark Office, *Manual of Patent Examining Procedure.* Washington, DC: Superintendent of Documents, U.S. Government Printing Office, 2000.

Automated Sequencer

The process of determining the order of nucleotides (A, C, G, and T) along a DNA strand is called DNA sequencing. Knowing the **nucleotide** sequence of a gene or region of DNA is important in studying relatedness between species and between individuals and for a better understanding of how genes function. Several techniques have been developed for "reading" the sequence of any particular DNA segment. One of these techniques was developed by Fred Sanger in 1977 and is called the chain termination method (or Sanger method). The essence of the technique is the creation of a set of DNA fragments that match the chain to be sequenced. Each fragment is one nucleotide longer than the last. By determining the identity of the final nucleotide in each fragment, the sequence of the whole chain can be determined.

> **nucleotide** the building block of RNA or DNA

The chain termination method makes use of special forms of the four nucleotides that, when incorporated at the end of a growing chain during DNA synthesis, stop (terminate) further chain growth. In four separate reactions, each containing a different terminator base (called a dideoxynucleotide), a collection of single-stranded fragments is made. These fragments all differ in length and all end in the dideoxynucleotide added to the particular reaction. **Gel electrophoresis** is then used to separate the fragments according to their length. By knowing which terminator base is associated with which fragment on the gel, the base sequence can be constructed.

> **gel electrophoresis** technique for separation of molecules based on size and charge

The Need for Automated Sequencing

When chain termination sequencing is performed manually, each of four reaction tubes contains a different type of terminator base as well as a

radioactive nucleotide for labeling the DNA fragments as they are made. Each of the four reactions is electrophoresed in a separate lane of a gel, and X-ray film is used to detect the fragments. Using this technique, a dedicated and skilled technician can determine the sequence of as many as 5,000 bases in a week. Demand for the ability to read more sequence in a shorter amount of time, however, led to the development of instruments that could, with the aid of computers, automate the DNA sequencing process.

primer short nucleotide sequence that helps begin DNA replication

The first step toward this goal was achieved in 1985, when Leroy Hood at the California Institute of Technology attached fluorescent dyes to the **primer** used in the sequencing reactions; each different color dye (blue, green, yellow, and red) was matched with a different terminator base. He and Michael Hunkapiller from Applied Biosystems, Inc. (ABI) built an instrument, dubbed the ABI Model 370, to read the sequence of the dye-labeled fragments. It was equipped with an argon ion laser for exciting the dyes, a flat gel laid between two glass plates (referred to as a "slab" gel) capable of sixteen-lane electrophoresis, and a Hewlett-Packard Vectra computer boasting 640 megabytes of memory for data analysis.

Using fluorescent dyes, all four sequencing reactions could now be loaded into a single gel lane. As the fragments electrophoresed, the beam of the laser focused at the bottom of the gel made the dye-labeled fragments glow as they passed. The color of each dye-labeled fragment was then interpreted by the computer as a specific base (A if green, C if blue, G if yellow, and T if red). Over 350 bases could be read per lane. With this new automated approach, a technician could read more sequence in a day than could be read manually in an entire week.

Refinements in Automation

Shortly after ABI placed its automated DNA sequencer on the market, the Dupont company introduced its own model, the Genesis 2000. Dupont had also developed a new method of labeling sequencing fragments: attaching the fluorescent dyes to the terminator bases. With this innovation, four separate sequencing reactions were no longer required; the entire sequencing reaction could be accomplished in a single tube. However, Dupont failed to effectively compete in the automated sequencer market and sold the rights to the dye terminator chemistry to ABI.

ABI continued to refine its automated sequencer. More powerful computers, increased gel capacity (to 96 lanes), improvement of the optical systems, enhancement of the chemistry, and the introduction of more sensitive fluorescent dyes increased the reading capacity of the instrument to over 550 bases per lane. The ABI PRISM Model 377 Automated Sequencer, introduced in 1995, incorporated these changes and could read, at maximum capacity, over 19,000 bases in a day. Even at this rate, however, the sequencing of entire genomes, as that of humans (3 billion bases in length), was still not practical. Genome sequencing awaited several further innovations.

Working with the Model 377 Automated Sequencer, a laboratory technician had to pour the slab gels and mount them on the instrument. This process alone was time-consuming and cumbersome. In addition, the technician had to add each sequencing reaction into each individual lane of the gel prior to the run. The MegaBase, developed by Molecular Dynamics, and the ABI Model 3700 Automated Sequencer, developed by ABI, addressed these limitations by using multiple capillaries, thin, hollow glass tubes filled with a gel polymer.

The ABI PRISM Model 3700 Automated Sequencer, developed with the Hitachi Corporation and having a price tag of $300,000, uses ninety-six capillaries, each not much wider than a strand of human hair. The capillaries are automatically cleaned and filled with fresh gel polymer between each electrophoresis run. The instrument is also equipped with a robot arm that automatically loads the sequencing reactions into the capillaries, greatly decreasing the amount of human labor required for its operation. The Model 3700 Automated Sequencer can read over 400,000 bases in a day, a greater than twenty-fold increase over the maximum capacity of the Model 377. Beginning in September 1999 and using 300 of these instruments, the Celera Corporation had sequenced the entire human genome five times over within four months. SEE ALSO CYCLE SEQUENCING; GEL ELECTROPHORESIS; NUCLEOTIDE; SANGER, FRED; SEQUENCING DNA.

Frank H. Stephenson and Maria Cristina Abilock

Bibliography

Smith, Lloyd M., et al. "Fluorescence Detection in Automated DNA Sequence Analysis." *Nature* 321 (1986): 674–679.

Bacteria *See Eubacteria, Archaea*

Balanced Polymorphism

Balanced polymorphism is a situation in which two different versions of a gene are maintained in a population of organisms because individuals carrying both versions are better able to survive than those who have two copies of either version alone. The evolutionary process that maintains the two versions over time is called balancing selection.

Genes are carried on chromosomes. Different versions of a gene are called alleles. The standard allele found in a population is referred to as the wild-type allele. Most plants and animals have at least two copies of each chromosome, one inherited from each parent. The copies of the genes found on these **homologous** chromosomes may be identical or different; that is, the organism may carry two copies of one allele, or one each of two different alleles. In the first case, the organism is called homozygous for that gene, and, in the second, it is called heterozygous.

homologous similar in structure

Alleles differ from each other in their sequence of **nucleotides**, which may change the structure and function of the protein the gene codes for. Because of this, different alleles may have different effects on an organism's appearance or ability to survive. These effects can be helpful, harmful, or neutral.

nucleotides building blocks of DNA or RNA, composed of sugars, phosphates, and bases

An example of balanced polymorphism can be illustrated with the set of **enzymes** in the liver that act like an assembly line (or, more accurately, a disassembly line) to detoxify poisons and other chemicals. Different alleles for these enzymes can affect how well an organism can protect itself from

enzymes proteins that control a reaction in a cell

exposure to harmful chemicals. An especially active form of a detoxifying enzyme, which is encoded by a specific allele, can cause accumulation of potentially harmful intermediates. If the other allele encodes an enzyme with low activity, the potential for this enzyme to cause harm is lessened, and the benefits of its activity will be felt by the organism. If an individual has two copies of the very active allele or two copies of the low-activity allele, it may not survive well. In the first case, too much enzyme activity will result in high levels of the harmful intermediate, and in the second case, too little enzyme activity will be present for detoxification. Therefore, the best situation for the organism is to have one copy of each allele. Because of this, both copies are maintained in the population.

The effects of alleles and whether they are maintained in a population can be influenced by the environment. A classic case of balanced polymorphism in humans that is influenced by the environment is the sickle-cell allele of the β-globin gene. This gene forms part of hemoglobin, which carries oxygen in red blood cells.

Individuals who have two copies of the β-globin sickle-cell allele develop sickle-cell disease and generally do not survive into adulthood without intensive medical care. Individuals with one copy of the β-globin sickle-cell allele and one β-globin wild-type allele have red blood cells that are functional and resistant to the organism that causes malaria. Because individuals with this combination of alleles tend to survive malaria better than those who carry only the wild-type allele, the combination is advantageous to those who live in areas where malaria is present. This is called "heterozygote advantage." As a result, the beta-globin sickle-cell allele will be maintained along with the wild-type allele in populations exposed to malaria—an example of balancing selection. SEE ALSO ALZHEIMER'S DISEASE; HEMOGLOBINOPATHIES; HETEROZYGOTE ADVANTAGE; POPULATION GENETICS.

R. John Nelson

Bibliography

Weaver, Robert F., and Philip W. Hedrick. *Genetics*, 2nd ed. Dubuque, IA: William C. Brown, 1992.

Barr Body *See Mosaicism*

Behavior

What is behavior? A dictionary definition reveals that behavior consists of our activities and actions, especially actions toward one another. As such definitions suggest, many behavioral terms have meaning only in social comparisons: We identify others as contentious, courteous, or conscientious only by their actions in social contexts. A long-standing question in science and in everyday affairs inquires about the causes of individual differences in behavior: Why are some people gregarious extroverts and others timid, shy introverts?

Behavior genetics is a **hybrid** area of science, at the intersection of human genetics and psychology. Its focus is on how genes and environments contribute to differences in behavior. It is a young discipline. A book that

β the Greek letter beta

hybrid combination of two different types

Identical twins share all their genes, and, if raised together, much of their environment. Together, genes and environment determine many aspects of behavior.

gave the field its name was published in 1960, and a decade later the Behavior Genetics Association was founded. For a time, most behavior genetics research was an effort to show that the term was not itself an oxymoron—that variations in genes do contribute to individual differences in behavior. Now, as a result of that research, the relevance and importance of genetic variation to individual differences in behavior are widely accepted, and the challenging task is to identify specific gene-behavior pathways. In this entry, we will review the methods used to identify such pathways and then focus on one set of behaviors, use and abuse of alcohol, as a model for the study of genetic and environmental influences.

Twin and Adoption Studies

To determine whether variation in some dimension of behavior is heritable (whether behavioral differences are, in some part, due to genetic differences between people), human researchers use family, twin, and adoption designs. The first step in determining whether a behavior is influenced by genes is to establish that it aggregates or "runs" in families. Similarities in behavioral characteristics among family members suggest that genes influence the trait, but they cannot conclusively demonstrate genetic influence, because family members share their experiences (i.e., their environments) as well as their genes.

Twin and adoption studies allow one to tease apart the effects of genes and environments. Twin studies compare the patterns of behavioral characteristics between identical, or monozygotic (MZ), and fraternal, or dizygotic (DZ), twins. MZ twins share 100 percent of their genetic information, whereas DZ twins share, on average, one-half, just like non-twin siblings. Thus, the presence of greater behavioral similarities between MZ twins than DZ twins suggests that genetic factors contribute to those behaviors.

Adoption studies compare whether an adopted child is more similar behaviorally to the child's adoptive parents (with whom environments, but not genes, are shared) or to the child's biological parents (with whom genes, but not environments, are shared). Twin and adoption techniques have been used to demonstrate that nearly all behavior is under some degree of genetic influence, and, in the context of the Human Genome Project, behavior genetics has attracted great interest and some controversy.

Complex Genetics

New techniques allow behavior geneticists to ask not just whether a behavior is under genetic influence, but also what specific genes are involved. To identify genes involved in behavior, investigators use genetic markers—stretches of DNA that differ among individuals. One can either use genetic markers that are evenly spaced on all chromosomes, to search for genes influencing the behavior that are located anywhere in the genome (called genomic screening), or one can test markers at a specific gene believed to be, on theoretical grounds, involved in the behavior (called the candidate gene approach).

The idea behind these analyses is that if a particular gene is involved in the behavior, then people who are more alike with respect to the behavior will be more likely to share the same stretch of DNA that is at or near the gene. The difficulty in searching for genes involved in behavior is that there is no one-to-one correspondence between carrying a particular gene and exhibiting a particular behavior. There are no genes *for* behavior; there are only genes that *influence* behavior. Any particular behavior is a complex trait that involves more than one gene and is influenced by the environment as well.

For example, having a particular gene may make a person *more likely* to have problems with alcohol, but it does not *determine* whether or not the person will be an alcoholic. Some individuals will carry genes predisposing them to alcohol abuse but will never exhibit any problems, because they choose to abstain from alcohol. Other individuals will exhibit obvious alcohol problems, but will not carry the particular genes known to be involved.

This is because a large number of genes are risk-relevant for use and abuse of alcohol, and each has only a very small effect. Different genes may be acting in different individuals. And genes interact with each other and with the environment. Thus, individual outcomes result from a complex and ill-understood mixture of both genetic and environmental risk factors. That very complexity creates the diverse nature of human behavior. Indeed, it is what makes us uniquely human, but it also makes finding genes involved in human behavior extraordinarily difficult.

Animal Models

Because of this complexity, some investigators use animal models to complement human studies. Like humans, mice and rats differ in a variety of behavioral characteristics, including levels of alcohol use and tolerance or sensitivity to its effects. Animal studies allow breeding strategies that cannot be performed in humans. One approach that is commonly used in animal studies takes advantage of natural variation in behavior. Different strains of mice differ not only in coat color but also in preference for alcohol.

Under one of the most commonly used breeding strategies, animals from each of the behaviorally different mouse lines are allowed to mate with each other. Assuming the parents from each strain have different versions of genes contributing to alcohol use, subsequent generations of offspring will have different combinations of the genes contributing to the alcohol use and will display wide variation in their alcohol use. Such samples can be used to perform genetic studies searching for genes involved in the behavior, much like those described in humans: Animals more alike in their drinking behavior should be more likely to have inherited common stretches of DNA involved in the behavior.

One advantage of using animals is that the factors contributing to alcohol use in mice and rats are thought to be much simpler than the processes contributing to abuse in humans. Another is that animals' experience with alcohol can be experimentally controlled. Other strategies that are used in animals include inducing mutations or **"knocking out"** particular genes and studying the resultant aberrant behavior. If altering a particular gene consistently causes an alteration in a given behavior, the gene is likely involved in that behavior.

knocking out deleting a gene or obstructing gene expression

Alcoholism in Humans

The techniques available for human research are more limited, and many questions remain. Although behavior geneticists now possess the techniques to identify genetic influence and to begin to identify specific genes, questions remain regarding which behaviors, actions, and activities of people are the best candidates for behavior-genetic study.

Again, alcohol use and abuse provide an illustration. Alcoholism is a major social and medical problem in the United States and in most of the world. It is estimated that 10 percent of men and 4 percent of women in the United States experience alcohol dependency, at a cost of billions of dollars and 100,000 lives annually. Because use of alcohol is typically part of social interactions, familial (and possibly genetic) factors would be expected to contribute to variation in drinking.

But where shall we begin its study? Perhaps with diagnosed alcoholism? Most adults in our society use alcohol, yet only a fraction of them ever experience clinical symptoms of alcoholism. Perhaps we should begin much earlier, studying the decision to begin drinking? Obviously, one cannot become alcoholic without initiating drinking and then drinking large quantities regularly and with high frequency. Or perhaps much earlier yet, for behavioral predictors of alcoholism can be identified years before alcohol is first consumed.

Such predictors are apparent in early childhood, in behaviors evident to the children's parents, teachers, and peers. Long-term (i.e., longitudinal) studies conducted in several countries suggest that, as early as kindergarten and elementary school, behavioral ratings made by parents, teachers, or classmates distinguish children who are more likely to abuse alcohol later, in adolescence and early adulthood. Children who were impulsive, exploratory, excitable, curious, and distractible—and those who were less

cautious, less fearful, less shy, and less inhibited—have a much greater risk of adult alcoholism than do children without those characteristics.

Twin studies have demonstrated that additive genetic variance, as well as familial-environmental influences, significantly contributes to the childhood behaviors that play a central role in the development of alcoholism risk. So, to understand the development of alcoholism, one must appreciate the complex developmental influences that affect children years before they first consume alcohol. Those influences reflect the interactions of dispositional differences in children's behavior with variations in their familial, social, and school environments.

Twin Studies of Alcoholism

That risk-related behaviors are evident early in life, remain stable into adolescence, and are associated with a family history of alcoholism suggests that those behaviors are, at least in part, of genetic origin. To establish that, researchers must use genetically informative study designs.

One approach is to study child or adolescent twins and their parents. Several such studies, which specifically assess the initiation of alcohol use and the transition to alcohol abuse, are being conducted throughout the world. We illustrate with two ongoing studies from Finland.

One, "FinnTwin12," is a study of approximately 2,800 twin pairs and their parents. The twins represent all pairs from five consecutive twin-birth cohorts (1983–1987) who were entered into the study as they reached age twelve (1995–1999), when behavioral ratings by teachers and parents were obtained on all participating pairs.

The ratings include multidimensional scales (i.e., scales that rate various characteristics) of behaviors associated with increased alcoholism risk. Two years later, at age fourteen, the twins were followed up, and, while most reported abstinence, about one-third were then using alcohol.

What predicts drinking or abstaining at age fourteen? Genetic factors played a role only among twin sisters, perhaps reflecting their more accelerated pubertal maturation, and environmental effects shared by twin siblings accounted for most of the variation in drinking or abstaining at this age. Differences that twins attributed to their home environments (e.g., in parental monitoring, support, and understanding) and differences in teachers' ratings of twins' behavior at age twelve (in problem behaviors of aggressiveness, impulsivity, and inattention) differentiated those who were drinking from those still abstaining at fourteen.

But once drinking is initiated, genetic effects become evident in individual differences in frequency and quantity of consumption and in behavioral problems that then result. "FinnTwin16," another study of five consecutive, complete birth cohorts of Finnish twins, illustrates. These twins were first studied as they reached age 16, with follow-up twelve and thirty months later, at ages 17 and 18½. At age 16, about 25 percent had remained abstinent.

Of 2,810 twin pairs, both twins in 459 pairs (16.3%) were abstaining, co-twins in 1,964 pairs (69.9%) had concordantly begun drinking by age sixteen, and only 387 pairs were discordant, with one twin drinking and the

other abstaining. Concordance is the co-occurrence of the behavior in the twin pair (e.g., both drinking or both not drinking). Overall concordance exceeded 85 percent, regardless of the twins' gender or zygosity.

There was extremely high familial aggregation for alcohol use or abstinence at age sixteen, additional evidence that genes play little role in abstinence or initiation. But thirty months later, individual abstinence had dropped to 10 percent, concordance among twin pairs had declined considerably, and genetic factors increasingly influenced the frequency and quantity of an adolescent's alcohol consumption. MZ twins were significantly more similar in drinking frequency than were DZ twins. The influence of genetic factors increases over time, with increasing experience with alcohol, and the differences between MZ and DZ twins becomes greater at each follow-up.

Regional residency moderates parental and sibling influences on adolescent drinking. Where abstinence is relatively rare, as in the large cities of Finland, siblings have greater effects on one another. Conversely, the protective effect of parental abstinence on that of their adolescent twin children was more evident in sparsely populated rural areas of the country, where abstinence was more prevalent. And, most interestingly, genetic factors exerted a larger role in urban settings than in rural settings from age 16 through the follow-up at age 18½. Common environmental factors assumed greater importance in rural settings.

Such results suggest that environments moderate the impact of genetic effects across many dimensions of behavior. But what aspects of the environment matter? In an analysis of results at age 18½, we demonstrated that specific characteristics of rural and urban environments moderate the effects of genes on drinking behavior. In areas with proportionately more young adults, genetic effects were nearly five times more evident than in communities with relatively few young adults. Thus, dramatic differences in the magnitude of genetic effects can be demonstrated across communities at environmental extremes of specific risk-relevant characteristics.

Complex Behaviors, Complex Causes

Thus, for use and abuse of alcohol, we know that the importance of genetic and environmental effects changes with sequencing in the use and abuse of alcohol, from abstinence or initiation to frequency of regular consumption, to problems associated with consumption, and ultimately, to diagnosed alcoholism and end-organ damage from the cumulative effects of alcohol. Similar stories could be told for many other behaviors of interest. Thus, for the major psychopathologies, from depression and schizophrenia in adults to attention deficit disorder in children or eating disorders in adolescents, genetic influences are invariably part of the story but never the whole story.

Genetic effects are always probabilistic and not deterministic. And the action of genes on behavioral outcomes is likely to be indirect. So we conclude with the same message with which we began: There are no genes *for* behavior, but behavioral development always represents an exquisite interplay between genes and environments. Gene-behavior correlations are modest and nonspecific; they alter risk but rarely determine outcome. Genes represent dispositions, not destinies.

Richard J. Rose and Danielle M. Dick

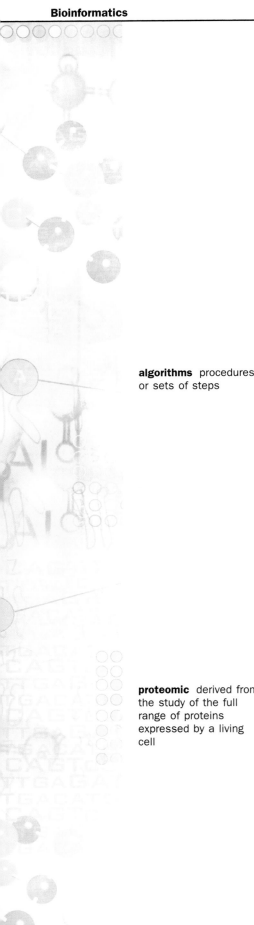

Bibliography

Dick, Danielle M., and Richard J. Rose. "Behavior Genetics: What's New? What's Next?" *Current Directions in Psychological Science* 11 (2002): 70–74.

Rose, Richard J. "A Developmental Behavior-Genetic Perspective on Alcoholism Risk." *Alcohol Health and Research World* 22 (1998): 131–143.

Rose, Richard J., et al. "Drinking or Abstaining at Age 14? A Genetic Epidemiological Study." *Alcoholism: Clinical and Experimental Research* 25 (2001): 1594–1604.

Bioinformatics

Bioinformatics is the use of mathematical, statistical and computer methods to analyze biological, biochemical, and biophysical data. Because bioinformatics is a young, rapidly evolving field, however, it also has a number of other credible definitions. It can also be defined as the science and technology of learning, managing, and processing biological information. Bioinformatics is often focused on obtaining biologically oriented data, organizing this information into databases, developing methods to get useful information from such databases, and devising methods to integrate related data from disparate sources. The computer databases and **algorithms** are developed to speed up and enhance biological research.

Bioinformatics can help answer such questions as whether a newly analyzed gene is similar to any previously known gene, whether a protein's sequence can suggest how the protein functions, and whether the genes turned on in a cancer cell are different from those turned on in a healthy cell.

Databases and Analysis Programs

A good deal of the early work in bioinformatics focused on processing and analyzing gene and protein sequences catalogued in databases such as GenBank, EMBL, and SWISS-PROT. Such databases were developed in academia or by government-sponsored groups and served as repositories where scientists could store and share their sequence data with other researchers. With the start of the Human Genome Project in 1990, efforts in bioinformatics intensified, rising to the challenge of handling the large amounts of DNA sequence data being generated at an unprecedented rate. By the mid- to late-1990s, much of the efforts in bioinformatics centered around genomic data, generated by the Human Genome Project and by private companies, and around **proteomic** data.

Early analysis of sequence information focused on looking for similarities between genes and between proteins. Algorithms were developed to help researchers rapidly identify similar gene or protein sequences. Such tools were extremely useful for determining whether a newly sequenced piece of DNA was at all similar to sequences already entered in a database. To determine how multiple sequences align and to view their similarities, multiple-alignment programs were developed. Such programs helped scientists compare the sequences of closely related genes or compare the sequence of a particular gene or protein as it appears in several species.

To better understand the functional roles of new nucleotide and amino acid sequences, researchers developed algorithms to look for particular sequence "domains." Domains are regions where a particular sequence of

algorithms procedures or sets of steps

proteomic derived from the study of the full range of proteins expressed by a living cell

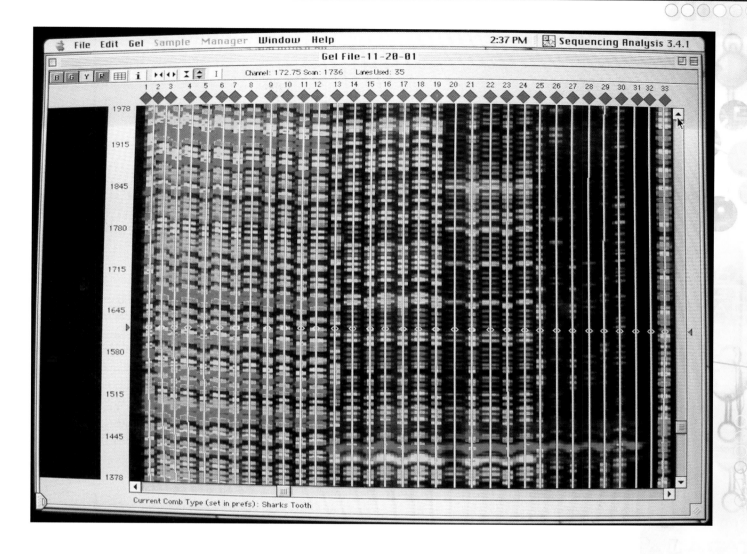

File   Edit   Gel   Sample   Manager   Window   Help   2:37 PM   Sequencing Analysis 3.4.1

Gel File-11-28-81

Channel: 172.75 Scan: 1736   Lanes Used: 35

Current Comb Type (set in prefs): Sharks Tooth

nucleotides or amino acids is indicative of function in the protein. For example, a protein may have a domain that binds to ATP or GTP, two important protein regulators.

In addition, these algorithms can detect sequences that denote a region involved in particular types of post-translational modifications, such as tyrosine **phosphorylation**. Tools such as prosite, blocks, prints, and Pfam can be used to detect and predict such protein domains in sequence data.

Structure is central to protein function, and another set of tools, including SWISS-MODEL, allows researchers to use gene and protein sequence data to predict a protein's three-dimensional structure. Such tools can help predict how mutations in a gene sequence could alter the three-dimensional structure of the corresponding protein. They accomplish such molecular modeling by comparing a novel sequence to the sequences of genes whose protein structures are known.

The majority of tools were developed as academic freeware distributed on the Internet. In the early- to mid-1990s, commercial companies began to develop their own **proprietary** algorithms and tools, as well as their own proprietary databases. Those databases were then marketed to pharmaceutical

The computer monitor of an automated gel sequencer displays a digital gel image. This data is more easily analyzed in this environment.

phosphorylation addition of the phosphate group PO_4^{3-}

proprietary exclusively owned; private

and biotech companies as well as to academic research groups. The most commercially viable and profitable businesses focused on the production and sale of proprietary DNA- and gene-sequence databases in the mid- to late-1990s. These databases primarily contained genetic information that were not in the public domain databases, such as GenBank, and they thus offered potential competitive advantages to the drug discovery groups of large pharmaceutical and biotech companies.

Applications of Bioinformatics to Drug Discovery

The application of bioinformatics to genomics data could be a huge potential boon for the discovery of new drugs. During the 1990s many pharmaceutical companies and biotech companies became convinced that they could speed up their drug-discovery pipelines by taking advantage of the data from the Human Genome Project as well as by funding their own internal genomics programs and by collaborating with third-party genomics companies.

The goal in such practical applications is to use such data as DNA sequence information and gene expression levels to help discover new drug targets. The vast majority of drugs target proteins, but there are a handful of drugs, such as some **chemotherapeutic** agents, that bind to DNA. In cases where the target is a protein, the drugs themselves are primarily small chemical molecules or, in some cases, small proteins, such as hormones, that bind to a larger protein in the body. Some drugs are therapeutic proteins delivered to the site of the disease.

The extent to which genomics will actually be able to help identify validated drug targets is uncertain. Genomics and bioinformatics are still young areas, and the drug development cycle can take up to ten years. As of 2001 relatively few of the drugs on the market or in the late stages of clinical trials were discovered via genomics or bioinformatics programs.

Specialists

Bioinformatics is applied to at least five major types of activities: data acquisition, database development, data analysis, data integration, and analysis of integrated data.

Data Acquisition. Data acquisition is primarily concerned with accessing and storing data generated directly off of laboratory instruments. Many of these instruments are either automated or semi-automated **high-throughput** instruments that generate large volumes of data. The Human Genome Project utilized hundreds of DNA sequencers, producing enormous amounts of data. The data had to be captured in the appropriate format, and it had to be capable of being linked to all the information related to the DNA samples, such as the species, tissue type, and quality parameters used in the experiments. This area of bioinformatics primarily relates to the use of "laboratory information management systems," which are the computer systems used to manage the information needs of a particular laboratory.

Database Development. Many laboratories generate large volumes of such data as DNA sequences, gene expression information, three-dimensional molecular structure, and high-throughput screening. Consequently, they must develop effective databases for storing and quickly accessing data. For each type of data, it is likely that a different database organization must be

chemotherapeutic use of chemicals to kill cancer cells

high-throughput rapid, with the capacity to analyze many samples in a short time

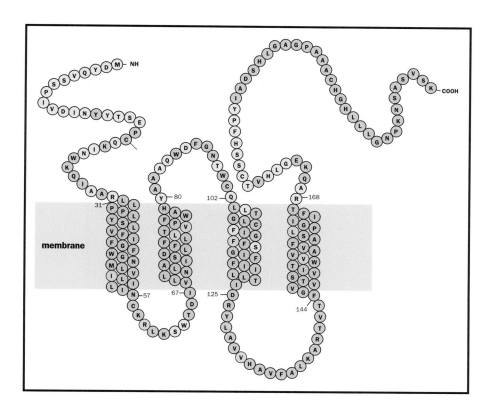

Predicted structure of a chemokine receptor in the membrane of a white blood cell. Knowing the amino acid sequence (represented by single letters) allows researchers to predict the three-dimensional structure of a protein. Adapted from <http://www.cdc.gov/ncidod/eid/vol3no3/smith.htm>.

used. A database must be designed to allow efficient storage, search, and analysis of the data it contains. Designing a high-quality database is complicated by the fact that there are several formats for many types of data and a wide variety of ways in which scientists may want to use the data. Many of these databases are best built using a relational database architecture, often based on Oracle or Sybase.

A strong background in relational databases is a fundamental requirement for working in database development. Having some background in the molecular biology techniques used to generate the data is also important. Most critical for the bioinformatics specialist is to have a strong working relationship with the researchers who will be using the database and the ability to understand and interpret their needs into functional database capabilities.

Data Analysis. Being able to analyze data efficiently requires having a good database design, allowing researchers to query the database effectively and letting them quickly obtain the types of information they need to begin their data analysis. If queries cannot be performed, or if performance is tediously slow, the whole system breaks down, since scientists will not be inclined to use the database. Once data is obtained from the database, the user must be able to easily transform it into the format appropriate for the desired analysis tools.

This can be challenging, since researchers often use a combination of publicly available tools, tools developed in-house, and third-party commercial tools. Each tool may have different input and output formats. Starting in the late 1990s, there have been both commercial and in-house efforts at pharmaceutical and biotech companies to reduce the formatting complexities.

Such simplification efforts focus on building analysis systems with a number of tools integrated within them such that the transfer of data between tools appears seamless to the end user.

Bioinformatics analysts have a broad range of opportunities. They may write specific algorithms to analyze data, or they may be expert users of analysis tools, helping scientists understand how the tools analyze the data and how to interpret results. A knowledge of various programming languages, such as Java, PERL, C, C++, and Visual Basic, is very useful, if not required, for those working in this area.

Data Integration. Once information has been analyzed, a researcher often needs to associate or integrate it with related data from other databases. For example, a scientist may run a series of **gene expression** analysis experiments and observe that a particular set of 100 genes is more highly expressed in cancerous lung tissue than in normal lung tissue. The scientist might wonder which of the genes is most likely to be truly related to the disease. To answer the question, the researcher might try to find out more information about those 100 genes, including any associated gene sequence, protein, enzyme, disease, metabolic pathways, or signal **transduction** pathway data.

gene expression use of a gene to create the corresponding protein

transduction conversion of a signal of one type into another type

Such information will help the researcher narrow the list down to a smaller set of genes. Finding this information, however, requires connections or links between the different databases and a good way to present and store the information. An understanding of database architectures and the relationship between the various biological concepts in the databases is key to doing effective data integration.

Analysis of Integrated Data. Once various types of data are integrated, users need a good way to present these various pieces of data so they can be interpreted and analyzed. The information should be capable of being stored and retrieved so that, over time, various pieces of information can be combined to form a "knowledge base" that can be extended as more experiments are run and additional data are integrated from other sources. This type of work requires skills related to database design and architecture. It also requires specific programming skills in various computer languages, as well as expertise in developing interfaces between a computer and its user. SEE ALSO Combinatorial Chemistry; Computational Biologist; Evolution of Genes; Genomics; Genomics Industry; High-Throughput Screening; Human Genome Project; Pharmacogenetics and Pharmacogenomics; Proteins; Proteomics; Sequencing DNA.

Anthony J. Recupero

Bibliography

Howard, Ken. "The Bioinformatics Gold Rush." *Scientific American* 283, no. 1 (2000): 58–64.

Internet Resources

"EID V3 N3: Host Genes and HIV." Centers for Disease Control and Prevention. <http://www.cdc.gov/ncidod/eid/vol3no3/smith.htm>.

EMBL Nucleotide Sequence Database. Release 69. December 2001. European Bioinformatics Institute. <http://www.ebi.ac.uk/>.

GenBank. National Center for Biotechnology Information. <http://www.ncbi.nlm.nih.gov/>.

SWISS-PROT. Swiss Institute of Bioinformatics. <http://www.expasy.org/sprot/>.

Biopesticides

Plants, growing in the wild or in cultivation, face numerous threats from insects, bacteria, viruses, and fungi, as well as from other plants. Biopesticides are inert substances or living organisms that can help protect plants from such threats. Chemical pesticides can offer similar protection but, by contrast, are neither alive nor made by living organisms.

Natural Chemical Defenses

A variety of chemicals produced by plants help ensure that parasites, predators, plant feeders, and **herbivores** seldom increase in number sufficiently to destroy the plant populations they prey upon. Chemicals found in very low concentrations in certain plants have been found to help keep locusts from feeding on those plants, and some trees produce nearly 1,000 different chemical compounds that help them resist herbivores and parasites.

herbivores plant eaters

Living Organisms Serving as Biopesticides

Plant predators are themselves subject to attack by predators, parasites, and microbes, all of which can indirectly help protect a plant and therefore are also considered biopesticides. An oak tree may have about 100 species of insect herbivores feeding on it. In turn, there can be up to 1,000 species of predators, parasites, and microbes feeding on the herbivores. The microbes, parasites, and predators attacking the herbivore populations are considered "biopesticides," as are any protective chemicals produced by the tree.

Such living biopesticides play a vital role in agriculture and nature, helping to control insect pests, plant pathogens, and weeds. Numerous organisms, including viruses, fungi, protozoa, bacteria, and **nematodes**, as well as insects, such as parasitic wasps, can attack pest insects and weeds. In some cases, biologists search around the world to find natural organisms to help control an insect, a plant **pathogen**, or weed populations.

nematodes worms of the Nematoda phylum, many of which are parasitic

pathogen disease-causing organism

The use of natural organisms as biopesticides is sometimes hampered by the presence of chemical pesticides, which can threaten populations of a pest insect's natural enemies. Pest outbreaks that result from chemical pesticides destroying a pest's natural enemies are estimated to cost the United States more than $500 million per year.

Genetically Modified Organisms as Biopesticide Producers

Since the 1980s, many crops have been genetically modified to produce biopesticides that will help protect them from insects and pathogens (including viruses). In 1998, 40 million hectares of engineered crops were planted throughout the world (though 74% of the modified cropland is in the United States). Globally, 20 percent of this area has been planted with herbicide-tolerant crops, 8 percent with insect-resistant crops, and 0.3 percent with insect- and herbicide-resistant crops.

Disease Resistance in Crops

More than 95 percent of all crops have some degree of pathogen resistance bred into them, with resistance to fungi, bacteria, and viruses being most common. Most of this resistance was either added by farmer selection or

plant breeder selection, rather than through genetic engineering. It is because of this natural resistance that has been bred into the crops that only 12 percent of the pesticides used in U.S. agriculture are fungicides.

Some viral resistance, however, has been bred into a number of crops through insertion of viral genes into the plant chromosomes. These genes may lead to the plant's producing viral proteins—biopesticides of a sort—that hamper a virus's own actions. This pathogen-derived resistance has been successfully used to protect Hawaii's papaya crop from the devastating papaya ringspot potyvirus. The viral gene was inserted into the papaya genome using a "gene gun," which shoots viral genes into papaya embryo cells.

Insect-Resistant Crops

About 8 percent of land covered by genetically modified crops is planted with insect-resistant crops. Although insect-resistant crops have not been employed as widely as disease-resistant crops, there are some notable examples. These include Hessian fly resistance in wheat and European corn borer resistance in corn. Resistance to corn borers has been provided by using a naturally occurring toxin produced by *Bacillus thuringiensis* (BT). This bacterium has traditionally been applied to corn and other crops. Genetic engineering has allowed the toxin to be manufactured by the corn plant itself.

The most serious problem with the use of BT toxin genes has been with STARLINK corn. This variety of genetically modified corn was approved for use only as animal feed, not as human food. STARLINK corn nevertheless found its way into the food-processing industry when farmers did not keep their STARLINK corn separate from corn to be used as human food. Some grain elevator operators as well did not keep the STARLINK corn separated. The STARLINK mix-up cost the United States an estimated $5 billion of processed food that had to be destroyed because of the BT toxin restriction.

Plants that have been genetically modified to make BT toxin produce it in every cell. BT corn pollen may harm nontarget moths and butterflies, like Monarch butterflies. Milkweed leaves dusted with drifting BT corn pollen are toxic to Monarch butterfly larvae, but some corn does not naturally pollinate at the same time of the year as butterflies are in their larval stage, so the effects may not be great. The full extent to which natural populations of butterflies are affected by BT corn is not known.

Herbicide-Resistant Crops

Some crops (e.g. corn) are being engineered to contain both herbicide tolerance and the BT toxin. Generally, the use of herbicide-tolerant crops will likely increase the use of herbicides. This has the potential to increase environmental pollution since it might increase the farmers' reliance on chemicals rather than mechanical and other means of weed control. SEE ALSO AGRICULTURAL BIOTECHNOLOGY; GENETICALLY MODIFIED FOODS; TRANSGENIC PLANTS.

David Pimentel

Bibliography

Paoletti, Maurizio G., and David Pimentel. "Genetic Engineering in Agriculture and the Environment." *BioScience* 46, no. 9 (1996): 665–673.

Pimentel, David, ed. *Techniques for Reducing Pesticide Use: Economic and Environmental Benefits*. Chichester, U.K.: John Wiley & Sons, 1997.

Pimentel, David, and Hugh Lehman, eds. *The Pesticide Question: Environment, Economics, and Ethics*. New York: Chapman and Hall, 1993.

Internet Resource

Religion and Ethics Newsweekly. Harvest of Fear. Public Broadcasting System. <http://www.pbs.org/wgbh/harvest/2001/>.

Bioremediation

Bioremediation is the use of organisms to break down and thereby detoxify dangerous chemicals in the environment. Plants and microorganisms are used as bioremediators. The technology can take advantage of a natural metabolic pathway or genetically modify an organism to have a particular toxic "appetite."

Natural Microbial Bioremediators

On March 24, 1989, an oil tanker called the Exxon Valdez crashed into a reef in the Prince William Sound in Alaska, spilling 11 million gallons of oil that devastated the highly populated **ecosystem**. Attempts to clean rescued animals and scrub oily rocks were of little help and actually killed some organisms. Bioremediation was more successful. Ten weeks after the spill, researchers from the U.S. Environmental Protection Agency applied phosphorus and nitrogen fertilizers to 750 oil-soaked sites. The fertilizer stimulated the growth of natural populations of bacteria that metabolize polycyclic aromatic hydrocarbons, which are organic toxins that were present in the spilled oil. Over the next few years, ecologists monitored and compared the areas that the bacteria had colonized to areas where they did not grow, and found that the level of polycyclic aromatic hydrocarbons fell five times faster in the bioremediated areas.

ecosystem an ecological community and its environment

Another environmental disaster being treated with natural bioremediation is the pollution of the Hudson River in New York with polychlorinated biphenyls (PCBs). General Electric Corporation deposited these compounds along a 40-mile stretch of the river between 1947 and 1977. PCBs were used to manufacture hydraulic fluids, capacitors, pigments, transformers, and electrical equipment. PCBs come in 209 different and interconverting forms, and the toxicity of a particular PCB depends upon the number of chlorine atoms it includes. Debate rages over whether it is better to remove and bury the most contaminated sediments, or to allow natural bacteria in the river to detoxify the PCBs.

The bioremediation of the Hudson River is occurring in three stages. First, buried **anaerobic** bacteria strip off chlorines. In the water column, aerobic bacteria cleave the two organic rings of the PCBs. Finally, other microorganisms degrade the dechlorinated, broken rings into carbon dioxide, water, and chloride. While the process effectively detoxifies the PCBs, it is a long-term process that can take up to two centuries.

anaerobic without oxygen or not requiring oxygen

Cleanup crews use hot and cold high-pressure water jets to clean the rocks following the Exxon Valdez oil spill in 1994. This cleanup strategy was followed by the introduction of fertilizers to some areas of the spill, helping to encourage the growth of bacterial colonies.

vacuoles cell structures used for storage or related functions

polypeptides chains of amino acids

Natural Plant Bioremediators

For many millions of years, plants have adapted to the presence of various metals in varying amounts in soils. Some metals, such as zinc, nickel, cobalt, and copper, function as nutrients when eaten by humans in small amounts, but are toxic when consumed in excess. Heavy metals that are toxic even in trace amounts include mercury, lead, cadmium, silver, gold and chromium. Human activities such as mining, municipal waste disposal, and manufacturing have increased heavy metal pollution to dangerous levels in some areas. These chemicals cause oxidative damage, which destroys lipids, DNA, and proteins.

Certain plants, called hyperaccumulators, cope with excess heavy metals in the environment by taking them in and sequestering them in **vacuoles**, which are bubble-like structures in their cells. Sometimes the plant combines a pollutant with another molecule, a process called chelation. Organic acids often serve this role. Citric acid, for example, surrounds and thereby detoxifies cadmium, and malic acid does the same for zinc. A class of **polypeptides** called phytochelatins can also bind metals and escort them to vacuoles. Yet a third strategy that plants use to control metal accumulation is to employ a class of small, metal-binding proteins called metallothioneins. The intentional use of plants that use any of these ways to take heavy metals from soil is termed phytoremediation. It is a form of bioremediation.

This petroleum-contaminated dirt was mixed with clean dirt and manure before being injected with microbes. The microbes fed on the petroleum's hydrocarbons, converting the harmful materials into carbon dioxide and oxygen.

Natural phytoremediators can be amazing. Consider *Sebertia acuminata*, a tree that lives in the tropical rain forest of New Caledonia, near Australia. Up to 20 percent of the tree's dry weight is nickel. If slashed, the bark oozes a bright green. This plant can perhaps be used to clean up nickel-contaminated soil. Soybeans also preferentially take up nickel from soil. Another phytoremediator is *Astragalus*, also know as locoweed. It accumulates selenium from soil to counteract toxic effects of phosphorus, which tends to be abundant in selenium-rich soils. Cattle that munch on locoweed stagger about from selenium intoxication. Some plants act as sponges for metals in their environment. For example, plants that grow near gold mines assimilate gold into their tissues, apparently without harm. Prospectors use the gold content of such plants to locate deposits of the precious metal. Plants that grow near highways take up lead from gasoline exhaust. Near nuclear test sites, plants absorb radioactive strontium.

Genetically Modified Bioremediators

Biotechnology can transfer the ability to manufacture detoxifying proteins from one type of organism to another. One organism that was so modified has earned the distinction of being the first micro-organism to be patented. Called the "oil eater," the microorganism was actually a naturally occurring bacterium that had been given four **plasmids** that were also naturally occurring (plasmids are rings of DNA that can be transferred from one cell to another). It was the combination of the four transferred plasmids in a single bacterial cell that was novel and therefore patent-worthy. The four plasmids in the oil eater gave the bacterium the ability to degrade four components of crude oil. It was invented by Ananda Chakrabarty at General Electric in 1980.

plasmids small rings of DNA found in many bacteria

Today, transgenic technology creates designer bioremediators. A transgenic organism contains a gene from another type of organism in all of its cells. The altered organism then manufactures the protein that the transgene encodes. The technology works because all organisms use the same genetic code. In other words, the same DNA and **RNA triplets** encode the same amino acids in all species.

RNA triplets sets of three nucleotides

catalyzes aids in the reaction of

Transgenic bioremediation can engineer microbial metabolic reactions into plants whose root cells then produce the needed proteins and distribute them in the soil. For example, transgenic yellow poplar trees can thrive in soil that has been heavily contaminated with mercury if they have been given a bacterial gene that encodes the enzyme called mercuric reductase. This enzyme **catalyzes** the chemical reaction that converts a highly toxic form of mercury in soil to a less toxic gas. The leaves of the tree then emit the gas to the atmosphere, where it dissipates.

Cleaning up munitions dumps is yet another target of transgenic plants, with some interesting biological participants. In one approach, a bacterial gene that breaks down trinitrotoluene (TNT, the major component of dynamite and land mines) is linked to a jellyfish gene that makes the protein glow green. The bacteria can be spread directly on soil that is thought to contain weapons residues, or the genes can be transferred to various types of plants, whose roots then glow when they are near buried explosives. In the future, plants that have been genetically modified in several ways will be able to detect a variety of pollutants or toxins. SEE ALSO EUBACTERIA; TRANSGENIC ORGANISMS: ETHICAL ISSUES.

Ricki Lewis

Bibliography

Bolin, Frederick. "Leveling Land Mines with Biotechnology." *Nature Biotechnology* 17 (1999): 732.

Eccles, Harry. *Bioremediation*. New York: Taylor and Francis, 2001.

Hooker, Brian S., and Rodney S. Skeen. "Transgenic Phytoremediation Blasts onto the Scene." *Nature Biotechnology* 17 (1999): 428.

Lewis, Ricki. "PCB Dilemma." *The Scientist* 15 (2001): 1.

Biotechnology

Biotechnology, broadly defined, refers to the manipulation of biology or a biological product for some human end. Before recorded history, humans grew selected plants for food and medicines. They bred animals for food, for work, and as pets. The ancient Egyptians learned how to maintain selected yeast cultures, which allowed them to bake and brew with predictable results. These are all examples of biotechnology. In more recent times, however, the term "biotechnology" has mainly been applied to specifically industrial processes that involve the use of biological systems. Today many biotechnology companies use processes that make use of genetically engineered microorganisms.

A Revolution in Biology

Following 1953, when Thomas Watson and Francis Crick published their famous paper on the double helix structure of DNA, a series of independent discoveries were made in chemistry, biochemistry, genetics, and microbiology, which together brought about a revolution in biology and led to the first experiments in genetic engineering in 1973. Because of this revolution, scientists learned to modify living microorganisms in a permanent, predictable way. Bacteria have been made to produce medical products, such as hormones, vaccines, and blood factors, that were formerly not available or available only at great expense or in limited amounts. Crop plants have been developed with increased resistance to disease or insect pests, or with greater tolerance to frost or drought. What has made all these things possible is the collection of bio-chemical and molecular biological techniques for manipulating genes, which are the basic units of biological inheritance. These are the techniques used in genetic engineering or **recombinant DNA** technology.

The fusion of traditional industrial microbiology and genetic engineer-ing in the late 1970s led to the development of the modern biotechnology industry. Using recombinant DNA technology, this industry has brought a long and steadily growing list of products into the marketplace. Human insulin produced by genetically engineered bacteria was one of the first of these products. It was followed by human growth hormone; an anti-viral protein called interferon; the immune stimulant called interleukin 2; a tis-sue plasminogen activator for dissolving blood clots; two blood-clotting fac-tors, labeled VIII and IX, which are administered to hemophiliacs; and many other products.

recombinant DNA DNA formed by combining segments of DNA, usu-ally from different types of organisms

Vitamin C

The production of certain chemicals has already become an important biotechnological industry. Vitamin C is a prime example. Humans, as well as other primates, guinea pigs, the Indian fruit bat, several species of fish, and a number of insects, all lack a key enzyme that is required to convert a sugar, glucose, into vitamin C.

No single bacterial genus or species is known that will carry out all of the reactions needed to synthesize vitamin C, but there are two (*Erwinia* species and *Corynebacterium* genus) that, between them, can perform all but one of the required steps. In 1985 a gene from one of these genus (*Corynebac-terium*) was introduced into the second organism (*Erwinia herbicola*), result-ing in a new bacterial form. This engineered organism can be used to produce a **precursor** to vitamin C that is converted via one chemical reac-tion into this essential vitamin. The engineering of many other microor-ganisms is being used to replace complex chemical reactions. For example, **amino acids**, needed for dietary supplements, are produced on a large scale using genetically modified microorganisms, as are antibiotics.

Before Captain James Cook, the famous English sailor and navigator, had his men drink lime juice (which contains vitamin C) during extended sea voyages, many sailors fell ill or died of the vitamin C deficiency known as scurvy.

precursor a substance from which another is made

amino acids building blocks of protein

Laundry Detergents

Another important class of compounds produced by biotechnology is **enzymes**. These protein **catalysts** are used widely in both medical and indus-trial research. Proteases, enzymes that break down proteins, are particularly important in detergents, in tanning hides, in food processing, and in the chemical industry. One of the most significant commercial enzymes of this type is subtilisin, which is produced by a bacterium. Because many stains con-tain proteins, the manufacturers of laundry detergents include subtilisin in their product. Subtilisin is 274 amino acids long, and one of these, the methio-nine at position 222, lies right beside the active site of the enzyme. This is the site on the enzyme's surface where the substrate is bound, and where the reaction that is catalyzed by the enzyme takes place. In this instance the sub-strate is a protein in a stain, and the reaction results in the breaking of a pep-tide bond in the backbone of the protein. Unfortunately, methionine is an amino acid that is very easily **oxidized**, and laundry detergents are often used

enzymes proteins that control a reaction in a cell

catalysts substances that speed up a reac-tion without being con-sumed

oxidized reacted with oxygen

in conjunction with bleach, which is a strong oxidizing agent. When used with bleach, the methionine in subtilisin is oxidized and the enzyme is inactivated, preventing the subtilisin from doing its work of breaking down the proteins present in food stains, blood stains, and the like.

To overcome this problem, genetic engineering techniques were used to isolate the gene for subtilisin, and the small part of the gene that codes for methionine 222 was replaced by chemically synthesized DNA fragments that coded for other amino acids. The experiment was done in such a way that nineteen new subtilisin genes were produced, and every possible amino acid was tried at position 222. Some of the altered genes gave rise to inactive versions of the enzyme, but others resulted in fully functional subtilisin. When these subtilisins were tested for their resistance to oxidation, most were found to be very good (except when cysteine replaced methionine: It too is easily oxidized). So now it is possible to use laundry detergent and bleach at the same time and still remove protein-based stains. This type of gene manipulation, which has been called "protein engineering," has already been used for making beneficial changes in other industrial enzymes, and in proteins used for medical purposes.

Other Examples

Biotechnology companies are continuing to produce new products at an impressive rate. Numerous clinical testing procedures for human disorders such as AIDS and hepatitis and for disease-causing organisms such as those responsible for malaria and Legionnaires' disease (a lung infection caused by the bacterium *Legionella pneumophila*), are based on diagnostic testing kits that have been developed by biotechnology companies. Many of these assays make use of recombinant **antibodies**, while others rely on DNA **primers** that are used in the polymerase chain reaction to detect DNA sequences present in an infecting organism, but not in the human genome.

Trangenic plants are now grown on millions of acres. Many of these plant species have been engineered to produce a protein, normally synthesized by the bacterium *Bacillus thuringiensis*, which is toxic to a number of agriculturally destructive insect pests but harmless to humans, most other non-insect animals, and many beneficial insects such as bees.

Ethical Issues

Like all industries, the biotechnology industry is subject to rules and regulations. Legal, social, and ethical concerns have been raised by the ability to genetically alter organisms. These have resulted in the establishment of governmental guidelines for the performance of biotechnology research, and specific requirements have been set to control the introduction of recombinant DNA products into the marketplace. General governmental guidelines for biotech research are published on the Internet at http://www.aphis.usda.gov/biotech/OECD/usregs.htm. Guidelines for plant genetic engineering and biotechnology are available at http://sbc.ucdavis.edu/Outreach/resource/US_gov.htm. SEE ALSO AGRICULTURAL BIOTECHNOLOGY; BIOTECHNOLOGY AND GENETIC ENGINEERING, HISTORY OF; CLONING GENES; CLONING ORGANISMS; GENE THERAPY; GENETIC TESTING; GENETICALLY MODIFIED FOODS; HEMOPHILIA; POLYMERASE CHAIN REACTION;

antibodies immune system proteins that bind to foreign molecules

primers short nucleotide sequences that help begin DNA replication

Recombinant DNA; Transgenic Animals; Transgenic Microorganisms; Transgenic Plants.

Dennis N. Luck

Bibliography

Glick, Bernard R., and Jack J. Pasternak. *Molecular Biotechnology: Principles and Applications of Recombinant DNA.* Washington, DC: ASM Press, 1998.

Marx, Jean L. *A Revolution in Biotechnology.* Cambridge, MA: Cambridge University Press, 1989.

Primrose, S. B. *Molecular Biotechnology.* Boston: Blackwell Scientific Publications, 1991.

Rudolph, Frederick B., and Larry V. McIntire, eds. *Biotechnology: Science, Engineering, and Ethical Challenges for the Twenty-first Century.* Washington, DC: Joseph Henry Press, 1996.

Biotechnology Entrepreneur

The biotechnology entrepreneur often starts with a technical background, most commonly including scientific laboratory research. It is quite common for scientists who have spent a number of years in an academic environment to start a new company based on a technology platform or a novel discovery they made during their tenure at a university or research institute. Alternatively, a scientist or a group of scientists from a rather large company may decide to leave and start their own venture based on some of the research and development work they had either performed or thought about doing for their previous employer. This type of business may or may not be a spin-off from the parent company. In some cases, a businessperson or two may join the scientists in the start-up of the new company. A less frequent but still relatively common phenomenon is the start-up of a new venture by someone with a stronger background in business than in science. All of these individuals may receive financial support from venture capitalists, people who provide funding for new businesses in return for a share in future profits.

Whatever the background of the entrepreneur, there are a few key initial elements that are critical to achieving success. First and foremost is funding. There are a number of ways to secure funding, including small business research grants from the government, personal funds and loans, private "angel" investors, and venture capitalists. Entrepreneurs must become proficient at marketing themselves, their ideas, and their company in order to raise capital for the business. Equally important is having an attractive and viable business plan and business model, preferably based upon a novel technology or combination of technologies that has or can develop a strong patent position. It is also extremely important to have a good understanding of the market that will be served by the new business. Excellent communication skills are also crucial, as is the ability to recognize and recruit talented individuals, to hire them in the right order and place them in the positions that will best help the young company to focus, execute its plans, and grow.

A common mistake of the technically oriented entrepreneur is to try to completely manage both the technical and business side of the company

Business-minded scientists seek support from their peers at web sites such as <http://www.bioe2e.org>.

without consulting an experienced administrator to manage the business affairs. There are, however, the occasional entrepreneurs who can navigate both the business and technical sides with equal success. These leaders will usually retain the position of chief executive officer (CEO) as the company continues to grow. The technical entrepreneur who recruits experienced business people to help foster the growth of the company quite often ends up becoming the firm's chief scientific officer or the chief technical officer. This is probably the most commonly encountered scenario for companies that end up being traded on the stock exchange by releasing an initial public offering (IPO) of shares in the business.

A successful entrepreneur must also be an effective leader, able to contribute with creative ideas as well as to motivate staff and colleagues. He or she is often an individual with unending enthusiasm, a strong vision, and the ability to convince others that this vision will be successful. Entrepreneurs must have the conviction to do what is necessary to successfully execute the company's vision, while being flexible enough to adjust to new or different opportunities when they present themselves.

For a successful entrepreneur in the field of biotechnology, the rewards can be enormous. There is the satisfaction that comes from starting with an original idea and, through hard work, making it a reality, but there are financial rewards as well. Biotechnology entrepreneurs may draw initial salaries of $150,000 or more, and potentially can claim very valuable stock options should their start-up companies eventually go public. A number of CEOs ultimately became multimillionaires on the strength of such options in the biotech boom of the late 1990s. SEE ALSO FINANCIAL ANALYST.

Anthony J. Recupero

Bibliography

Pappas, Michael G. *The Biotech Entrepreneur's Glossary*, 2nd ed. Shrewsbury, MA: M. G. Pappas & Company, 2002.

Robbins-Roth, Cynthia. *From Alchemy to IPO: The Business of Biotechnology*. Cambridge, MA: Perseus Publishing, 2001.

Werth, Barry. *Billion-Dollar Molecule: One Company's Quest for the Perfect Drug*. New York: Simon & Schuster, 1994.

Biotechnology: Ethical Issues

Biotechnology is the use of organisms or their parts or products to provide a valuable substance or process. **Fermentation** using microorganisms in brewing, baking, and cheese production are biotechnologies that date back centuries. Production of human insulin in bacteria to treat type I diabetes mellitus without causing allergic reactions is a more modern example of biotechnology. Two widely used biotechnologies that manipulate genes are recombinant DNA technology, which endows single-celled organisms with novel characteristics using genes from other organisms, and transgenic technology, which creates multicellular organisms that bear genes from other types of organisms. Genetically modified (GM) fruits and vegetables, such as a type of corn that manufactures a bacterial insecticide, are transgenic plants.

Ethical issues that arise from modern biotechnologies include the availability and use of privileged information, potential for ecological harm,

fermentation biochemical process of sugar breakdown without oxygen

An environmental activist opposed to the further development of genetically modified foods demonstrates outside Bangkok's UN headquarters during the opening session of the "New Technology Food and Crops: Science, Safety and Society" conference held July 10, 2001.

access to new drugs and treatments, and the idea of interfering with nature. Applications include agriculture and health care.

Agriculture

In agriculture, GM crops have been in the food supply in the United States for several years. Foods containing GM ingredients are not usually labeled to indicate their origin. This is because regulatory agencies determine food safety based on its similarity to existing foods, its chemical composition, and effects on the digestive systems of test animals, not on whether the plant variant arose from traditional agriculture or transgenic technology. If a food is found to include a chemical that could cause an allergy or is a toxin, it is

not marketed. As of early 2002, there have been no reports of harm coming from the consumption of GM foods.

Still, individuals who object to genetic modification would like the opportunity to select plant foods that were not produced in this manner. Labeling would solve this problem, and perhaps with continued consumer pressure it may come to pass. Some argue that at times, those who object to GM foods have acted in unethical ways. In several instances, protesters destroyed what they erroneously thought were fields of GM plants. Companies have behaved in ethically questionable ways in the GM food debate too. Before consumer outrage put an end to it, certain agrichemical companies sold GM crops that did not produce viable seed, forcing farmers to purchase new seed each year.

Another concern arising in agricultural biotechnology is the unintended spread of **transgenes** to other organisms. When a crop is grown in the field, its DNA, including the transgene, can theoretically be spread to other organisms in several ways. Certain types of plant viruses can transfer DNA from the host chromosome to a wild relative as well. Bacteria take up genes from the environment in the process known as transformation, and pass genes among different types of plants through **conjugation**. It is not yet known whether any of these latter processes have occurred with GM plant DNA, and detection may be difficult. It is likely to be a question not of whether but of when, however, given the large acreages devoted to GM plants.

Again the question arises of whether the consequences of such gene transfers are qualitatively different from the same process occurring on crop plants modified through traditional breeding. Opponents of GM crops say yes, since the potential exists to transfer genes from sources that would otherwise never be found in the agricultural environment. For instance, jellyfish genes are used in some agricultural research. In addition, the potential for harm from "escaped genes" may be greater precisely because the gene is so useful agriculturally. The gene for a natural insecticide may help grow safer corn, for example, but it could also allow a wild plant to escape its natural controls and become a serious forest weed. While such scenarios are hypothetical for the moment, opponents say that so little is known about the intricacies of ecology that caution is the only safe policy.

Health Care

In health care, genetic testing presents several ethical challenges. Legislation is in place or is being developed to limit access to genetic information, so that employers or insurers cannot discriminate against individuals because of their genotypes. Testing for a genetic disease presents a complication not seen in other types of illnesses, because the diagnosis of one individual immediately reveals the risk that other family members may be affected, based on the rules of inheritance. For example, a young woman learned that she is a carrier of Wiskott-Aldrich syndrome, which causes severe immune deficiency that is lethal in childhood. Because the mutant gene is carried on the X chromosome, each of her sons faces a 50 percent chance of inheriting the illness. Knowing that the same would be true for other carriers in her family, the young woman chose to inform all of her relatives who might also carry the gene. The decision of whether or not to be tested rests with the individuals.

transgenes genes introduced into an organism

conjugation a type of DNA exchange between bacteria

Another ethical dilemma in health care that arises from biotechnology is cost and access to new treatments. Such drugs as tissue plasminogen activator, used to break up clots that cause heart attacks and strokes, and erythropoietin and colony-stimulating factors, used to restore blood supplies in cancer patients being treated with chemotherapy, are extremely expensive. Although insurers often cover the costs in the United States, people in many other nations cannot take advantage of these drugs.

Because biotechnology is a rapidly evolving field, experiments and **clinical trials** are ongoing. Participation in a clinical trial of a recombinant DNA-derived drug, or of a gene therapy, requires informed consent. A case in 1999 provoked reevaluation of the care with which such participants are screened, and of the adequacy of informed consent protocols. Jesse Gelsinger was 19 when he received experimental gene therapy to treat ornithine transcarbamylase deficiency. In this disorder, lack of an enzyme that metabolizes protein leads to buildup of ammonia, which damages the brain. Most affected individuals die within days of birth, but survivors can usually control symptoms with diet and drugs, as Jesse had been doing. This is why his death five days after receiving the gene therapy was especially tragic. An underlying and undetected medical condition may have contributed to his death.

clinical trials tests performed on human subjects

Medication usually controlled Gelsinger's condition, but he chose to join the clinical trial so that he might help babies who died of a more severe form of the illness. Although Gelsinger clearly stated that he realized he might die, questions arose about the extent of his knowledge, largely because he had been healthy. This issue does not arise in the more common situation in which a participant in a gene therapy trial has exhausted conventional treatments, because he or she has little to lose at that point. Clinical trials of gene therapy are now conducted with much greater care.

New Challenges

Another objection to biotechnology is that it interferes with nature, but so do traditional agriculture and medicine. However, the changes that biotechnology can introduce are usually quite unlikely to occur naturally, such as a tobacco plant that glows thanks to a firefly protein, or cloning a human. We place limits on some biotechnologies, but not on others, based on our perceptions and on the intents of the interventions. The glowing tobacco plant was done as an experiment to see if a plant could express a gene from an animal, but many countries ban human cloning because it is seen as unnecessary, dangerous, and unethical. Still, time can change minds. When Louise Joy Brown, the first baby conceived using in vitro fertilization, was born in 1980, objection to "test tube baby" technology was loud. The procedure is now routine. In general, it seems that a biotechnology will eventually be considered ethical if evidence accumulates demonstrating that it does no harm.

A biotechnology that by its very definition causes harm is bioterrorism, especially when genetic manipulation is used to augment the killing power of a naturally occurring pathogen. Bioterrorism dates back to the Middle Ages, when Tartan warriors hurled plague-ridden corpses over city walls to kill the inhabitants. The British used a similar approach in the eighteenth century, when they intentionally gave Native Americans blankets that carried smallpox virus. Efforts in the former Soviet Union to create bio-weapons

from the 1970s until the 1990s introduced genetic modifications. For example, they engineered plague bacteria to be resistant to sixteen different antibiotic drugs and to produce a toxin that adds paralysis to the list of its effects. International efforts to ban bio-weapon development in the wake of the attacks on the World Trade Center in New York City on September 11, 2001, might put an end to this subversion of biotechnology. SEE ALSO AGRICULTURAL BIOTECHNOLOGY; CLONING: ETHICAL ISSUES; GENE THERAPY: ETHICAL ISSUES; GENETIC DISCRIMINATION; GENETIC TESTING: ETHICAL ISSUES; METABOLIC DISEASE; REPRODUCTIVE TECHNOLOGY: ETHICAL ISSUES; TRANSGENIC ORGANISMS: ETHICAL ISSUES.

Ricki Lewis

Bibliography

Burgess, Michael M. "Beyond Consent: Ethical and Social Issues in Genetic Testing." *Nature Reviews Genetics* 2 (Feb., 2001): 147–151.

Dale, Philip. "Public Concerns over Transgenic Crops." *Genome Research* 10 (Jan., 2000): 1.

Stolberg, Sheryl Gay. "The Biotech Death of Jesse Gelsinger." *The New York Times Magazine* (Nov. 28, 1999): 17.

Biotechnology and Genetic Engineering, History of

The term "biotechnology" dates from 1919, when the Hungarian engineer Karl Ereky first used it to mean "any product produced from raw materials with the aid of living organisms." Using the term in its broadest sense, biotechnology can be traced to prehistoric times, when hunter-gatherers began to settle down, plant crops, and breed animals for food. Ancient civilizations even found that they could use microorganisms to make useful products, although, of course, they had no idea that it was microbes that were the active agents. About B.C.E. 7000, the Sumarians and Babylonians discovered how to use yeast to make beer, and winemaking dates from biblical times. In about B.C.E. 4000, the Egyptians found that the addition of yeast produced a light, fluffy bread instead of a thin, hard wafer. At the same time, the Chinese were adding bacteria to milk to produce yogurt.

Genetic Engineering versus Biotechnology

For many, the term "biotechnology" is often equated with the manipulation of genes, but, as Ereky's definition suggests, this is only one aspect of biotechnology. For the more specific technique of gene manipulation, the term "genetic engineering" is more appropriate. Genetic engineering dates from the 1970s. At that time molecular biologists devised methods to isolate, identify, and clone genes as well as to mutate, manipulate, and insert them into other species. One of the key elements in such research was the discovery of **restriction enzymes**. These enzymes are able to cleave DNA at a limited number of sequence-specific sites and often leave "sticky ends." Isolated DNA from any organism could be cleaved with a restriction enzyme and then mixed with a preparation of a **vector** that had been cleaved with the same restriction endonuclease. By virtue of the "sticky ends," a **hybrid**

restriction enzymes enzymes that cut DNA at a particular sequence

vector carrier

hybrid combination of two different types

Egyptian artwork, dating from between B.C.E. 1550 and 1295, depicts the harvest of the grapes and subsequent counting of the jars of wine. This art suggests that ancient civilizations fermented grape juice to make wine, establishing the basics of a process still used in wineries today.

molecule could be created that contained the gene of interest, which could then be inserted into such a cloning vector. The importance of restriction endonucleases was recognized in 1978 by the awarding of the Nobel Prize in physiology or medicine to Werner Arber, Daniel Nathans, and Hamilton Smith for their discovery of these enzymes.

Further Advances and Ethical Concerns

The first experiment to combine different DNA molecules was performed in 1972 in the laboratory of Paul Berg (who shared the 1980 Nobel Prize in chemistry for this work). The following year Stanley Cohen and Herbert Boyer combined some viral DNA and bacterial DNA in a **plasmid** to create the first recombinant DNA organism.

plasmid a small ring of DNA found in many bacteria

Realizing the potential dangers of moving genes from one organism to another, approximately ninety prominent scientists, whose laboratories were poised to start cloning experiments, met in 1975 at the Asilomar Conference Center in California to discuss the potential dangers of gene manipulation. This meeting, wherein scientists recognized and openly discussed the ramifications and potential dangers of their research before that research was actually begun, was unprecedented. The result of the Asilomar Conference was to call for and agree upon a one-year moratorium before any cloning experiments were to be done. This provided time to develop guidelines for the physical and biological isolation of recombinant organisms, to ensure that they not escape into the environment, and, if they did, to make sure that they would be so weakened as not to survive competition with naturally occurring organisms. By 1976, then, gene cloning was in full swing around the world.

recombinant DNA DNA formed by combining segments of DNA, usually from different types of organisms

hormone molecule released by one cell to influence another

Key Technical Developments

Advances in biotechnology were marked by the development of key research techniques. In 1976, Herbert Boyer and Robert Swanson founded Genentech, the first biotechnology company to use **recombinant DNA** technology in developing commercially useful products such as drugs. The year 1977 is considered the "dawn of modern biotechnology," for it was in that year that the first human protein was cloned and manufactured using genetic engineering technology: Genentech reported the cloning of the human **hormone** somatostatin. This year was also important for the development of the technique of DNA sequencing, achieved by Fred Sanger and Walter Gilbert (who, with Paul Berg, shared the 1980 Nobel Prize in chemistry).

In 1978 Genentech was able to isolate the gene for human insulin and begin clinical trials that resulted in the approval and marketing of the first genetically engineered drug for human use. This was a major accomplishment. Diabetes, the seventh leading cause of death in the United States, affects millions of Americans. In the past, insulin was extracted from the pancreases of cows or pigs, then used to treat diabetics. Although insulin from these species is very similar to human insulin and was effective in humans, the small differences between human and animal insulin were enough to cause problems for some patients. Often patients developed immunological reactions to the foreign protein, reducing its effectiveness. With the availability of genetically engineered human insulin, these problems were eliminated.

Patents and the Rise of Biotechnology Companies

In 1980 the U.S. Supreme Court provided an important incentive for the development of biotechnology companies. In the case of *Diamond* v. *Chakrabarty*, the court ruled that biological materials may be patented. Thus, private companies could look forward to making substantial profits from therapies that they developed through genetic engineering techniques.

Among the new companies to take advantage of the court ruling was the Chiron corporation, which cloned the protein that formed the outer coat of the human hepatitis B virus. This protein, which could now be produced without the virus that it normally enclosed, provided the material for the development of the first human vaccine using recombinant DNA technology. The hepatitis **vaccine** has been available since 1987.

vaccine protective antibodies

In the same year, the Food and Drug Administration (FDA) approved Genentech's drug tPA (tissue plasminogen activator). This is a human blood protein that helps to dissolve fibrin, the major protein involved in forming blood clots at the site of an injury. After the healing process is complete and clotting is no longer required at the site of the injury, the body normally releases tPA to activate an enzyme called plasmin, which dissolves fibrin. However, it was discovered that tPA could also be used as a powerful drug in the treatment of certain heart attacks. Sometimes a blood clot forms spontaneously in the body. If the clot forms or lodges in the coronary arteries of the heart, the clot blocks blood flow to the heart muscle, resulting in what is commonly called a heart attack. If tPA is given to such patients within four hours of onset, the recovery is truly remarkable. Such patients are able to leave the hospital the next day with little or no after-effects of the heart attack. A patient not treated with tPA often remains in the hospital for a week or longer and can not resume normal activities until after a long recovery period. The drug has subsequently been approved for use with patients suffering a stroke from a blood clot in the brain with similar success.

Biotechnology has also been sucessful in development of other useful products. Today many laundry detergents contain proteases, enzymes that remove stains by digesting the protein components of the stain. However, such enzymes are inactivated by bleach. In 1988 the biotechnology company Genecor received approval for a bleach-resistant protease. This had been accomplished by isolating the gene for protease and then, using site-directed mutagenesis, changing the gene such that the corresponding protein was no longer sensitive to inactivation by bleach.

Biotechnology has also made a great impact in agriculture. The first genetically engineered plant was patented in 1983. The first genetically engineered food was produced by a company called Calgene, in 1987. Calgene, now a part of Monsanto, produced a tomato that could be ripened on the vine and transported ripe to market. Tomatoes are normally shipped green to market and left to ripen at their destination because they are easily bruised and damaged if shipped when fully ripe. Today there is a new "green revolution" under way, in which genetically modified food will provide greater nourishment and higher yields, while simultaneously reducing the use of fertilizers and herbicides. Although there is considerable controversy surrounding these foods (sometimes referred to as "Frankenfood"), there have been no documented cases of anyone being hurt by eating them. In 1990 the biotech firm GenPharm created a transgenic dairy cow into which the genes for human milk proteins were inserted. The milk from such cows will be used for producing infant formula.

Biotechnology and the Law

Biotechnology has also made important contributions to the field of law. Most notably, scientists have developed exquisitely sensitive methods for identifying DNA. Indeed, with the invention of the **polymerase** chain reaction in 1988, enough DNA can be extracted from a drop of blood, a tiny shred of skin, a single hair, or a small semen sample to identify the individual from whom it originated. Such "genetic fingerprinting" was developed in 1984 and first used in a trial in 1985.

polymerase enzyme complex that synthesizes DNA or RNA from individual nucleotides

Perhaps the most famous case involving DNA-based evidence was the O. J. Simpson murder trial in 1995. During the 1990s however, genetic evidence in the courtroom became commonplace and accepted by trial lawyers, judges, and juries alike. In fact, several innocent people were released from prison as a result of the reexamination of evidence using DNA fingerprinting.

The Human Genome Project

In 1990 molecular biologists around the world began working on what ranks as perhaps the greatest achievement of biotechnology, the Human Genome Project, in which the more than 3 billion **nucleotides** of DNA in the human nucleus were ultimately sequenced. Although DNA sequencing began in 1977, it was the development in the 1990s of automated DNA sequencers and powerful computers to store and analyze the data that made this project

nucleotides the building blocks of RNA or DNA

feasible. The first draft of the human genome was completed in 2001. With the complete sequence available, scientists will be able to "mine the genome" to find important gene products and to design specific drugs to target gene product. The twenty-first century will see tremendous new advances using biotechnology. SEE ALSO AGRICULTURAL BIOTECHNOLOGY; BIOINFORMATICS; BIOTECHNOLOGY; BIOTECHNOLOGY ENTREPRENEUR; CLONING GENES; CLONING ORGANISMS; GENETICALLY MODIFIED FOODS; GENOMICS INDUSTRY; HUMAN GENOME PROJECT; MUTAGENESIS; PATENTING GENES; PLANT GENETIC ENGINEER; RESTRICTION ENZYMES; SANGER, FRED.

Ralph R. Meyer

Bibliography

Alcamo, I. Edward. *DNA Technology: The Awesome Skill*, 2nd ed. San Diego: Academic Press, 2001.

Bud, Robert, and Mark F. Cantley. *The Uses of Life: A History of Biotechnology*. Cambridge, U.K.: Cambridge University Press, 1983.

Weaver, Robert F. *Molecular Biology*, 2nd ed. New York: McGraw-Hill, 2002.

Birth Defects

A birth defect is an anomaly that is congenital, or present from birth. Birth defects are the leading cause of infant mortality, causing 22 percent of all infant deaths. Approximately 3 to 4 percent of all live births are affected by a birth defect; the causes of most of them are unknown. Some birth defects are considered to be physical, while others are thought of as functional. Physical birth defects result in the malformation of a physical organ or limb, whereas functional birth defects are those that cause primarily functional, rather than physical, problems. Functional birth defects include mental retardation, congenital hearing loss, early-onset vision impairment, and numerous other health concerns. They may be caused by single gene mutations, or they may be due to polygenic or multifactorial inheritance.

Birth defects may be found in isolation or they may occur in combination in one child, as part of a larger syndrome. A syndrome is usually defined as the presence of three or more birth defects due to one underlying cause, that characterize a particular disease or condition. An example of this is Down syndrome.

Various Causes, Various Treatments

etiologies causations of disease

There are several **etiologies** of birth defects, including single gene mutations, polygenic and multifactorial conditions, chromosomal abnormalities, and teratogens, which cause growth or developmental abnormalities. These etiologies may cause both physical and functional birth defects. The diagnosis, treatment, and management of birth defects often involves a team of professionals and specialists. Among these specialists are clinical geneticists, medical geneticists, and genetic counselors. These are all genetic service providers who are trained to help make a diagnosis and identify whether a birth defect is isolated or part of a syndrome. An accurate diagnosis is of utmost importance in being able to treat the condition and anticipate any future health concerns that may arise.

LEADING CATEGORIES OF BIRTH DEFECTS	
Birth Defects	**Estimated Incidence**
Structural/Metabolic	
Heart and circulation	1 in 115 births
Muscles and skeleton	1 in 130 births
Club foot	1 in 735 births
Cleft lip/palate	1 in 930 births
Genital and urinary tract	1 in 135 births
Nervous system and eye	1 in 235 births
Anencephaly	1 in 8,000 births
Spina bifida	1 in 2,000 births
Chromosomal syndromes	1 in 600 births
Down syndrome (Trisomy 21)	1 in 900 births
Respiratory tract	1 in 900 births
Metabolic disorders	1 in 3,500 births
PKU	1 in 12,000 births
Congenital Infections	
Congenital syphilis	1 in 2,000 births
Congenital HIV infection	1 in 2,700 births
Congenital rubella syndrome	1 in 100,000 births
Other	
Rh disease	1 in 1,400 births
Fetal alcohol syndrome	1 in 1,000 births

Adapted from "Leading Categories of Birth Defects." March of Dimes Perinatal Data Center, 2000.

Single-Gene Mutations

Single-gene mutations are defects or changes in genes that may be passed on from generation to generation. The most common forms of inheritance include **autosomal** dominant, autosomal recessive, X-linked recessive, and new dominant mutations. Single-gene mutations may result in both structural and functional birth defects. These defects may be found in isolation or as part of a known syndrome and may display **phenotypic** variation, in which the same mutant gene leads to variable clinical problems. This often occurs with dominant mutations such as Marfan syndrome. A genetic disorder may also display reduced penetrance, wherein not everyone with the genetic mutation will have the disorder or trait. These are also known as "silent" disorders.

autosomal describes a chromosome other than the X and Y sex-determining chromosomes

phenotypic related to the observable characteristics of an organism

Autosomal Dominant Disorders

An example of an autosomal dominant disorder is achondroplasia, the most common form of short-limbed dwarfism in humans. Achondroplasia displays complete penetrance (everyone with the genetic defect also has the disorder), and it occurs in 1 out of 25,000 births. Most cases are **sporadic** rather than inherited.

Achondroplasia is a growth disorder caused by a mutation of the gene that encodes the fibroblast growth factor receptor 3 (*FGFR3*), and it is

sporadic caused by new mutations

characterized by short limbs, malformed hands, a disproportionately large head, and abnormal facial features. Medical problems are due to abnormally configured bones and related structures, leading to hydrocephalus, problems of the spine, frequent sinus and ear infections, and orthopedic problems. If one parent is affected with achondroplasia, there is a 50 percent risk that an offspring will also be affected. If both parents are affected, there is a 25 percent chance that an offspring will inherit two gene copies and develop severe, life-threatening features. Two known mutations in *FGFR3* account for 98 percent of all achondroplasia cases. This makes early identification, even prenatal diagnosis, relatively easy.

Another autosomal dominant condition is Marfan syndrome, which results from a defect in the synthesis, secretion, or utilization of the protein fibrillin, an important component of connective tissue throughout the body. The gene for Marfan syndrome is fibrillin 1 (*FBR1*). Marfan syndrome features are variable, including cardiovascular, skeletal, and ocular defects.

Marfan syndrome's most serious medical complication is the risk of sudden death from aortic dissection, a tear in the inner wall of the major artery leading from the heart. Approximately 75 percent of individuals with Marfan syndrome have a family history of the disease, with the rest occurring as new mutations. Because the syndrome is autosomally dominant, affected indviduals have a 50 percent risk of passing the mutated gene to their offspring. The condition has full penetrance; therefore, all individuals who inherit this mutation will express some features of Marfan syndrome.

Functional Birth Defects

base pair two nucleotides (either DNA or RNA) linked by weak bonds

Fragile X syndrome is the most common cause of inherited mental retardation, occurring in one out of 1,000 births. It is caused by expansion of a "triplet repeat" section of nucleotides in the *FMR-1* gene on the X chromosome. Triplet repeats are three-**base-pair** sequences in a gene that are abnormally repeated, sometimes dozens or even hundreds of times, causing abnormal protein sequence and structure. Because it is carried on the X chromosome, it affects males more often than females. In males, an FMR-1 gene with greater than 200 repeats is always associated with the syndrome. Inactivated *FMR-1* gene causes impaired mental function. The FMR-1 protein is thought to help shape the connections between **neurons** that underlie learning and memory. Affected individuals may also have large testes, abnormal facial features, seizures, and emotional and behavior problems. DNA testing allows for detection of carriers as well as affected individuals, enabling the use of genetic counseling and prenatal testing.

neurons brain cells

Another common functional genetic disorder is hearing loss, which can result from a defect in any one of more than fifty different genes. One gene, connexin 26, may be responsible for a large portion of inherited hearing loss. Some forms of congenital hearing loss may be due to prenatal exposure to infectious agents such as rubella. Genetic screening or screening for hearing loss at birth may be the most important test for hearing impairment yet to be developed, as early recognition and treatment can lead to dramatic improvements in hearing and, consequently, in the development of language in early childhood.

Other genetic disorders that cause functional birth defects include those involved in various aspects of the immune system. The most severe form is severe combined immune deficiency (SCID), in which a major type of immune cell, the lymphocyte, is absent. People with SCID suffer life-threatening infections beginning in infancy and may require complete physical isolation. This was the case for David Vetter, who became known to the world as the "bubble boy."

Multifactorial and Polygenic Inheritance

Many traits and diseases are caused by the interaction of inherited genes and the environment. These are known as "multifactorial" traits. While all genes interact with the environment, the impact of the environment in multifactorial traits and diseases is usually greater than in single-gene traits and diseases. Prenatal environmental influences are inevitably filtered through the maternal-placental system and include factors such as infections, drugs, tobacco or alcohol use, diabetes, and industrial toxins.

Polygenic traits and diseases are due to the cumulative effect of multiple genes, working together. Many congenital birth defects are thought to be multifactorial, such as pyloric stenosis (narrowing of the passage from stomach to intestine), cleft lip and palate, clubfoot, and neural tube defects. When found as isolated birth defects, these conditions are thought to be explained by a "multifactorial threshold model."

The multifactorial threshold model assumes the gene defects for multifactorial traits are normally distributed within the population. This means that almost everyone has some genes involved with these conditions, with most individuals having too few of them to cause disease. Individuals will not become affected with the condition unless they have a genetic liability that is significant enough to push them past the threshold, moving them out of the unaffected range and into the affected range (Figure 1).

Examples of Multifactorial and Polygenic Effects

Cleft lip with or without cleft palate (CL/P) is a heterogeneous disorder (those children affected may have somewhat different abnormalities) occurring in 1 out of 1,000 births. Some CL/P cases occur as isolated birth defects, while others occur as part of a larger syndrome. The majority of CL/P cases are associated with multifactorial inheritance. The risk to relatives of affected individuals can be anywhere from 0.5 to 15 percent, depending on the severity of the clefting and the degree of relationship to the affected individual, with risks highest for first-degree relatives. Some unique cases of CL/P may be associated with genetic syndromes that are due to single-gene mutations or chromosomal abnormalities.

Clubfoot is another primarily multifactorial defect and occurs in 1 out of 10,000 Caucasian newborns. The estimated risk to relatives of inheriting this defect is between 2 to 20 percent, depending upon the family history. Clubfoot can also have genetic causes such as chromosomal abnormalities or single-gene disorders, or it may have an environmental origin, such as problems caused by amniotic fluid or structural abnormalities of the uterus that restrict fetal growth and mobility. Clubfoot can also be due to autosomal recessive as well as autosomal dominant inheritance, and it may also occur as part of a larger syndrome.

Another class of multifactorial disorders is known as neural tube defects (NTDs). The neural tube is the embryonic structure that develops into the

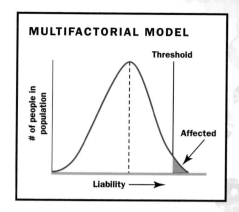

Figure 1. The threshold model for multifactorial traits. The vertical axis represents the number of people with a given level of overall liability. On the horizontal axis, individuals below the threshold do not have the disease, while those above it do.

Thawiphop Phimthep waits for his surgery to fix his cleft lip and palate in Si Saket Hospital in Sisaket Province, Thailand. Thawiphop's surgery will be performed by a group of plastic surgeons from Bangkok.

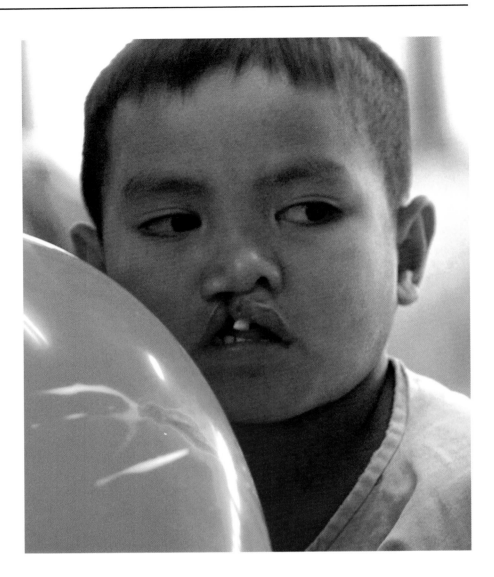

brain and spinal cord. Failure of the neural tube to close, which normally occurs during in the fourth week of gestation, results in an NTD, usually spina bifida or anencephaly. Spina bifida ("open spine") is a defect of the spine. The most common form of spina bifida causes some degree of leg paralysis, impaired bladder and bowel control, and sometimes mental retardation. Anencephaly is a rapidly fatal condition in which a baby is born with a severely underdeveloped brain and skull.

While most NTDs are inherited as multifactorial disorders, a few result from single-gene disorders, chromosomal abnormalities, or teratogens. NTDs currently have an incidence of 1 per 2,000 births. This rate has fallen dramatically over the past thirty years, due to the remarkable effects of NTD-prevention efforts. Maternal deficiency in folate (a B vitamin) greatly increases the risk of NTDs, but taking multivitamins containing folic acid before conception and early in pregnancy is highly effective in preventing these disorders. High doses of folic acid are needed to help protect the fetuses of women with pregnancies previously affected by NTDs, and for those who need to take certain medications that interfere with folate **metabolism**.

metabolism chemical reactions within a cell

Chromosome Disorders

Humans normally have twenty-two pairs of **autosomes** and two sex chromosomes, XX or XY, making forty-six chromosomes in total. Chromosomal abnormalities occur in about 0.5 percent of all live births and are usually due to an abnormal number of chromosomes. These are nearly always an addition or deletion of a single autosome or sex chromosome in a pair. One extra copy of a chromosome is called a trisomy, while one missing copy of a chromosome is called a monosomy. Sometimes only a segment of a chromosome is duplicated or lost.

Chromosomal disorders are diagnosed by **karyotype** analysis and can be done on adults by testing blood, skin, or other tissue. Karyotypes can also be performed on a fetus through specialized testing such as amniocentesis. Prenatal maternal blood tests are routinely used to screen for some trisomies, though accurate diagnosis requires fetal karyotyping.

Chromosomal abnormalities can occur in offspring of mothers of all ages, but the frequencies of these disorders increase with maternal age, rising exponentially after the maternal age of thirty-five. Advanced paternal age has far less impact. Chromosomal abnormalities can result in either physical or functional birth defects. The severity of these birth defects is highly variable and depends upon the exact chromosome problem.

Chromosomal defects include such problems as Down syndrome, Klinefelter's syndrome, and Turner's syndrome. The majority of Down syndrome cases are due to an extra chromosome 21. Trisomy 21 usually occurs as an isolated event within a family. It results in characteristic facial features, lax muscle tone, cardiac and intestinal anomalies, and mild or moderate mental retardation. Additional medical complications may also include recurrent ear and respiratory tract infections, vision problems, hearing difficulties, and short stature.

Klinefelter's syndrome is a sex chromosome abnormality that occurs in 1 of 600 males, with a karyotype of 47. Individuals with Klinefelter's syndrome possess an extra X chromosome: XXY. Clinical characteristics are variable and include some learning and developmental disabilities, **hypogonadism**, small testes, and **gynecomastia** occuring in puberty. The condition can be managed by administering testosterone supplements beginning in adolescence. As with some other sex chromosome abnormalities, adults with Klinefelter's syndrome are usually infertile.

Turner's syndrome is another sex chromosome disorder, with a karyotype of 45. In this condition, one X chromosome is missing. Turner's syndrome occurs in one out of 4,000 live births. Most females with Turner's syndrome are short and have webbing of the neck, a broad chest, and a lack of ovarian development, with a consequent lack of pubertal development and infertility. Female hormone therapy is often used to induce breast development and menstruation. The majority of conceptions resulting in a fetus with this condition end in a miscarriage, as Turner's syndrome is highly lethal in early fetal development.

Teratogen Exposure

A teratogen is any agent that can cause birth defects if a fetus is exposed to it. Teratogens are usually drugs or infectious agents such as bacteria or

autosomes chromosomes that are not sex determining (not X or Y)

karyotype the set of chromosomes in a cell, or a standard picture of the chromosomes

The average IQ of an individual with Down syndrome is 55. Most tests cite an IQ of 100 as "average."

hypogonadism underdeveloped testes or ovaries

gynecomastia excessive breast development in males

Clubbed feet (*talipes equinovarous*) result from abnormal development of the muscles, tendons and bones of the feet in utero. Heredity is thought to play a role when such deformities appear.

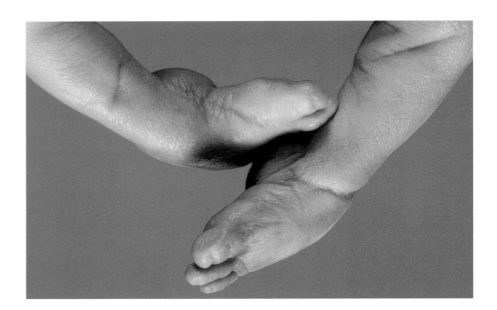

viruses, and can affect a fetus from as early as the first few weeks after conception through the second trimester. For this reason, ideally, women should avoid all medications during pregnancy. However, in some circumstances there are medical risks and benefits that must be weighed, particularly if a medication is important to the health of the prospective mother.

To properly assess the potential danger of a teratogen, information is required about its effect on embryonic development, its ease of passage across the placenta, and the dosage and timing of fetal exposure to the teratogen. There are, however, limitations on the ability of doctors to predict the risk of birth defects arising from fetal exposure to any particular drug because of the lack of information on the effects of multiple medication use and possible drug interactions, the inability to control for other exposures women may have during pregnancy, and the unique genetic susceptibilities of each person. In addition, there are limited clinical studies that address this problem.

Risks for birth defects or adverse pregnancy outcomes associated with any type of exposure are in addition to the 3 percent background risk for birth defects in all pregnancies. There is no evidence linking paternal exposures to teratogenicity for the developing fetus, though exposure to some agents can reduce male fertility.

teratogenic causing birth defects

One well-documented teratogen is the drug thalidomide, which was taken by tens of thousands of women in the 1950s and early 1960s to treat nausea during pregnancy, before its potent **teratogenic** effects were recognized. Even a single dose caused severe birth defects, including amelia (absence of limbs), phocomelia (short limbs), incomplete or absent bone growth, ear and eye abnormalities, congenital heart defects, and others.

microcephaly reduced head size

Another potent teratogen is a substance called isotretinoin, marketed under the brand name of Accutane, used to treat severe acne. Birth defects following prenatal exposure include serious central nervous system defects such as hydrocephalus, **microcephaly**, and mental retardation, as well as cleft lip and palate and anomalies of cardiovascular, limb, eye, and other systems. For women who have taken this substance, it is recommended to delay pregnancy at least one month after they have stopped using it.

Like medications, "recreational" drugs such as alcohol and cocaine can act as teratogenic agents. Fetal alcohol syndrome (FAS) and fetal alcohol effects (FAE) are the most common, completely preventable, and potentially devastating disorders caused by alcohol use during pregnancy. FAS is one of the leading causes of mental disabilities in children. Fetal exposure to alcohol creates additional problems in children born with FAS/FAE, such as characteristic facial features, growth retardation, central nervous system difficulties, learning disabilities, and behavioral problems. No amount of alcohol is thought to be safe during pregnancy; however, some of its effects may be prevented by stopping the exposure during or shortly after the first trimester.

Cocaine use during pregnancy is known to increase the risk of miscarriages and premature labor and delivery. Disturbances in the behavior of exposed newborns have been reported, such as irritability, irregular sleeping patterns, muscular rigidity, and poor feeding. Some birth defects associated with the use of this drug include urinary and genital malformations, as well as defects of the limbs, intestines, and the skull.

Conditions arising from infectious teratogenic agents include toxoplasmosis, syphilis, and rubella. In each of these cases, the mother is exposed to the infectious agent, then transmits it to the fetus. In toxoplasmosis, the parasite *Toxoplasma gondii* can be transmitted from cats to humans through contact with cat feces (cleaning litter box or gardening), or through consumption of undercooked meats, poorly washed fruits and vegetables, goat's milk, or raw eggs. Mother-to-fetus transmission is more likely if maternal infection occurs in the last few weeks before delivery, but early fetal exposure is generally associated with greater severity of defects in the child. Overall, 20 to 30 percent of untreated, infected newborns have birth defects, including seizures, microcephaly, and other severe effects on the nervous system. Treatment of the mother with antibiotics during pregnancy is safe for the fetus, and significantly reduces the likelihood of fetal infection.

Syphilis is an infection caused by the spirochete *Treponema pallidum*. This bacterium crosses the placenta and may result in fetal infection. If untreated, the pregnancy may end in miscarriage, stillbirth, or neonatal death. Signs of congenital infection include jaundice, joint swelling, rash, anemia, and characteristic defects of bone and teeth. Maternal treatment of this condition may help prevent the transmission to the fetus and its ill effects.

Rubella is the scientific name for the disease commonly known as German measles. **Congenital** rubella syndrome (CRS) results from the exposure of an unprotected pregnant woman to the rubella virus, and can lead to major birth defects, including serious malformations of the heart, blindness, deafness, and mental retardation. CRS has been virtually eradicated in the United States because of the near-universal vaccination against rubella, now part of the standard childhood immunizations program. Unfortunately, this vastly improved situation is not as common in much of the rest of the world.

congenital from birth

Maternal Conditions

Birth defects can also result from physical conditions affecting the health of the mother. One common maternal condition associated with birth defects is diabetes mellitus, a multifactorial disorder. Mothers with diabetes have a two- to three-fold times greater risk of having a child with birth defects than the general population if their condition is not well controlled. However, good glucose control has been shown to correlate with a decreased risk of congenital malformations. Characteristic diabetic malformations include cardiovascular, craniofacial, genitourinary, gastrointestinal, and neurological abnormalities. The risk that the child born of a mother with diabetes mellitis will also develop diabetes as an adult is 1 to 3 percent.

Another maternal condition giving rise to birth defects is maternal phenylketonuria (PKU). This is an autosomal recessive disorder in which an **enzyme** called phenylalanine hydroxylase is defective. This enzyme normally converts a substance in the blood called phenylalanine to another substance called tyrosine. As a result, phenylalanine levels are high, resulting in mental retardation, microencephaly, growth retardation, cardiac problems, seizures, vomiting, and hyperactivity. Other traits associated with PKU are fair hair and skin and blue eyes. PKU can be effectively managed through changes in the diet, and women who have appropriately managed their diet can have pregnancies with healthy offspring. SEE ALSO CHROMOSOMAL ABERRATIONS; CLINICAL GENETICIST; COMPLEX TRAITS; DIABETES; DOWN SYNDROME; FRAGILE X SYNDROME; GENETIC COUNSELING; GROWTH DISORDERS; SEVERE COMBINED IMMUNE DEFICIENCY; TRIPLET REPEAT DISEASE.

Nancy S. Green and Terri Creeden

enzyme a protein that controls a reaction in a cell

Bibliography

Batshaw, Mark L. *When Your Child Has a Disability: The Complete Sourcebook of Daily and Medical Care.* Baltimore, MD: Paul H. Brooks Publishing, 2001.

"Leading Categories of Birth Defects." March of Dimes Perinatal Data Center, 2000.

Internet Resource

National Organization of Rare Disorders. <http://www.rarediseases.org>.

Blood Type

Blood has two main components: **serum** and cells. In 1900 Karl Landsteiner, a physician at the University of Vienna, Austria, noted that the sera of some individuals caused the red cells of others to **agglutinate**. This observation led to the discovery of the ABO blood group system, for which Landsteiner received the Nobel Prize. Based on the reactions between the red blood cells and the sera, he was able to divide individuals into three groups: A, B, and O. Two years later, two of his students discovered the fourth and rarest type, namely AB.

serum fluid portion of the blood (plural sera)

agglutinate clump together

Antigens and Antibodies

To understand blood typing, it is necessary to define antigen and antibody. An antigen is a substance, usually a protein or a **glycoprotein**, which, when injected into a human (or other organism) that does not have the antigen,

glycoprotein protein to which sugars are attached

RELATIONSHIPS BETWEEN BLOOD TYPES AND ANTIBODIES

Blood Type	Antigens on Red Blood Cell	Can Donate Blood To	Antibodies in Serum	Can Receive Blood From
A	A	A, AB	Anti-B	A, O
B	B	B, AB	Anti-A	B, O
AB	A and B	AB	None	AB, O
O	None	A, B, AB, O	Anti-A and anti-B	O

will cause an **antibody** to be produced. Antibodies are a specific type of immune-system proteins known as immunoglobulins, whose role is to fight infections by binding themselves to antigens. In the case of the ABO blood groups, the antigens are present on the surface of the red blood cell, while the antibodies are in the serum. These antibodies are unique to the ABO system and are termed "naturally occurring antibodies." The table shows the relationships between blood types and antibodies.

This aspect of the ABO blood group system is very important in transfusion. Blood group O individuals are said to be universal donors, because their blood can be used for transfusion in individuals who have any one of the four blood types. On the other hand, individuals with blood type A can only donate to either type A or type AB, and individuals with blood type B can only donate to B or AB types. AB individuals can only donate to type AB. However, before any transfusions, donor blood is mixed with serum from the recipient (a process called cross matching) to ensure that no agglutination will occur after transfusion.

Multiple Alleles

The genetic basis of the ABO blood group system is an example of multiple **alleles**. There are three alleles, A, B, and O, at the ABO **locus** on chromosome 9. The expression of the O allele is recessive to that of A and B, which are said to be co-dominant. Thus, the genotypes AO and AA express blood type A, BO and BB express blood type B, AB expresses blood type AB, and OO expresses blood type O. In the past, ABO blood group typing was used extensively both in forensic cases as well as for paternity testing. More recently, DNA testing, which is much more informative, has superseded these tests.

The ABO blood group substances are glycoproteins, the basic molecule of which is known as the H substance. This H substance is present in unmodified form in individuals with blood type O. Adding extra sugar molecules to the H substance produces the A and B substances. The frequency of the ABO blood types varies widely across the globe. For example, blood group B has a frequency of 25 percent in Asians, 17 percent in Africans, but only 8 percent in Caucasians. The frequency of blood group O in Europe increases as one travels from southern to northern countries.

Alleles at a locus independent of the ABO blood group locus, known as the secretor locus, determine an individual's ability to secrete the ABO blood group substances in saliva and other body fluids. There are two genes, *Se* and *se*, where *Se* is **dominant** to *se*. In other words, an individual with at

antibody immune-system protein that binds to foreign molecules

alleles particular forms of genes

locus site on a chromosome (plural, loci)

dominant controlling the phenotype when one allele is present

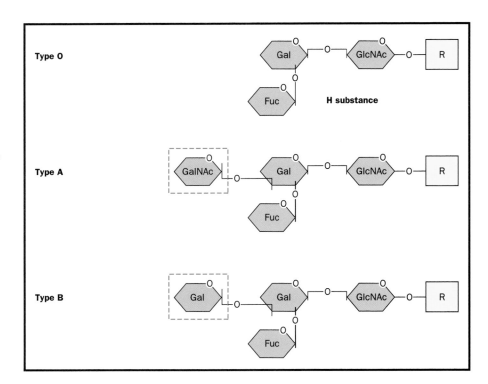

Molecular structure of blood type. A, B, and O antigens differ in the presence and type of the terminal sugar on a common glycoprotein base. The genes for A, B, and O blood type code for enzymes that add these sugars. Adapted from <http://www.indstate.edu/thcme/mwking/abo-bloodgroups.gif>.

least one *Se* gene is a secretor. Approximately 77 percent of Europeans are secretors. This frequency is rarely less than 50 percent and sometimes as high as 100 percent in other populations.

An interesting aspect of the ABO blood groups is their association with disease. Among individuals with stomach and peptic ulcers, there is an excess of type O individuals, whereas among those with cancer of the stomach, there is an excess of type A individuals. Not all type O individuals have an increased risk for peptic or stomach ulcers, however. If type O individuals are secretors, they are protected against ulceration, whereas non-secretors have a two-fold increased risk. Thus the presence of ABO blood group substances act as a protective agent against the development of stomach and peptic ulcers.

The Rh System

The second most important blood group in humans is the Rhesus (Rh) system. Landsteiner and Wiener discovered the Rh blood group in 1940. They found that when they injected rabbits with Rhesus monkey blood; the rabbits produced antibodies against the Rhesus red cells. These antibodies reacted with red blood cells taken from 85 percent of Caucasians in New York City, who were thus said to be Rh positive, while the remaining 15 percent were Rh negative.

One year earlier (1939), Levine and Stetson published a paper describing the mother of a stillborn infant who had a severe reaction when transfused with her husband's blood. They tested the woman's serum and found that it reacted with 77 percent of blood donors. They postulated that the mother had been exposed to blood from her fetus and produced an antibody that reacted with it. The same antigen was present in the baby's father, explaining the woman's reaction to his blood. Their con-

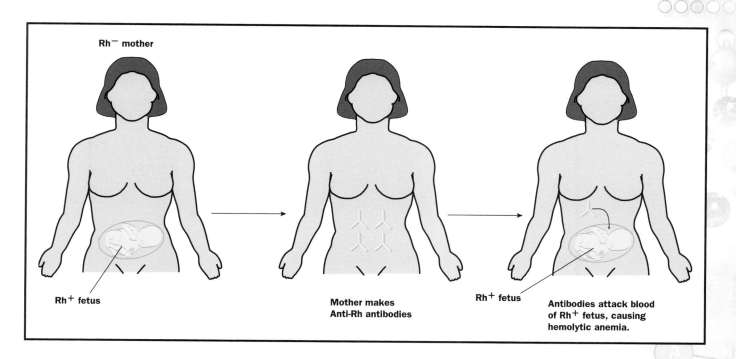

Rh⁻ mother

Rh⁺ fetus

Mother makes Anti-Rh antibodies

Rh⁺ fetus

Antibodies attack blood of Rh⁺ fetus, causing hemolytic anemia.

clusion was correct, and later they realized that they had discovered the same antigen (Rh) that was discovered in the following year. The antibody found in the mother of the stillborn child was shown to be identical to the anti-Rh antibody produced in the rabbit by Landsteiner and Wiener.

The Rh blood group system is the major cause of **hemolytic anemia** in the newborn. A fetus who is Rh⁺ and whose mother is Rh⁻ is at high risk for this disorder, because the mother will produce antibodies against the fetal antigen. The first such fetus is usually not at risk since the fetal cells do not enter the mother's circulation until the time of birth. Only at this time does the mother produce anti-Rh⁺ antibodies. This complicates future pregnancies, because her antibodies will enter the fetal circulation system and react with fetal blood, causing **hemolysis**.

A treatment for Rh⁻ women at risk to have an Rh⁺ fetus is now widely used. Anti-Rh⁺ antibody is injected into the mother soon after her first delivery. This antibody coats the fetal Rh⁺ cells in the mother's circulation, which prevents them from causing antibody production in the mother and, therefore, her next child will not be at risk for hemolytic anemia.

The precise genetics of the complex Rh system has been in dispute since the early discoveries. The Rh blood group system is, in fact, much more complex than simply Rh⁺ and Rh⁻. There are two genes, one of which has four possible alleles, giving six antigens of which five are commonly tested. The first is D, which is the dominant gene that determines whether one is Rh⁺ or Rh⁻. Individuals with genotypes DD and Dd are Rh⁺ and those who are dd are Rh⁻. The DD and Dd genotypes cannot be distinguished from one another, since there is no "anti-d" antibody. The remaining four antigens are C, c, E, and e. The Rh locus is on the short arm of chromosome 1 and consists of two tandem genes. The first, RHCE, codes for non-

Incompatability in Rh type can cause hemolytic anemia in a second child.

hemolytic anemia blood disorder characterized by destruction of red blood cells

hemolysis breakdown of the blood cells

polypeptide chain of amino acids

RhD proteins while the second codes for the RhD protein. The Rh **polypeptide** has been sequenced. It contains 417 amino acids. Thus the molecular genetics conferring different antigenic Rh types is now clear. SEE ALSO GENOTYPE AND PHENOTYPE; IMMUNE SYSTEM GENETICS; INHERITANCE PATTERNS; MULTIPLE ALLELES.

P. Michael Conneally

Bibliography

Cavalli-Sforza, L. L., and W. F. Bodmer. *The Genetics of Human Populations.* San Francisco: W. H. Freeman and Company, 1971.

Huang, Cheng-Han, Philip Z. Liu, and Jeffrey G. Cheng. "Molecular Biology and Genetics of the Rh Blood Group System." *Seminars in Hematology* 37, no. 2 (2000): 150–165.

Race, R. R., and Ruth Sanger. *Blood Groups in Man*, 6th ed. Oxford, U.K.: Blackwell Scientific Publications, 1975.

Internet Resource

"Blood Types." Indiana State University. <http://www.indstate.edu/thcme/mwking/abo-bloodgroups.gif>.

Blotting

Blotting is a common laboratory procedure in which biological molecules in a gel matrix are transferred onto nitrocellulose paper for further scientific analysis. The biological molecules transferred in this process are DNA fragments, RNA fragments, or proteins. Because the isolation and characterization of these types of materials is at the center of much molecular biology research, blotting is one of the most useful techniques in the molecular biology laboratory.

The blotting procedure is named differently depending on the type of the molecules being transferred. When the molecules to be transferred are DNA fragments, the procedure is called a **Southern blot**, named for the man who first developed it, Edward Southern, a molecular biologist at Oxford University. The **Northern blotting** procedure, which transfers RNA molecules, was developed shortly thereafter and, since it was patterned after Southern blotting, its name was a humorous play on words inspired by the name of the first procedure. Western blotting got its name in a similar fashion. All three blotting methods are relatively easy to carry out, can be conducted in a short period of time, and provide answers to many questions that are commonly raised in the field of molecular biology.

The Procedure

Southern blot a technique for separating DNA fragments by electrophoresis and then identifying a target fragment with a DNA probe

Northern blotting separating RNA molecules by electrophoresis and then identifying a target fragment with a DNA probe

gel electrophoresis technique for separation of molecules based on size and charge

All blotting procedures begin with a standard process called **gel electrophoresis**. During this step, DNA, RNA, or proteins are loaded on to an agarose or acrylamide gel (that functions like a molecular sieve) and are then run through an electric field. Two types of gels are commonly used: agarose gels and acrylamide gels. Agarose gels are based on a meshwork of agar filaments and are most often used to analyze DNA and RNA. Acrylamide gels are based on a meshwork formed from the chemical acrylamide and used most often to analyze proteins. Gels are loaded with a mixture of many differently sized molecules. When pulled through the gel by the electric current, they will separate into separate pools on the basis of their size; smaller molecules migrate farther through the gel than larger molecules. These separate pools of molecules will appear as bands on the gel if they are stained with an appropriate dye. After the molecules have been fractionated on the gel, they are ready for transfer to the nitrocellulose paper.

Transfer is initiated when the gel is retrieved from the electrophoresis apparatus and the nitrocellulose paper is carefully laid on top of the gel. The objective now is to transfer the bands of molecules found in the gel over to the nitrocellulose paper. Here they become immobilized, and will reflect the pattern seen on the gel. The paper now serves as a type of permanent record of the gel's banding pattern that can be used for further analysis.

There are two basic ways the actual transfer, or blotting, is carried out. One method takes a "sandwich" of gel and nitrocellulose paper and places it in a special apparatus that sets up an electric field running perpendicular to the band as preserved in the gel. This pulls the bands of molecules out of the gel, and they are immediately absorbed onto the nitrocellulose paper. This method is most commonly employed in Western (protein) blots.

The other method, commonly employed with Southern and Northern blots, lays the gel on top of a platform that in turn is placed in a tray containing a buffer solution. Underneath the gel is a strip of blotting paper that is folded down on each side of the platform, so that it dips into the buffer to serve as a wick. On top of the gel are placed, first, the strip of nitrocellulose paper, then several pieces of blotting paper, and finally a small stack of paper towels. A weight is then placed on top of the paper towels. The buffer flows up the blotting paper "wick" by capillary action, then through the gel, through the nitrocellulose paper, and ultimately into the paper towels. The DNA or RNA in the gel moves with the buffer but sticks to the nitrocellulose paper on contact. The paper towels soak up the transfer buffer, but only after it has passed through the gel and deposited the DNA or RNA on the nitrocellulose paper.

After transfer has been completed, the nitrocellulose paper can be examined by using probes. Short fragments of DNA that have a **nucleotide** sequence complementary to the molecule being analyzed are normally used as probes in Southern and Northern blots. **Antibodies** that react with the protein being analyzed are used as probes in a Western blot. In either case, the probe is "labeled," usually by making it radioactive, so that it is easy to identify. In all blotting experiments, the nitrocellulose paper is placed in a chamber full of buffer and mixed with the probe, which then binds to the molecule that is being studied. This is called the **hybridization** step. Detection of the probe indirectly detects the molecules being studied.

Illustrative Examples

Blotting is perhaps best understood with illustrative examples. Suppose a student was studying a newly identified gene, X, from cows. The student then asks three basic questions as part of a research project: (1) Do pigs also have gene X on their chromosomes? (2) Do cows express gene X in their brain tissue? (3) Is the protein product of gene X found in the cow's blood plasma? Blotting experiments can answer all three of these questions.

nucleotide the building block of RNA or DNA

antibodies immune-system proteins that bind to foreign molecules

hybridization (molecular) base-pairing among DNAs or RNAs of different origins

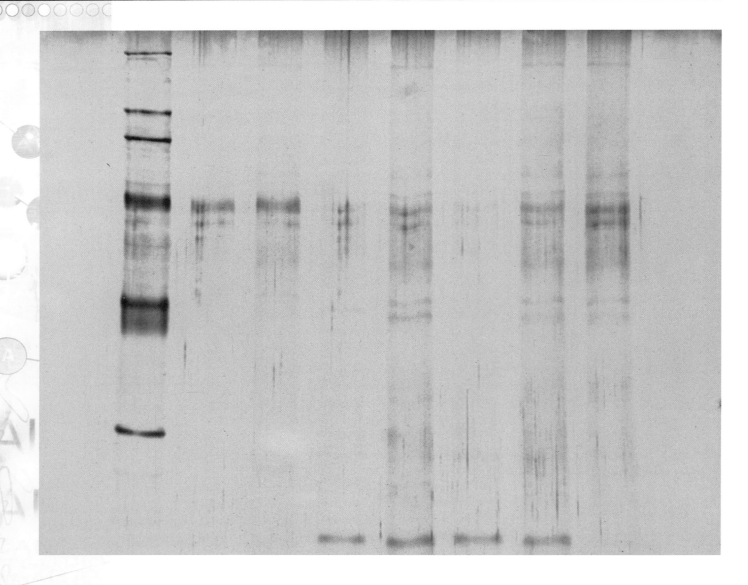

Southern blots, such as this one, are useful in forensic examinations of DNA evidence.

restriction enzyme an enzyme that cuts DNA at a particular sequence

probe molecule used to locate another molecule

complementary matching opposite, like hand and glove

A Southern (DNA) blot will answer the first question. The student obtains DNA from a pig, uses a **restriction enzyme** to cut the DNA into a large pool of fragments of different sizes, and then fractionates the DNA fragments using gel electrophoresis. The contents of the gel are then chemically treated so that the double-stranded DNA molecule "unzips" and exists in a single-stranded form, which is then blotted onto nitrocellulose paper. At this point, the student can take gene X (or a portion of the gene) from the cow, label it, make it single-stranded, and use it as a **probe** to analyze the pig's DNA. The labeled probe is then added to the nitrocellulose blotted with the pig DNA. If the pig's DNA also contains gene X, there should be a fragment on the nitrocellulose with a nucleotide sequence sufficiently **complementary** to the probe such that the probe will bind. In other words, the labeled probe will bind to any fragment from the blotted pig DNA that contains gene X, allowing the student to detect the presence of gene X in pigs.

To answer the second question, a Northern (RNA) blot would be used. The procedure is essentially the same as with the Southern blot, except that the student would isolate RNA from the cow's brain tissue and run it out on the gel. The same DNA probe described above would then be used to detect whether the RNA that represents gene X expression is present in the brain.

To answer the third question, the student would use a Western (protein) blot. This requires the use of an antibody that specifically reacts with the protein coded for by gene X. The student first obtains plasma from the cow and uses standard biochemical techniques to isolate the proteins for analysis. These proteins can then be run out on a gel and transferred to nitrocellulose. The proteins can then be probed with the labeled antibody. If the product of gene X is in the plasma, it will bind with the labeled antibody and can thus be detected. SEE ALSO GEL ELECTROPHORESIS; *IN SITU* HYBRIDIZATION; RESTRICTION ENZYMES; SEQUENCING DNA.

Michael J. Bumbulis

Bibliography

Bloom, Mark V., Greg A. Freyer, and David A. Micklos. *Laboratory DNA Science: An Introduction to Recombinant DNA Techniques and Methods of Genome Analysis.* Menlo Park, CA: Addison-Wesley, 1996.

Russell, Peter. *Genetics,* 5th ed. Menlo Park, CA: Benjamin Cummings, 1998.

Watson, James D., et al. *Recombinant DNA,* 2nd ed. New York: Scientific American Books, 1992.

Breast Cancer

Breast cancer remains the most common cause of cancer among women in the United States, and it results in more deaths from cancer among women than any other type of cancer, except lung cancer. Over 40,000 women die from breast cancer in the United States each year. A long history of research, now coupled with the new information emerging from the field of molecular genetics, is beginning to explain the basic steps leading to breast cancer, and it will enable the development of novel treatment and prevention strategies.

Almost all breast cancers begin in the glandular structures in the breast that, during lactation, produce milk. These mammary glands are under the control of reproductive hormones that stimulate the monthly cycle of gland expansion and shrinkage, which is a feature of the regular menstrual cycle. Many of the factors associated with the development of breast cancer appear to have their effect through interaction with the hormonal stimulation of these glands.

The risk of developing breast cancer increases throughout a woman's lifetime, and the disease is relatively rare in very young women. The overall association of breast cancer incidence with increasing age may be explained by a model of breast cancer in which a progressive and cumulative series of genetic changes within the cells of the glands is necessary for the initiation of cancer. The longer a woman lives, the more opportunities there are for these genetic changes to accumulate and reach a stage where cells can become cancerous.

One of the most consistent epidemiological observations is the association of reproductive events with risk of breast cancer. Women who have

MALE BREAST CANCER

According to the National Cancer Institute, male breast cancer is most common among males between 60 and 70 years of age. Two of the major risk factors for men include: exposure to radiation, and having a family history of breast cancer (especially the *BRCA2* gene). The survival rate for men with breast cancer almost equals that for women.

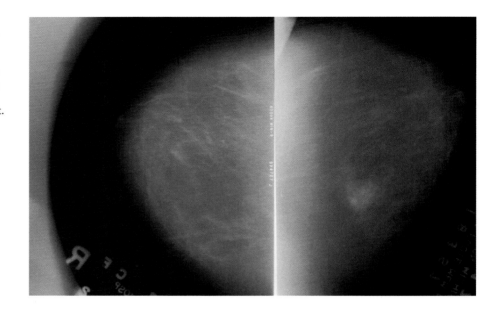

This composite of two mammograms compares healthy breast tissue (left) to partly cancerous breast tissue (right). The malignant lump is visible as a lighter patch in the mammogram on the right.

one or more full-term pregnancies have a *lower* risk for breast cancer, especially if they are pregnant before age twenty. Pregnancy at an early age may help to stabilize the mammary glands and make them less vulnerable to genetic changes later in life. The risk for breast cancer is also significantly decreased among women undergoing surgical removal of the ovaries, particularly if the surgery is performed before age thirty-five. This surgery removes the major source of reproductive hormones and therefore results in less stimulation of the glands in the breast.

Conversely, the greater number of years a woman has regular menstrual cycles, the higher the risk of breast cancer. There is also a modest increase in risk associated with postmenopausal estrogen replacement therapy (especially when used more than 15 years), and with exposure to the synthetic estrogen diethylstilbestrol during pregnancy. Studies have found a significant correlation between breast cancer and levels of hormones—estradiol, estrone, estrone sulfate, prolactin, and dehydroepiandrosterone sulfate. A drug used to treat breast cancer, tamoxifen, blocks **estrogen** receptors.

Taken together, a significant body of research shows that reproductive hormones—produced internally and taken as medicines—are major determinants of breast cancer risk. Other factors—including genetic predisposition, environmental exposure, and lifestyle choices—may increase cancer risk via hormone regulation.

There are striking racial and ethnic differences in breast cancer incidence and resulting deaths. Overall, rates are highest for Caucasian women and lowest for Native American and Korean women. The general international pattern of breast cancer incidence reveals higher rates for Western, industrialized nations, and lower rates for less industrialized and Asian countries. Even within the United States, there is significant geographic diversity in breast cancer rates, with mortality rates highest in the Northeast and lowest in the South. Much of this variation is thought to be due to regional differences in reproductive events, such as the age when women start having children and their use of hormone medications.

estrogen female horomome

There is also considerable evidence from international comparisons, migration studies, and time trends to support an important role for dietary fat in the causation of breast cancer. However when the diets of specific population groups are followed over time, no definite causal link can be demonstrated. The data on fiber and vitamins and minerals is also contradictory. Dietary studies also show a fairly consistent but weak increase in breast cancer risk with moderate to heavy alcohol consumption. Alcohol may act by stimulating the production of more internal hormones. Among postmenopausal women, body weight has also been positively correlated with both breast cancer incidence and mortality. Although exposure to large amounts of radiation is associated with an increased risk for breast cancer, there does not appear to be any risk associated with routine diagnostic imaging, such as chest X rays and mammograms.

Finally, there is limited data to support a protective role for physical activity, both during leisure time and at work, in terms of breast cancer risk. The effect is most pronounced among premenopausal and younger postmenopausal women. The known association of vigorous physical activity with decreased circulating levels of ovarian hormones may explain this finding, which could have significant public health implications.

Women undergoing breast **biopsies** whose tissue shows no evidence of cancer, but whose cells have **atypical** features or faster-than-normal rates of growth have an increased risk of breast cancer, with risks up to eightfold higher in some cases. It is thought that these atypical cells may be a precursor to the development of breast cancer, or they may act as markers for genetic instability within the glandular cells.

biopsies removal of tissue samples for diagnosis

atypical irregular

Population studies have documented that a history of breast cancer in first-, second-, or third-degree relatives increases cancer risk between twofold and fourfold. Recently two genes, *BRCA1* and *BRCA2*, have, when inherited in a mutated form, been associated with a hereditary form of breast cancer. This form is characterized by early age at onset (5 to 15 years earlier than noninherited cases), cancer in both breasts, and association in the family with tumors of other organs, particularly of the ovary in women and prostate gland in men. Among the normal functions of these genes are the control of the cell cycle and the maintenance of stability of the genes. Both genes are tumor suppressor genes whose proteins help both to control the cell cycle and to repair damaged DNA. Mutations interfere with this vital function, causing damaged cells to reproduce and become cancerous.

The frequency of mutations in *BRCA1* in the general population has been estimated to be 1 in 800. Carrier rates are not distributed evenly, however, and mutations tend to concentrate in families with multiple cases of breast or ovarian cancer. Different ethnic groups have unique *BRCA1* and *BRCA2* mutations. Most notably, three specific mutations are common in Ashkenazic Jews. Additional **founder** mutations have been described in Sweden and Iceland.

founder population

Individuals who have inherited a mutated *BRCA1-2* gene face an estimated 36 percent to 85 percent lifetime risk for breast cancer and an estimated 16 percent to 60 percent lifetime risk for ovarian cancer. Among female *BRCA1* carriers who have already developed a primary breast cancer, estimates for a second breast cancer in the opposite breast are as high as 64 percent by age seventy. Men who test positive for a mutation in the *BRCA2* gene also have a higher lifetime risk for breast cancer.

The identification and location of these breast cancer genes will now permit further investigation of the precise role they play in cancer progression and, specifically, how they interact with reproductive hormones. SEE ALSO CANCER; CELL CYCLE; COLON CANCER; ONCOGENES; TUMOR SUPPRESSOR GENES.

Mary B. Daly

Bibliography

Brody, Larry, and Barbara Biesecker. "Breast Cancer Susceptibility Genes *BRCA1* and *BRCA2*." *Medicine* 77 (1998): 208–226.

Kelsey, Jennifer, and Leslie Bernstein. "Epidemiology and Prevention of Breast Cancer." *Annual Review of Public Health* 17 (1996): 47–67.

Weber, Barbara L. "Genetic Testing for Breast Cancer." *Scientific American Science and Medicine* 3, no. 1 (1996): 12–21.

Caenorhabditis elegans *See Roundworm: Caenorhabditis elegans*

Cancer

Cancer is a number of related diseases that are characterized by the uncontrolled proliferation and disorganized growth of cells. Tumor cells invade and destroy normal tissues and may spread throughout the body via the circulatory systems.

A Genetic Disease

Cancer is the result of changes in the genetic material of a cell that cause the cell to gradually lose the ability to grow in a regulated fashion. These changes can be brought about by contact with harmful environmental agents or by inheritance of genes leading to a **genetic predisposition**.

genetic predisposition increased risk of developing diseases

mutations changes in DNA sequences

Cancer risk increases with age, as the probability of accumulating **mutations** in the DNA increases with time. Environmental factors include lifestyle (e.g., smoking), diet (e.g., saturated fats from red meat), and exposure to certain chemicals (e.g., asbestos, benzopyrenes), ionizing radiation (e.g., X-rays, radon gas), ultraviolet radiation (e.g., sun, tanning beds), and certain viruses (e.g., human papillomavirus, Epstein-Barr virus). Heredity also plays a role in **oncogenesis**, as mutations in certain genes increase the probability of developing certain types of cancer. For instance, women who inherit a mutated copy of the *BRCA1* or *BRCA2* gene have a greatly increased probability of developing breast cancer at a young age.

oncogenesis the formation of cancerous tumors

Classification of Cancer Types

The term "cancer" is general, in that it represents a large group of related diseases that arise from **neoplasms**. A neoplasm is classified by the type of tissue in which it arises and the stage to which it has progressed. Neoplasms are also called tumors. Not all tumors are cancerous. A tumor that grows in one place and does not invade surrounding tissue is called benign. In contrast, invasive tumors are called malignant. These are cancerous.

neoplasms new growths

ESTIMATED NEW CANCER CASES AND DEATHS IN THE UNITED STATES 2000

Site of Origin	New Cases*		Deaths*	
	Male	Female	Male	Female
Breast	1,400	182,800	400	40,800
Colorectal	63,600	66,600	27,800	28,500
Esophagus	9,200	3,100	9,200	2,900
Kidney & Bladder	57,100	27,300	15,400	8,700
Leukemia	16,900	13,900	12,100	9,600
Liver	10,000	5,300	8,500	5,300
Lung	89,500	74,600	89,300	67,600
Lymphoid	35,900	26,400	14,400	13,100
Ovary	-	23,100	-	14,000
Pancreas	13,700	14,600	13,700	14,500
Prostate	180,400	-	31,900	-
Skin	34,100	22,800	6,000	3,600
Stomach	13,400	8,100	7,600	5,400
Testis	6,900		300	-
Uterine	-	48,900	-	11,100

*(the American Cancer Society's Clinical Oncology, Lenhard R.E., Osteen R.T., Gansler T., 2001)

Estimated new cancer cases and deaths in the United States in 2000. Adapted from Lenherd, 2001.

Benign or Malignant Tumor

Whether a tumor is benign or malignant determines how potentially life-threatening it is. Benign tumors are usually harmless, although their location may be serious (if surgery to remove the tumor would carry significant risk). These tumors are not considered cancerous, are relatively slow-growing, and usually are encased within a fibrous capsule.

Malignant tumors (cancers) have great potential to spread, or metastasize, to other sites in the body. These tumors are fast-growing and aggressive, and they invade neighboring healthy tissue. They therefore are considered life threatening.

Type of Tissue

The body consists of many different organs, which in turn are composed of several different types of tissues. There are three major categories of tissue-related tumor types: carcinoma, sarcoma, and leukemia/lymphoma. There are also other specialized tumor categories, such as those of the central nervous system (e.g., brain tumors).

Carcinoma. This is the largest category, containing about 90 percent of all cancers, and it consists of neoplasms derived from epithelial cells. Epithelial cells make up the outer layers of the skin. They also line the inner structures of organs such as the lungs, intestines and testes, as well as complex tissue such as the breast.

Sarcoma. These are solid tumors derived from all connective tissues except the blood-forming tissues (these are the leukemias and lymphomas). These tumors account for about 2 percent of all cancers. They occur in such tissues as muscle, bone, and cartilage.

Leukemia and Lymphoma. This group contains about 8 percent of all cancers, including blood cancers that originate from the marrow (leukemias) and from the lymphatic system (lymphomas). This group also includes other nonsolid tumors of the bone marrow and lymphatic system, such as myeloma, which affects plasma cells—a type of white blood cell found in the marrow and in other tissues.

Type of Cell

Classifying a tumor by the type of cell from which it is derived is slightly more complex than classifying it by the type of tissue, since there are so many cell types. The main cell types include adenomatous cells (which are ductal or glandular cells), basal cells (found at the base of the skin), myeloid blood cells (granulocytes, monocytes, and platelets), lymphoid cells (lymphocytes or macrophages), and squamous cells (flat cells). Therefore it is possible for a cancer classified by its site of origin to be broken up into one of several cell types. For example, a skin cancer could be either a squamous cell carcinoma, a basal cell carcinoma, or a melanoma (from a pigment-producing cell).

Site of Origin

Solid tumors are firm masses that develop from a neoplasm's originating organ, such as the brain, esophagus, kidney, liver, lung, ovary, pancreas, prostate, or testis. Tumors of the blood-forming tissues and lymphatic systems are not solid and tend to remain free and circulating even when malignant. Some of the common forms of cancer are listed in the table above.

Cancer Progression

metastasis breaking away of cancerous cells from the initial tumor

There two main steps in cancer progression: the initial growth of the cancer and the subsequent spread via **metastasis**. Solid tumors are subject to the physiological constraints of biological systems: Without nutrients and oxygen, they will die. Therefore a solid tumor is initially limited in size to no larger than 1 to 2 millimeters in diameter (about the size of a small pea).

angiogenesis growth of new blood vessels

For a tumor to become aggressive, it needs to be able to nourish the cells at the center of its mass that are too far away from blood vessels. This is achieved by **angiogenesis**. Through mutation, a few cancer cells may gain the ability to produce angiogenic growth factors. These growth factors are proteins that are released by the tumor into nearby tissues, where they stimulate new blood vessels to grow into the tumor. This allows the tumor to rapidly expand in mass and invade surrounding tissue. It also provides a route for the cancer cells to escape into the new blood vessels and circulate throughout the body, where they can lodge in other organs forming metastases.

The most common way for a cancer to metastasize is through the lymphatic system. The lymphatic system is a network of channels throughout the body that carry a tissue fluid called lymph.

When a primary neoplasm metastasizes to another location, its cell type does not change. If leukemia metastasizes to the liver and develops a tumor, the tumor will display the characteristics of the leukemia, not those of a liver cancer. In some cases this can help physicians determine the original site of a tumor.

Genes Altered in Tumors

Although each cell in the body maintains itself and carries out its specific function, it is part of a large colony of collaborating cells that constitute the whole organism. A cell communicates with its surrounding cells by releasing chemical messages, in a process called signal transduction. These

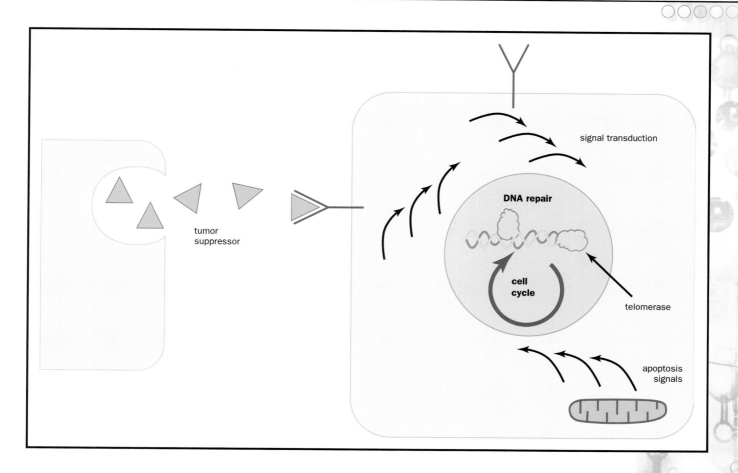

messages bind to specific receptor proteins on the surface of the surrounding cells. The gene expression of these cells is changed as a result of the messages.

A **hyperplastic cell** or a cancerous cell will stimulate neighboring cells to grow by secreting growth factors. Several types of genes can be mutated in tumor cells: oncogenes, tumor suppressor genes, DNA repair genes, and genes involved in cell mortality.

Oncogenes. These genes are involved in signal transduction, and some are involved in the various phases of the cell cycle. Mutations in cell-cycle regulation or signal transduction can "push" the cell into dividing rapidly and without regard to its surroundings. Over 100 oncogenes have been identified so far. They include genes such as *ABL1* (Abelson murine strain leukemia viral homolog) and *EGFR* (Epidermal Growth Factor Receptor).

Tumor Suppressor Genes. These genes inhibit cell division, working in a manner opposite to that of the oncogenes. Surrounding cells secrete growth-inhibitory signals that help prevent proliferation. These growth-inhibitory signals work in conjunction with tumor suppressor genes. If a tumor suppressor gene is mutated, proliferating cells can ignore these inhibitory messages. This group includes the genes *p53*, *BRCA1*, and *BRCA2*.

DNA Repair Genes. These are the genes that provide the cell with the ability to sense and correct damage to the DNA. Damage to the DNA can be caused by radiation, chemicals, ultraviolet light, or errors in transcription. If these errors are not corrected, they accumulate in the genome and can

Multiple systems interact to control the cell cycle, ensuring that cell division occurs only when it is advantageous for the organism and when undamaged DNA is availale for replication. Cancer may occur when any one of these systems is disrupted.

hyperplastic cell cell that is growing at an increased rate compared to normal cells, but is not yet cancerous

Inheritance of a mutated retinoblastoma gene (Rb*) greatly increases the likelihood of developing the disease.

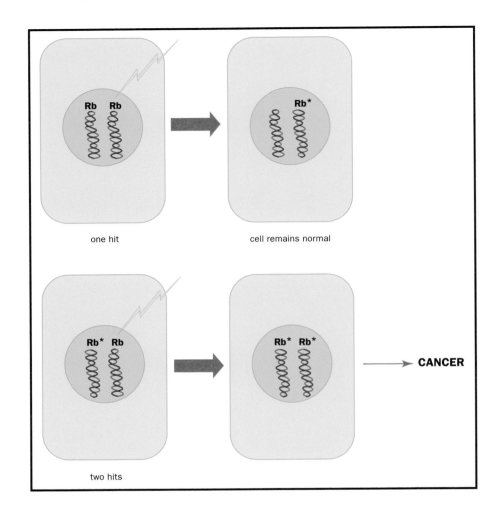

one hit

cell remains normal

two hits

CANCER

senescence a state in a cell in which it will not divide again, even in the presence of growth factors

apoptosis programmed cell death

quickly increase the chance that a cell will become cancerous. Repair genes include those in the DNA-ligase and excision-repair gene families.

Genes Involved in Cell Mortality. A normal cell can only undergo about forty divisions, after which it dies or enters **senescence**. If a tumor had this limitation it would be very limited in its size, as it would reach its forty divisions relatively quickly. This process is controlled by the enzyme telomerase, which maintains the telomeres (repetitive DNA sequences at the ends of chromosomes that shorten after each round of DNA replication, until they reach a length that causes the cell to die) by not allowing them to shorten. Some cancer cells become immortal as a result of mutations in the telomerase gene, causing the telomeres to be extended indefinitely, allowing the cell to continue dividing without limit. Other mutations affect the process of **apoptosis**.

Cancer does not usually arise by a single event. Instead, two or more "hits" are needed to convert a well-regulated cell to a cancer cell. This is the case because each cell contains two copies of each gene, one inherited from each parent. Most cancer-causing mutations cause a loss of function in the mutated gene. Often, having only one functional copy is enough to prevent disease. Thus, two mutations are needed.

This can be illustrated by looking at retinoblastoma, a common cancer of the retina. The affected gene (called the *retinoblastoma* gene) is a tumor suppressor. **Spontaneous** mutations are rare, but since there are many millions of cells in the retina, several will develop the appropriate gene mutation over the course of a lifetime. It would be very unlikely, though, for a single cell to develop two spontaneous mutations (at least in the absence of prolonged exposure to carcinogens), and thus spontaneous retinoblastoma is very rare.

spontaneous non-inherited

If, however, a person inherits one copy of an already-mutated gene from one parent, every cell in the eye starts life with one "hit." The chances are very high that several cells will suffer another hit sometime during their life, and so the chances are very high that the person will develop retinoblastoma. Since inheriting a single copy of the mutated gene is so likely to lead to the disease, the gene is said to show a dominant inheritance pattern.

Future Directions in Diagnosis and Treatment

The increased knowledge of cancer at the biochemical and genetic level has led to many advances toward better diagnosis and treatment of cancer, including the design of more specific drugs that are less toxic to normal tissue. This includes the use of antisense molecules, which are nucleic acid sequences that are complementary to the **mRNA** of a target gene. As the two sequences are complementary, they anneal and thus the mRNA is blocked from being translated into a protein, resulting in less of that particular protein being produced (such as growth factor receptors). Drugs specific in blocking angiogenesis are able to control the growth and spread of tumors, especially when used in combination with other treatments. SEE ALSO AMES TEST; ANTISENSE NUCLEOTIDES; APOPTOSIS; BREAST CANCER; CARCINOGENS; CELL CYCLE; COLON CANCER; DNA REPAIR; MUTAGEN; MUTATION; ONCOGENES; SIGNAL TRANSDUCTION; TELOMERE; TUMOR SUPPRESSOR GENES.

mRNA messenger RNA

Giles Watts

Bibliography

Greider, C. W., and E. H. Blackburn. "Telomeres, Telomerase and Cancer." *Scientific American* 274 (1996): 80–85.

Kiberstis, Paula, and Jean Marx. "The Unstable Path to Cancer." *Science* 297, no. 5581 (2002): 543.

Lenherd, Raymond E., et al. *Clinical Oncology*. Atlanta, GA: American Cancer Society, 2001.

Rosenberg, S. A., and B. M. John. *The Transformed Cell: Unlocking the Mysteries of Cancer*. New York: Putnam, 1992.

Weinberg, R. A. *Racing to the Beginning of the Road: The Search for the Origin of Cancer*. New York: Putnam, 1998.

Weinberg, R. A. *One Renegade Cell: How Cancer Begins*. New York: Basic Books, 1999.

Carcinogens

Carcinogens are agents that cause cancer, and include chemicals, radiation, and some viruses. While avoiding contact with carcinogenic agents is wise, it is virtually impossible to steer clear of them completely. Ultraviolet radiation from the sun, substances in food, and even oxygen can induce **malignancies**. In spite of the pervasive nature of carcinogens, however, not all individuals develop cancer, which suggests that mere contact

malignancies cancerous tissues

Breathing second hand smoke increases a person's heart rate and blood pressure, and delivers dangerous amounts of carbon monoxide to the blood. It is estimated that 2 percent of lung cancer deaths are caused by passive smoking.

with a carcinogenic agent is insufficient to produce this lethal disease. That is because organisms have evolved protective mechanisms to prevent cancer, and some of these defenses work by thwarting the potentially harmful effects of carcinogens.

Cancer-Causing Chemicals

References to cancer have been found in the annals of human disease since ancient times, but the disease's association with carcinogen exposure is a relatively new concept. Sir Percival Potts, a British physician who lived in the eighteenth century, was the first to suggest that the induction of cancer might be linked to agents in the environment. Potts had observed high rates of scrotal and nasal cancer among England's chimney sweeps, men who were exposed to accumulated fireplace soot during their work. After some careful studies, Potts suggested correctly that exposure to soot caused the high cancer rates, providing the impetus for identifying other carcinogens present in the environment.

In retrospect, it was fortuitous that soot was acknowledged as one of the first carcinogenic agents. Soot is a complex mixture of chemicals that arises from the combustion of organic material. As scientists and physicians separated soot's individual components, it became clear that chemicals called polycyclic aromatic hydrocarbons (PAHs) were among its principal carcinogenic compounds. The story became even more intriguing when it was shown that many PAHs behave as **procarcinogens**. Procarcinogens do not cause cancer per se, but they can be converted to active carcinogens by **enzymes** located in organs like the liver and lung. The implications of this discovery are noteworthy. For example, cigarette smoke contains a wide variety of procarcinogenic PAHs that are turned into active carcinogens in lung cells. Since smokers draw these PAHs deep into their lungs with each inhale on a cigarette, one reason that cigarette smoking correlates so highly with the induction of lung cancer becomes very clear.

procarcinogens substances that can be converted into carcinogens, or cancer-causing substances

enzymes proteins that control a reaction in a cell

Oncogenes and Tumor Suppressors

How do carcinogens cause cancer? Answering this question still forms the core of much basic research, but a common feature of many carcinogens, particularly chemicals and **radiation**, is that they act as mutagens. Mutagens are agents that generate changes in DNA, sometimes by reacting with the DNA building blocks, guanine, adenine, thymine, and cytosine, which results in damaged DNA. When such damage remains in chromosomes, genes are often mutated in a way that impairs their normal function and enhances cancer induction. Cells try to prevent such mutations by repairing DNA damage, but they are not always successful. In fact, some individuals are susceptible to hereditary skin and colon cancers because they lack the ability to remove damaged DNA from chromosomes.

radiation high energy particles or waves capable of damaging DNA, including X rays and gamma rays

There are two general classes of genes that contribute to malignant tumor formation when they are mutated by carcinogens: oncogenes and tumor suppressor genes. **Oncogenes** (the prefix "onco-" meaning "tumor") are altered versions of normal genes called proto-oncogenes. Proto-oncogenes encode proteins that are often involved in regulating normal cell growth and division. When a proto-oncogene is mutated by exposure to a carcinogen, the protein it encodes may lose its ability to govern cell growth and division, often giving rise to the rapid, unrestrained cell proliferation that is characteristic of cancer. In such a case, the mutations in the proto-oncogene convert it into an actual oncogene.

oncogenes genes that cause cancer

While many oncogenes have been identified, numerous cancers are associated with mutations in one particular proto-oncogene, called *ras*, which is an abbreviation for "rat sarcoma." "Ras" is written as Ras when biologists refer to the protein, and as *ras* when they refer to the gene that encodes the protein. The *ras* gene encodes Ras protein, which acts to regulate cell growth. Normally, Ras protein cycles between an "off" and "on" form. Many carcinogens induce **mutations** in the *ras* proto-oncogene, converting it to a *ras* oncogene, which encodes a form of the Ras protein that is locked in the "on" state. By abolishing Ras protein's regulatory off/on cycle, the accumulated mutations in the *ras* gene contribute to the formation of malignancies.

mutations changes in DNA sequences

Not all oncogenes arise from mutations in normal cellular proto-oncogenes. In the early twentieth century, Peyton Rous discovered a

carcinogenic virus that now bears his name, the Rous sarcoma virus. This virus harbors a gene called v-*src* (viral-sarcoma) that is a mutant form of a normal cellular proto-oncogene called c-*src* (cell-sarcoma). Like Ras protein, c-Src protein helps to regulate cell growth. When cells are infected by Rous sarcoma virus, the v-*src* gene, which is classified as an oncogene, is expressed in those cells. High amounts of mutant v-Src protein encoded by the v-*src* oncogene are made in the cell, and they dominate the normal cellular c-Src protein, an event that contributes to abnormal cell growth and proliferation, eventually leading to cancer.

Tumor suppressor genes encode proteins that tend to repress cancer formation. When tumor suppressor genes are mutated by carcinogens, they often lose their ability to stem tumor formation, resulting in cancer. Some hereditary forms of breast cancer are linked to mutations in a tumor suppressor gene called *BRCA-1*. BRCA is derived from BReast CAncer. The *BRCA-1* gene encodes BRCA-1 protein, which participates in controlling cell division, preventing cells from growing out of control, thus contributing to the suppression of tumor formation. Mutations in the *BRCA-1* gene result in altered BRCA-1 protein that no longer functions correctly in cell-growth regulation, contributing to the formation of tumors, particularly in breast tissue.

Reducing Exposure

Decreased carcinogen contact along with improved methods for treating cancer provide two important means for curtailing the suffering, expense, and death associated with the disease. The documented existence of carcinogens has prompted a worldwide effort to detect additional cancer-causing agents. A variety of toxicological assessments, including the Ames test, are used to identify potential mutagens and carcinogens. When possible, established carcinogens, such as asbestos, are removed from the environment, home, and workplace.

Exposure can also be reduced if the population is provided with protective warnings, like those advising the use of sunblock to shield skin from the cancer-causing effects of ultraviolet radiation in sunlight. The cost and manpower of such efforts are enormous, but carcinogen identification is critical for ensuring that exposure is minimized. A great challenge is reducing exposure to the carcinogens to which people actively expose themselves, most notably cigarette smoke. Prolonged education programs have helped cut down the use of cigarettes, but continued education is needed for each new generation. SEE ALSO AMES TEST; BREAST CANCER; CANCER; CELL CYCLE; DNA REPAIR; MUTAGEN; ONCOGENES; TUMOR SUPPRESSOR GENES.

David A. Scicchitano

Bibliography

Lodish, Harvey, et al. *Molecular Cell Biology*, 4th ed. New York: W. H. Freeman, 2000.

Tomatis, Lorenzo. "The Identification of Human Carcinogens and Primary Prevention of Cancer." *Mutation Research* 462 (2000): 407–421.

Trichopoulos, Dimitrios, Frederick P. Li, and David J. Hunter. "What Causes Cancer?" *Scientific American* 275 (1996): 80–87.

Weinberg, Robert. "How Cancer Arises." *Scientific American* 275 (1996): 62–71.

Cardiovascular Disease

Cardiovascular disease is a set of diseases affecting the heart and blood vessels. As with most chronic diseases whose incidence increases with age, it involves both inherited and environmental contributors and is therefore classified as a complex genetic disease. Most researchers believe that all major risk factors for cardiovascular disease have been identified. It is estimated that cigarette smoking, **hypertension**, abnormal serum cholesterol (low-density lipoprotein cholesterol or high-density lipoprotein cholesterol), obesity, lack of physical exercise, and diabetes account for 50 percent of the variability of risk in high-risk populations. The remaining risk is likely composed of a large number of yet-to-be identified minor risk factors or genetic influences that account for the development of disease in most individuals. Investigators who have attempted to estimate the overall contribution of genetics to the development of cardiovascular disease have proposed numbers ranging from 20 to 60 percent, based upon the analysis of large epidemiologic studies.

hypertension high blood pressure

Finding Genes for Cardiovascular Disease

Genetics studies of cardiovascular disease involve searches for genes in two general classes: causative genes and disease-susceptibility (or disease-modifying) genes. These are sought through gene-linkage analysis or candidate-gene studies, respectively. Identifying causative genes for this disease is likely several years away at best. Before that time, however, a new understanding will have been reached regarding the relationship between inherited risks and outcomes in cardiovascular disease. With the development of new technology, we also have the promise of a detailed catalogue of disease-modifying genes that may open the door to therapeutic advances.

Gene-linkage analyses involve the study of families that express the cardiovascular trait of interest. In such studies, it is important also to establish the relative risk. Relative risk is defined as the probability of developing a condition (such as cardiovascular disease) if a risk factor (such as a gene) is present, divided by the probability of developing the condition if the risk factor is absent. A relative risk greater than 4.0 (that is, a four-fold greater risk due to presence of a gene or genes) will be associated with a reasonable likelihood of success in finding associated genes, given a study of 200 sibling pairs demonstrating the condition.

One of the best-studied types of cardiovascular disease is early-onset (or premature) coronary artery disease, which has a particularly strong genetic or inherited component. The coronary arteries are those around the heart that supply it with blood. Early-onset is defined as disease presentation (as reversible heart pain, heart attack, or cardiovascular surgery) before the age of 50. Approximately 8 to 10 percent of the U.S. population with cardiovascular disease presents before age 50, according to most surveys. Based upon a number of relatively small **epidemiologic** studies and several genetics studies in twins, a conservative estimate of the relative risk ratio contributed by genetics to the development of early-onset cardiovascular disease is between 4.0 and 8.0. Despite the fact that it has an inherited component, the actual genes responsible for familial predisposition to early-onset coronary artery disease have been incompletely investigated and remain obscure.

epidemiologic the spread of diseases in a population

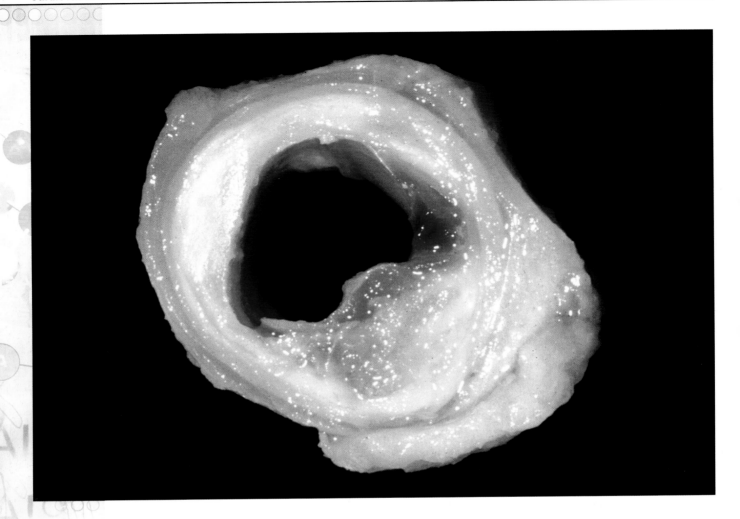

Coronary artery plaque is a symptom of cardiovascular disease. The buildup of excessive plaque can be attributed to both one's hereditary predisposition to coronary disease and one's lifestyle.

While population-level relative risk for developing cardiovascular disease can be known with a great deal of accuracy, therefore, this knowledge cannot be used to counsel or direct therapy for an individual in any given family. In fact, it has become clear to most practicing cardiologists that even when we know which cardiovascular risk factors are present, we have a very limited ability to predict the development of disease in most individuals.

Ongoing Studies

The ongoing studies of the genetics of cardiovascular disease consist of two general types: those that accumulate individual cases with the goal of performing association candidate-gene studies, and those that collect data from families (sibling pairs or extended families) with the idea of performing gene-linkage studies. Candidate-gene studies examine variations in genes that code for proteins that are likely to be involved in a disease or its prevention, such as genes controlling cholesterol metabolism or blood pressure. Linkage studies look for chromosome regions that are co-inherited with risk for disease, and then look carefully at the region to determine what genes are present.

Patients for both types of studies may be located in similar ways. Disease registry databases contain information on patients with particular conditions, which may have been collected by hospitals, charitable organizations, or research organizations. Clinical trials databases are generated during the testing of a new drug or other treatment. Population-based longitudinal studies collect data on a large number of randomly selected people (not just those with disease) and follow them over many years, to determine what factors lead to development of disease. Each study has its own contribution to make, and only through the combined efforts of multiple studies and approaches will we discover and understand the genetic contributions to the development of cardiovascular disease.

Goals of Genetic Studies

Many of the promises of genetics investigations have probably been grossly overstated. The immediate potential of the ongoing and planned investigations into the genetics of cardiovascular disease is more promising for gene-directed therapy (the use of genetic information to guide the judicious use of medical interventions) than for somatic gene therapy (the use of a gene or gene product which, when introduced into a human organ, changes the function of the organ).

The realistic promises of current genetics studies include the elucidation of disease mechanisms; the identification of new targets for the development of therapeutic pharmacologic agents; and the use of genetic markers to identify individuals for whom a particular agent is either effective or unusually hazardous. This approach, called pharmacogenomics, improves the safety and efficacy of treatments, and enhances the ability to preferentially select subjects for clinical trials based upon genetic predispostion and for gene-directed therapy. In the latter case, for example, a genetic contributor to the development of early-onset cardiovascular disease might be used as an additional risk factor whose identification could focus the allocation of preventive resources, whether educational, behavioral, or pharmacologic, to populations at particularly high risk for the disease. SEE ALSO COMPLEX TRAITS; GENE AND ENVIRONMENT; PHARMACOGENETICS AND PHARMACOGENOMICS; PUBLIC HEALTH, GENETIC TECHNIQUES IN; STATISTICS.

Bill Kraus

Bibliography

Lander, E. S., and N. J. Schork. "Genetic Dissection of Complex Traits." *Science* 265 (1994): 2035–2048.

Cell Cycle

The cell cycle is the process by which a cell grows, duplicates its DNA, and divides into identical daughter cells. Cell cycle duration varies according to cell type and organism. In mammals, cell division occurs over a period of approximately twenty-four hours.

In multicellular organisms, only a subset of cells go through the cycle continuously. Those cells include the **stem cells** of the **hematopoietic** system, the basal cells of the skin, and the cells in the bottom of the **colon crypts**. Other cells, such as those that make up the endocrine glands, as well as liver cells, certain renal (kidney) tubular cells, and cells that belong to connective tissue, exist in a nonreplicating state but can enter the cell cycle after receiving signals from external stimuli. Finally, postmitotic cells are

stem cells cells capable of differentiating into multiple other cell types

hematopoietic blood-forming

colon crypts part of the large intestine

Cyclin-dependent kinases (CDK) trigger transition between cell-cycle phases. Adapted from Robinson, 2001.

replication duplication of DNA

incapable of cell division even after maximal stimulation, and include most neurons, striated muscle cells, and heart muscle cells.

The cell cycle is functionally divided into discrete phases. During the DNA synthesis (S) phase, the cell replicates its chromosomes. During the mitosis (M) phase, the duplicated chromosomes are segregated, migrating to opposite poles of the cell. The cell then divides into two daughter cells, each having the same genetic components as the parental cell. Mammalian cells undergo two gap, or growth, phases (G_1 and G_2). G_1 occurs prior to the S phase, and G_2 occurs before the M phase.

Control of the Cycle

During the G_1 and G_2 phases, cells grow and make sure that conditions are proper for DNA replication and cell division. During the G_1 phase, cells monitor their environment and determine if conditions, including the availability of nutrients, growth factors and hormones, justify DNA replication. The decision to initiate replication is made at a specific "checkpoint" in G_1 called the "restriction point."

The processes of DNA **replication** and mitosis, and intervening events during the cell cycle, occur in a highly ordered and specific manner. A complex network of proteins ensures that these events occur at the proper times. Intracellular and extracellular signals block cell-cycle progression at checkpoints if certain events have not yet been completed. After the restriction point, the cell is committed to replicating its genome and dividing, completing one round of the cell cycle. If, prior to the restriction point, cells sense inadequate growth conditions or receive inhibitory signals from other cells, they enter G_0 (G-zero) phase, also called quiescence. In the G_0 phase, they are maintained for prolonged periods in a nondividing state. If cells sense such conditions after the restriction point, they complete the current

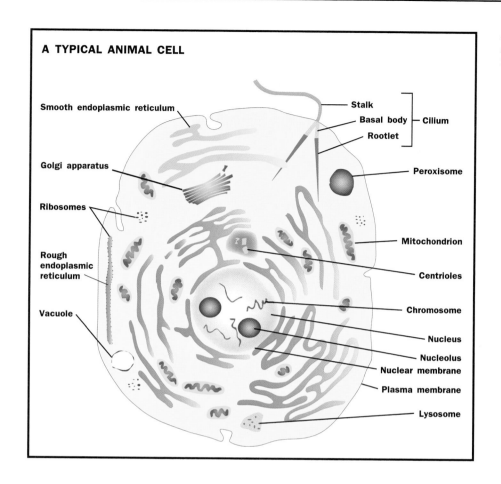

A TYPICAL ANIMAL CELL

Smooth endoplasmic reticulum

Golgi apparatus

Ribosomes

Rough endoplasmic reticulum

Vacuole

Stalk
Basal body ⎤ Cilium
Rootlet ⎦

Peroxisome

Mitochondrion

Centrioles

Chromosome

Nucleus

Nucleolus
Nuclear membrane
Plasma membrane

Lysosome

Animal cell parts.
Adapted from Robinson,
2001.

round of the cell cycle and exit to G_0 during the subsequent G_1 phase. The G_2 phase is shorter than G_1, but it, too, consists of important mechanisms that control the completion and fidelity of DNA replication and that prepare the cell for entry into mitosis. Whereas some conditions cause cells to enter the G_0 phase, others trigger **apoptosis**. One such signal that may trigger apoptosis is if a cell's DNA has undergone significant damage.

After the restriction point, at the transition from the G_2 to the M phase, another checkpoint occurs. Mitosis is prevented if DNA damage has occurred or if genomic replication is not complete. The final key checkpoint occurs at the end of mitosis, when the cycle stops if chromosomes are not properly attached to the mitotic spindle.

Proteins That Regulate the Cycle

The mammalian cell cycle control system is regulated by a group of protein **kinases** called c̲yclin-d̲ependent k̲inases (CDKs). These proteins catalyze the attachment of **phosphate groups** to specific serine or threonine amino acids in a target protein. The phosphate groups alter the target protein's properties, such as its interaction with other proteins. (The alteration of protein activity by the attachment of phosphate groups occurs frequently in cells.)

CDKs are called "cyclin-dependent" because their activity requires their association with activating subunits called cyclins. While the number of CDKs in a cell remains constant during the cell cycle, the levels of cyclins

apoptosis programmed
cell death

kinases enzymes that
add a phosphate group
to another molecule,
usually a protein

phosphate groups
PO_4^{3-} groups, whose
presence or absence
often regulates protein
action

Figure 1.

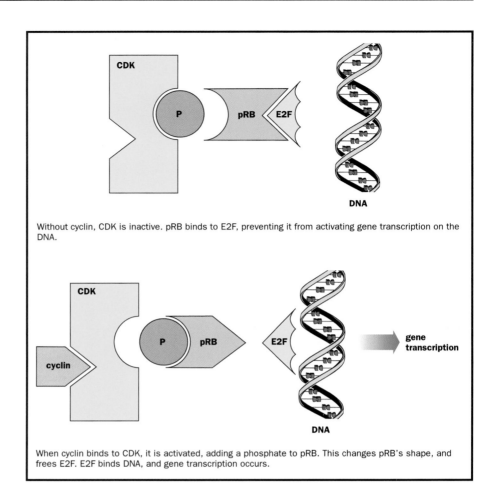

Without cyclin, CDK is inactive. pRB binds to E2F, preventing it from activating gene transcription on the DNA.

When cyclin binds to CDK, it is activated, adding a phosphate to pRB. This changes pRB's shape, and frees E2F. E2F binds DNA, and gene transcription occurs.

oscillate. There are G_1 cyclins, S-phase cyclins, and G_2/M cyclins, each of which interact differently with CDK subunits to regulate the various phases of the cell cycle. CDKs can also associate with inhibitory subunits called CDK inhibitors (CKIs). In response to signals that work against proliferation, such as growth factor deprivation, DNA damage, cell-cell contact inhibition and lack of cell adhesion, CKIs cause the cell cycle to halt.

By the end of 2001, many structurally related cyclins (A1, A2, B1, B2, B3, B4, B5, C, D1, D2, D3, E1, E2, F, G1, G2, H, I, L, and T) and nine CDKs (CDK1 to CDK9) were identified in mammalian cells. Complexes of cyclin D and CDK4, as well as complexes of cyclin D and CDK6, operate during the G_1 phase. Complexes of cyclin A and CDK2, as well as complexes of cyclin E and CDK2, act during the transition from the G_1 to the S phase. Complexes of cyclin A and CDK1, as well as cyclin B and CDK1, function during the transition from the G_2 to the M phase.

Active complexes of cyclins and CDKs exert their biological effects by phosphorylating proteins. During the G_1 phase, a major target of cyclin/CDK complexes is the retinoblastoma protein (pRb). pRb is a growth-suppressing protein whose activity is controlled by whether or not it is phosphorylated.

When pRb is in the dephosphorylated form, during the G_0 phase and early in the G_1 phase, it is active. pRb exerts its growth-suppressing effects by binding to many cellular proteins, including the **transcription factors** of the E2F family (Figure 1). E2F transcription factors regulate the expression of numerous genes that are expressed during G_1, or at the transition from the G_1 to the S phase, to initiate DNA replication.

pRb that is bound to an E2F transcription factor inhibits the transcription factor's activity. Following phosphorylation by cyclin/CDK complexes, pRb dissociates from E2F, allowing the transcription factor to bind DNA sequences and activate the expression of genes necessary for the cell to enter the S phase. Cyclin D1/CDK4 complexes phosphorylation of pRb during the middle of the G_1 phase. They allow for subsequent phosphorylation of pRb by additional cyclin/CDK complexes that act later in the cell cycle.

Two families of CKIs have been identified, based on their **amino acid** sequence similarity and the specificity of their interactions with CDKs. One of the families of CKIs, the INK family, includes four proteins (p15, p16 p18 and p20). These CKIs exclusively bind complexes of cyclin D and CDK4, as well as complexes of cyclin D and CDK6, to block cells that are in the G_1 phase of the cell cycle. The other family of CKIs, the Cip/Kip family, consists of three proteins (p21, p27, and p57). These inhibitors bind to all complexes of cyclins and CDKs that function during the G_1 phase and during the transition from the G_1 to the S phase. They act preferentially, however, to block the activity of complexes containing CDK2.

Deregulation and Cancer

Deregulation of cell cycle control proteins plays a key role in the development of cancer. Overactivation of proteins that favor cell cycle progression, namely cyclins and CDKs, and the inactivation of proteins that impede cell cycle progression, such as CKIs, can result in uncontrolled cell proliferation.

In human **tumors**, it is genes encoding the proteins that control the transition from the G_1 to the S phase that are most commonly altered. These genes include those for cyclins, CKIs, and pRb. Such mutations overcome the inhibitory effects of pRb on the cell cycle, causing cells to have a growth advantage. In some cancers, this occurs after the direct mutation of the pRb gene, resulting in the protein's loss of function. In a larger set of cancers, pRb is indirectly inactivated by the hyper-activation of CDKs. This may result from over expression of cyclins, from an activating mutation in CDK4, or from inactivation of CKIs.

There is much evidence to suggest that cyclins can act as **oncogenes** to induce cells to become cancerous. In particular the G_1 cyclins, cyclin D1, and cyclin E have been implicated in the development of cancer. Overexpression of the cyclin D1 protein is frequently detected in human breast cancer, and increasing evidence suggests that cyclin E overexpression plays an important role in the **pathogenesis** of breast cancer.

CKIs antagonize the function of cyclins, and considerable evidence suggests that these proteins function as tumor suppressors. CKI function is often altered in cancer cells. The gene encoding p16, a protein that belongs to the INK family of CKIs, is mutated, deleted, or inactivated in a large number of human malignancies and tumors. Such alterations

transcription factors proteins that increase the rate of gene transcription

amino acid a building block of protein

tumors masses of undifferentiated cells; may become cancerous

oncogenes genes that cause cancer

pathogenesis pathway leading to disease

prevent the inhibition of cyclin D/CDK4 and cyclin D/CDK6 complexes during G_1.

Decreased expression of p21 and p27, proteins that belong to the Cip/Kip family of CKIs, also has been demonstrated in numerous human tumors. In contrast to the genetic mutations observed with p16, the decrease in p27 levels in tumors is due to enhanced degradation of the p27 protein. One of the proteins required for the degradation of p27, Skp2, has oncogenic properties. Skp2 over expression is observed in several human cancers and likely contributes to the uncontrolled progression of the cell cycle by increasing the degradation of p27. Understanding of the fine details of cell cycle regulation is likely to lead to specific cancer therapies targeting one or more of these important proteins. SEE ALSO APOPTOSIS; CANCER; CELL, EUKARYOTIC; MEIOSIS; MITOSIS; ONCOGENES; REPLICATION; SIGNAL TRANSDUCTION; TUMOR SUPPRESSOR GENES; TRANSCRIPTION FACTORS.

Joanna Bloom and Michele Pagano

Bibliography

Goldberg, Alfred L., Stephen J. Elledge, and J. Wade Harper. "The Cellular Chamber of Doom." *Scientific American* 284, no. 1 (2001): 68–73.

Gutkind, J. Silvio, ed. *Signaling Networks and Cell Cycle Control*. Totowa, NJ: Humana Press, 2000.

Murray, Andrew, and Tim Hunt. *The Cell Cycle: An Introduction*. Oxford, U.K.: Oxford University Press, 1993.

Pagano, Michele, ed. *Cell Cycle Control*. New York: Springer-Verlag, 1998.

Weinberg, Robert A. "How Cancer Arises." *Scientific American* 275, no. 3 (1996): 62–70.

Cell, Eukaryotic

Archaea one of three domains of life, a type of cell without a nucleus

All living organisms are composed of cells. A eukaryotic cell is a cell with a nucleus, which contains the cell's chromosomes. Plants, animals, protists, and fungi have eukaryotic cells, unlike the Eubacteria and **Archaea**, whose cells do not have nuclei and are therefore termed prokaryotic. In addition to having a nucleus, eukaryotic cells differ from prokaryotic cells in being larger and much more structurally and functionally complex. Eukaryotic cells contain subcompartments called organelles, which carry out specialized reactions within their boundaries. A eukaryotic cell may be an individual organism, such as the amoeba, or a highly specialized part of a multicellular organism, such as a **neuron**.

neuron nerve cell

Physical Characteristics

micrometer 1/1000 of a meter

A typical eukaryotic cell is about 25 **micrometers** in diameter, but this average hides a large range of sizes. The smallest cell is a type of green algae, *Ostreococcus tauri*, with a diameter of only 0.8 micrometers, about the size of a typical bacterium. The human sperm is about 4 micrometers wide, but 40 micrometers long, while the egg is about 100 micrometers in diameter. Single neurons can be a meter or more in length. While schematic diagrams often picture cells as simple cubes or spheres, most cells have highly individual shapes. Human red blood cells are flattened disks indented on either side; muscle cells are highly elongated; neurons are long and thin with many branches on each end; and white blood cells constantly change their shapes as they crawl through the body.

Cells are also often depicted as a bag of fluid with a smattering of structures within, but this is far from the truth. Instead, the interior of the cell is a dense network of structural proteins, collectively termed the cytoskeleton, within which is embedded a large collection of organelles. The material within the cell except for the nucleus is called the cytoplasm. The nonorganelle portion of the cytoplasm is called the cytosol. The consistency of the cytoplasm is much like egg white, and not at all like freely flowing water.

Membranes

Eukaryotic cells include large amounts of membrane, which enclose the cell itself and surround each of the organelles. The membrane surrounding the cell is termed the plasma membrane. Membranes are bilayered structures, made of two layers of phospholipid molecules, built from phosphoric acids and fatty acids. One end of the phospholipid molecules (the exterior head) is **hydrophilic**, and it is oriented to the outer side of the membrane; the other end (the interior tails) are **hydrophobic**. Despite this, water molecules can pass freely through the bilayer, as can oxygen and carbon dioxide. Ions such as sodium or chloride cannot pass through, however, and neither can larger molecules such as sugars or **amino acids**. Instead, these materials must pass through the membrane via specialized proteins. This selective permeability allows the membrane to control the flow of materials in and out of the cell and its organelles.

hydrophilic water-loving

hydrophobic "water hating," such as oils

amino acids building blocks of protein

Proteins and Membrane Transport

Proteins are long chains of amino acids. They have unique shapes and chemical properties that dictate their diverse functions. Proteins govern the range of materials that enter and leave the cell, relay signals from the environment to the interior, and participate in many metabolic reactions, harvesting or harnessing energy to transform raw materials into the molecules needed by the cell for growth, repair, or other functions. Cytoskeleton proteins give the cell its structure. Approximately half the weight of a membrane is due to the proteins embedded in it. Proteins give each organelle, and the cell as a whole, its unique character.

As noted, ions cannot pass freely through the cell's phospholipid membrane. Instead, most ions flow through special channels built from multiple protein subunits that together form a pore from one side of the membrane to the other. Some channels are gated, fitted with proteins that act as hinged doors, blocking the opening until stimulated to swing out of the way. Neurons, for instance, have gated sodium channels that open to allow an electrical impulse to pass and then close to recharge the cell for another firing. Molecules can also cross the membrane attached to protein pumps that are powered by **ATP**. Transport of scarce molecules such as sugars can also be powered indirectly, by coupling their movement to the flow of another substance. In addition to traversing the membrane directly, water passes through special channels formed by a protein called aquaporin.

ATP adenosine triphosphate, a high-energy compound used to power cell processes

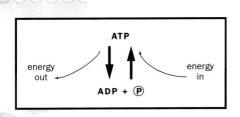

Energy is released when ATP is broken down, and required when it is formed.

hormones molecules released by one cell to influence another

transcription factor protein that increases the rate of transcription of a gene

glucose sugar

Signal Transduction

Proteins, including membrane proteins, also play critical roles in signal transduction, or relay. Signals can include **hormones**, ions, environmental changes such as odors or light, or mechanical disturbances such as stretching. A hormone is a small molecule released by one cell in the body to influence the behavior of another. A hormone exerts its influence by binding to a protein receptor in the target cell either on the membrane or within the cytoplasm. Cells that do not make receptors for a particular hormone are not susceptible to its effects. Adrenaline and testosterone are examples of hormones that illustrate two major modes of hormone action.

Adrenaline binds to a membrane-spanning receptor that projects both to the outside and the inside of the cell. The binding of adrenaline to the exterior portion changes the shape of the receptor, which in turn sets in motion other changes within the cell. The result is the production of a molecule called cyclic AMP (adenosine monophosphate), another form of the adenosine nucleotide. This "second messenger" binds to a variety of enzymes within the cell, activating them and leading to production of a variety of products. The exact set of enzymes turned on by cyclic AMP and the exact set of consequences depend on the particular cell. Kidney cells, for instance, increase their permeability to water, while liver cells release sugar into the bloodstream. The unique set of proteins within each cell is determined by the genes it has expressed, which in turn is determined by its own history and the hormones and other influences to which it has been exposed.

Testosterone's effects come on more slowly than adrenaline's, but last much longer. Testosterone passes through the plasma membrane and binds to a receptor in the cytosol. Once this occurs, the receptor-hormone complex is transported to the nucleus. Here, it binds to DNA, altering the rate of gene expression for a wide variety of genes. Thus, testosterone acts as a **transcription factor**. The prolonged action of testosterone is in part because it stimulates the production of new, long-lasting proteins that alter the cell's function for much longer than the very rapid and short-lived effects of adrenaline.

Cells continually respond to signals, and they influence other cells through the signals they release. Signaling pathways within the cell control the rate of cell division, the development and differentiation of the cell, the secretion of proteins and other molecules, and the response to injury, among many other reactions.

Metabolism

Metabolism refers to the entire set of reactions within the cell. Most reactions can be classified as either anabolic or catabolic. Anabolic reactions use stored energy to build more complex molecules from simpler ones. Protein synthesis is an example. Catabolic reactions break down complex molecules to simpler ones, releasing energy in the process that may be harvested and stored by the cell. **Glucose** breakdown is an example.

The energy transfer in each type of reaction almost always involves the interconversion of ATP and ADP (adenosine diphosphate). Energy is released when ATP loses a phosphate to become ADP, while energy is required to make ATP from ADP and phosphate. ATP can also be con-

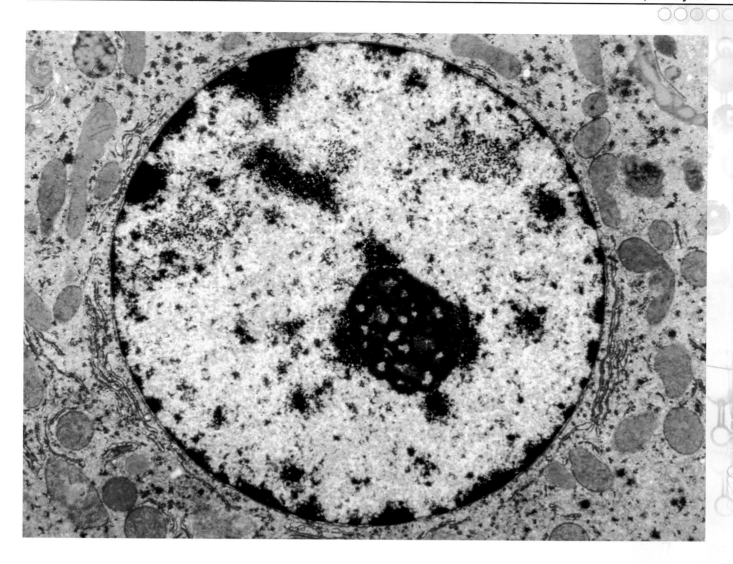

verted to AMP by the loss of two phosphates. This reaction, which releases even more energy, is used in replication of DNA and synthesis of RNA (transcription).

Mitochondrion

Glucose breakdown begins in the cytosol, but the majority of the process occurs in the mitochondrion, the energy-harvesting organelle of the cell. In addition to participating in the breakdown of glucose (and making ATP in the process), the mitochondrion is also involved in breaking down fats and amino acids. All these fuels are processed in two major steps, termed the Krebs cycle and the electron transport chain. In the Krebs cycle, the carbon skeletons are broken apart to make CO_2, while the hydrogen atoms are removed on special nucleotide carriers. In the electron transport chain, the hydrogens are stripped of their energy in a series of steps to make ATP, and in the end are reacted with oxygen to form water. The mitochondrion consumes virtually all the oxygen used by the cell. The mitochondrion also participates in many anabolic reactions, using the intermediates of the Krebs cycle as a source of carbon skeletons for creating and modifying nucleotides, amino acids, and other building blocks of the cell.

The nucleus of this liver cell, magnified nearly 3,000 times, has been stained to show some of its components: DNA (dark purple), nucleolus (burgundy). Mitochondria (red), and cytoplasm (pink), are visible outside the nucleus.

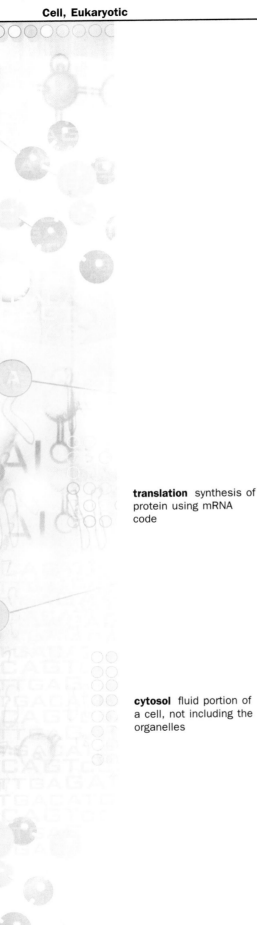

The mitochondrion is the descendant of a once free-living bacterium that took up residence inside an ancient cell, probably to take advantage of high-energy molecules the host could not metabolize. Mitochondria retain their own DNA on their own bacteria-like chromosome, although over time most of the original mitochondrion's genes were transferred to the host and now reside in the nucleus.

Chloroplast

The cells of plants and some protists possess chloroplasts, whose green chlorophyll gives plant leaves their characteristic color. Embedded in an internal membrane, chlorophyll absorbs sunlight and funnels it to a complex set of proteins nearby. Light energy is used to split water into oxygen (released as a waste product) and hydrogen, which is attached to nucleotide carriers. The hydrogen is then reacted with CO_2 from the air to form sugars, the essential high-energy product that powers all of life. Like the mitochondrion, the chloroplast is a relic of a former free-living bacterium, and has its own DNA on its own chromosome.

Nucleus

The nucleus contains the chromosomes. Chromosomes contain the genes, which are DNA sequences used to create RNA. The nucleus is bounded by a double membrane, called the nuclear envelope. Numerous large pores provide channels through which materials enter and exit. One of the chief exports of the nucleus is messenger RNA, which is used in the cytoplasm for protein construction.

translation synthesis of protein using mRNA code

Translation occurs in the cytoplasm at ribosomes, large complexes made of protein and RNA. Ribosomes are assembled in the nucleus, in the region called the nucleolus. RNA is synthesized by the enzyme RNA polymerase, which unwinds DNA and transcribes short portions, known as genes. These RNA molecules are processed further before being exported as messenger RNA. Other RNAs made in the nucleus include the RNA used in ribosomes (ribosomal RNA), RNAs that carry amino acids to the ribosome (transfer RNA), and a host of small RNAs that mostly function in the nucleus to modify other RNAs.

Protein Synthesis, Modification, and Export

cytosol fluid portion of a cell, not including the organelles

Messenger RNA exported from the nucleus binds to a ribosome in the **cytosol**, which then proceeds to translate the genetic message into a protein. Some proteins, with their ribosomes, remain free in the cytosol throughout translation, but others do not. Those that do not remain free carry a special sequence of amino acids at their leading end, called a signal peptide. This sequence directs the growing protein with its ribosome to the surface of the endoplasmic reticulum (ER), the most extensive organelle in the cell. Here, the ribosome attaches and extrudes the growing protein into the interior, or lumen, of the ER. Attachment of numerous ribosomes gives portions of the ER a rough appearance under the electron microscope. The ER also synthesizes most of the lipids used in the cell's many membranes. Lipid-synthesizing ER does not have ribosomes attached, and so appears smooth.

Many of the proteins entering the ER lumen are destined for other compartments in the cell, and contain organelle-specific targeting sequences that direct them to their final destination. Most of these proteins are first modified by the addition of branched sugar groups to make "glycoproteins." Most proteins in the plasma membrane, for instance, are glycoproteins. The full range of functions of these sugar groups is unknown, but they may help the protein to fold correctly after synthesis, act in cell-cell recognition and adhesion, and promote appropriate interactions with other proteins.

Proteins are further modified and sorted in the Golgi apparatus, a set of flattened membrane disks that is continuous with the ER. Here proteins and lipids are packaged in **vesicles** that bud off and travel along the cytoskeleton to their final destination. Fusion of the vesicle membrane with the target membrane delivers the contents to the target organelle. Proteins and other materials that the cell exports travel to the plasma membrane via vesicles. Fusion of the vesicle with the **plasma membrane** delivers the contents to the exterior of the cell.

vesicles membrane-bound sacs

plasma membrane outer membrane of the cell

Cell Cycle

Cells must reproduce in order for the organism to grow or repair damage. For single-celled organisms, cellular reproduction creates a new organism. Each new cell must get a complete set of chromosomes, which therefore must be duplicated and evenly divided between the two daughter cells.

The orderly series of events involving cell growth and division is termed the cell cycle. Immediately following a division, the cell grows by taking up and metabolizing nutrients, and by synthesizing the many proteins, lipids, nucleic acids, sugars, and other molecules it needs. DNA replication occurs next, making duplicate chromosomes, followed by a short period in which the cell synthesizes the numerous proteins specific for cell division itself.

Cell division includes two linked processes: mitosis, or chromosome division, and **cytokinesis**, or cytoplasm division. Triggered by specific protein changes, the chromosomes begin to coil up tightly and become visible under the microscope. Cytoskeleton fibers attach to them, and position the chromosomes in pairs along the cell's imaginary equator. At the same time, the nuclear envelope breaks down into numerous small vesicles. The cytoskeleton fibers (termed the spindle) pull the chromosome duplicates apart, segregating one member of each pair to opposite sides of the cell. Other cytoskeleton proteins pinch the membrane along the equator (in animal cells) or build a wall across it (in plant cells) to separate the two cell halves, ultimately forming two daughter cells. Finally, the nuclear envelope re-forms and the chromosomes uncoil, starting a new round of the cell cycle.

cytokinesis division of the cell's cytoplasm

SEE ALSO ARCHAEA; CELL CYCLE; EUBACTERIA; INHERITANCE, EXTRANUCLEAR; MEIOSIS; MITOCHONDRIAL GENOME; MITOSIS; NUCLEUS; PROTEINS; RIBOSOME; RNA PROCESSING; SIGNAL TRANSDUCTION; TRANSCRIPTION FACTORS; TRANSLATION.

Richard Robinson

Bibliography

Alberts, Bruce, et al. *Molecular Biology of the Cell*, 3rd ed. New York: Garland Publishing, 1994.

Centromere

During mitosis in a typical plant or animal cell, each chromosome divides **longitudinally** into two sister chromosomes that eventually separate and travel to opposite poles of the mitotic **spindle**. At the beginning of mitosis, when the sister chromosomes have split but are still paired, every chromosome attaches to the spindle at a specific point along its length. That point is referred to as the centromere or spindle attachment region.

longitudinally length-wise

spindle football-shaped structure that separates chromosomes in mitosis

Images from an electron microscope show that each sister is attached to fibers emanating from only one pole of the spindle. This allows the sisters to be pulled to opposite poles during mitosis. The electron microscope images also show that the spindle fibers do not terminate on the chromosomes themselves but rather on separate structures, known as kinetochores. Kinetochores are **trilaminar** bodies that assemble at the centromeres during the early stages of mitosis and disappear after the chromosomes have separated.

trilaminar three-layer

The budding yeast *Saccharomyces cerevisiae* has the simplest known centromeres consisting of a DNA segment only 110 bases in length. The DNA segment in the yeast centromere binds to specific proteins, which, like the kinetochores in higher organisms, link the chromosome to spindle fibers during mitosis. Centromeres of higher plants and animals are much larger, consisting of thousands or millions of bases of DNA and numerous proteins. For reasons that are unknown, centromeres are often flanked by long segments of DNA that do not contain functional genes. These nonfunctional DNA segments, called pericentric **heterochromatin**, vary in length in different organisms. In some cases they constitute more than half of the whole chromosome.

heterochromatin condensed portion of chromosomes

The crucial role of centromeres in the orderly behavior of chromosomes can be demonstrated by using X rays or other treatments that cause chromosomes to fragment. Pieces of chromosomes that lack centromeres (acentric fragments) do not attach to the spindle and are not pulled to the poles during mitosis. They generally are not included in the nuclei that are newly formed after cell division, and they usually degenerate in the cytoplasm.

Conversely, two fragments that each contain a centromere sometimes fuse, producing a dicentric chromosome. If the two centromeres happen to attach to the same pole at mitosis, the chromosome may move intact to that pole. However, if the centromeres attach to opposite poles, the chromosome will be stretched during mitosis and will eventually break. In general, therefore, only chromosomes with one centromere are stable.

Some organisms, including hemipteran insects and nematode worms, have *holocentric* or *holokinetic* chromosomes. In these organisms, spindle fibers attach all along one side of each sister chromosome, and the chromosomes are pulled more or less sideways to the pole.

Centromeres also play an important role during meiosis, in which the number of chromosomes is halved. The first meiotic division differs from a typical mitotic division in two respects:

1. In the first meiotic division, chromosomes derived from the organism's maternal and paternal parents pair at the beginning of meiosis. As in

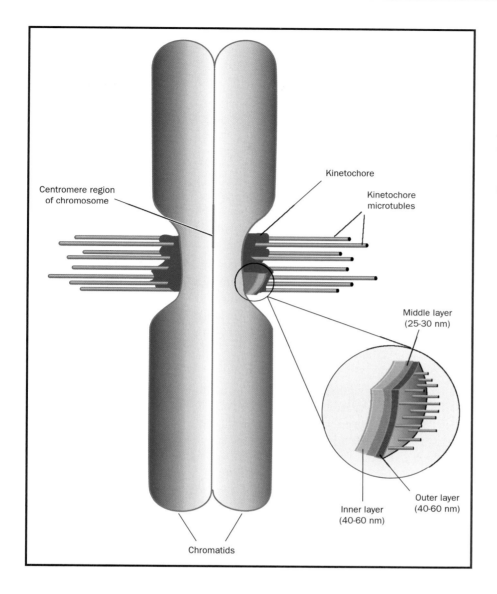

Centromere region of chromosome

Kinetochore

Kinetochore microtubles

Middle layer (25-30 nm)

Outer layer (40-60 nm)

Inner layer (40-60 nm)

Chromatids

The kinetochore is a protein structure that forms at the centromere before mitosis. Microtubules attach to it, and pull apart the two chromatids (not shown to scale). Adapted from <http://www.sus.mcgill.ca/bio202/00b/lec2/sld028.htm and http://www.esb.utexas.edu/dr325/Supplements/kinetchr.htm>.

a typical mitosis, each of these chromosomes has split into two sisters, so after pairing there are four chromosomes in a group.

2. When these four chromosomes attach to the spindle, sister chromosomes attach to the *same* pole, not to opposite poles as occurs in mitosis.

As a result, both maternal chromosomes move to one pole, while both paternal chromosomes move to the opposite pole. It is this unique behavior of the centromeres at meiosis that accounts for the separation of maternal and paternal genes during formation of sperm and eggs, which in turn is the basis of Mendelian genetics. SEE ALSO CELL CYCLE; CHROMOSOME, EUKARYOTIC; MEIOSIS; MENDELIAN GENETICS; MITOSIS.

Joseph G. Gall

Bibliography

Alberts, Bruce, et al. *Molecular Biology of the Cell*, 3rd ed. New York: Garland Publishing, 1994.

Internet Resources

Kinetochore Function. McGill University. <http://www.sus.mcgill.ca/bio202/00b/lec2/sld028.htm>.

Kinetochore Structure. University of Texas. <http://www.esb.utexas.edu/dr325/Supplements/kinetchr.htm>.

Chaperones

Molecular chaperones are proteins and protein complexes that bind to misfolded or unfolded **polypeptide** chains and affect the subsequent folding processes of these chains. All proteins are created at the ribosome as straight chains of amino acids, but must be folded into a precise, three-dimensional shape (conformation) in order to perform their specific functions. The misfolded or unfolded polypeptide chains to which chaperones bind are said to be "non-native," meaning that they are not folded into their functional conformation. Chaperones are found in all types of cells and cellular compartments, and have a wide range of binding specificities and functional roles.

polypeptide chain of amino acids

Discovery of Chaperones

Chaperones were originally identified in the mid-1980s from studies of protein folding and assembly in plant chloroplasts. A new protein was identified that was required for correct folding of a large enzyme complex in chloroplasts, yet the mysterious protein was not associated with the final assembled complex. It was quickly determined that this "chaperone" protein directing correct assembly was identical to one of the many proteins expressed at high levels when cells are grown at high temperatures (hence the common alternative name, "heat-shock protein," or Hsp).

It was later discovered that chaperones recognize the non-native, partially misfolded states of proteins that accumulate during high temperature stress. Most chaperones are also abundantly expressed under normal cell growth conditions, where they recognize non-native conformations occurring during both protein synthesis (prior to correct polypeptide chain folding), and later misfolding events.

Recognizing and Correcting Mistakes

in vivo "in life"; in a living organism, rather than in a laboratory apparatus

non-polar without charge separation; not soluble in water

Careful study, both **in vivo** and in the test tube, has demonstrated that molecular chaperones bind to their non-native substrate proteins by recognizing exposed **non-polar** surfaces ("non-polar" means that they are not attracted to water). In correctly folded proteins, these surfaces are usually buried away from the watery environment surrounding the protein. Chaperones promote correct folding of their substrate proteins by unfolding incorrect polypeptide chain conformations, and, in some cases, by providing a sequestered environment in which correct protein folding can occur. The activity of chaperones often requires the binding and hydrolysis of adenosine triphosphate (ATP).

Although only 20 to 30 percent of polypeptide chains require the assistance of a chaperone for correct folding under normal growth conditions, molecular chaperones are absolutely required for cell viability. Discussed below are a few of the most common classes of molecular chaperones and their effects on protein folding in the cell.

Two Common Chaperone Systems: Hsp70 and Hsp60

Hsp70 chaperones (so called because their size is approximately 70,000 daltons, or atomic mass units) are a very large family of proteins whose amino acid sequences are very similar, indicating how important their structure is to their function. A single cell or cellular compartment may contain multiple Hsp70 chaperones, each with a specific function. In addition, the Hsp70 chaperones often work in concert with one or more smaller co-chaperone proteins, which serve to modulate the activity of the chaperone.

Some of the well-studied Hsp70 chaperones include DnaK from the bacterium *Escherichia coli*, the Ssa and Ssb proteins from yeast, and BiP (for "binding protein") from the mammalian **endoplasmic reticulum**. Hsp70 chaperones are often located where unfolded polypeptide chains typically appear. For example, Ssb chaperones associate with ribosomes, so that they are close to newly synthesized, unstructured polypeptide chains. It is thought that the binding of Hsp70 chaperones to these unfolded chains prevents inappropriate partial folding until the entire polypeptide chain is available for correct folding.

Hsp60 chaperones (also called "chaperonins") are barrel-shaped structures composed of fourteen to sixteen subunits of proteins that are approximately 60,000 daltons in size. Each subunit has a patch of non-polar amino acid groups lining the inner surface of the barrel; this patch recognizes the exposed non-polar amino acids of misfolded proteins. The binding and **hydrolysis** of ATP triggers conformational changes within the barrel, which (1) unfolds the misfolded conformation and releases the unfolded chain into the center of the barrel, (2) closes the top of the barrel with the binding of a co-chaperone "cap," and thereby (3) provides a protected environment in which correct folding can occur. Upon dissociation of the co-chaperone, the fully or partially folded protein is released into the general cellular environment.

The most extensively studied Hsp60 chaperones include GroEL from *E. coli* and TRiC/CCT from eukaryotic cells. GroEL appears to function as a general chaperone and interacts with 10 to 15 percent of all *E. coli* polypeptide chains, with a definite bias toward proteins that are small enough to fit within its central cavity. TRiC/CCT recognizes a much smaller set of proteins, and appears to play an additional role in the assembly of multiprotein complexes.

Other Chaperone Systems

While Hsp70 and Hsp60 chaperones are the most extensively studied chaperone systems, there are many other chaperones with distinct cellular functions. These functions include modifying polypeptides after formation by altering the bonds within and between chains. It appears that some chaperones, in addition to attempting to rescue partially misfolded proteins, also alert the protein degradation system of the cell to the presence of substrate proteins that are too misfolded for rescue. It is expected, with the explosion of information provided by genome sequencing efforts, that many additional chaperones will be identified in the near future.

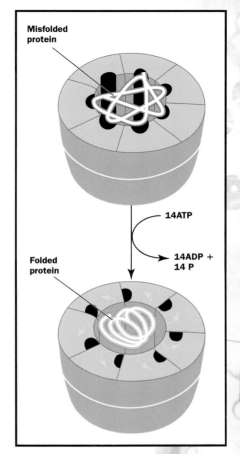

Chaperones use energy from ATP to help misfolded proteins refold properly.

endoplasmic reticulum network of membranes within the cell

hydrolysis splitting with water

Heat-shock proteins and chaperonins cooperate to fold newly synthesized proteins. The correct three-dimensional conformation is essential for protein function. Adapted from <http://www.nurseminerva.co.uk/chaperon.htm>.

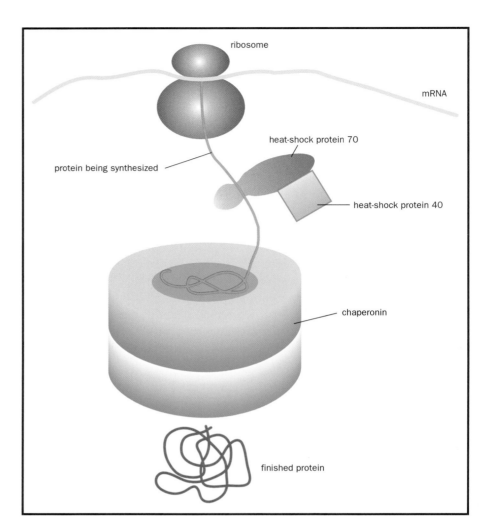

Chaperones and Human Disease

It is clear that molecular chaperones assist with the folding of newly synthesized proteins and correct protein misfolding. Recent studies now suggest that defects in molecular chaperone/substrate interactions may also play a substantial role in human disease. For example, mutations linked to Alzheimer's disease have been shown to disrupt the expression of chaperones in the endoplasmic reticulum. In addition, several genes linked to eye degeneration diseases have recently been identified as putative molecular chaperones. SEE ALSO CELL, EUKARYOTIC; POST-TRANSLATIONAL CONTROL; PROTEINS; RIBOSOME; TRANSLATION.

Patricia L. Clark

Bibliography

Frydman, Judith. "Folding of Newly Translated Proteins In Vivo: The Role of Molecular Chaperones." *Annual Review of Biochemistry* 70 (2001): 603–647.

Wickner, Sue, Michael R. Maurizi, and Susan Gottesman. "Posttranslational Quality Control: Folding, Refolding, and Degrading Proteins." *Science* 286 (1999): 1888–1893.

Internet Resources

"Chaperone." Nurse Minerva. <http://www.nurseminerva.co.uk/chaperon.htm>.

"Innovations." Environmental Health Perspectives. National Institutes of Health. <http://ehpnet1.niehs.nih.gov/docs/1994/102-6-7/innovations.html>.

"Molecular Chaperones." Federation of American Societies for Experimental Biology. <http://www.faseb.org/opar/protfold/molechap.html>.

Chromosomal Aberrations

Chromosomal aberrations are abnormalities in the structure or number of chromosomes and are often responsible for genetic disorders. For more than a century, scientists have been fascinated by the study of human chromosomes. It was not until 1956, however, that it was determined that the actual diploid number of chromosomes in a human cell was forty-six (22 pairs of **autosomes** and two sex chromosomes make up the human genome). In 1959 two discoveries opened a new era of genetics. Jerome Lejeune, Marthe Gautier, and M. Raymond Turpin discovered the presence of an extra chromosome in Down syndrome patients. And C. E. Ford and his colleagues, P. A. Jacobs and J. A. Strong first observed sex chromosome anomalies in patients with sexual development disorders.

autosomes chromosomes that are not sex determining (not X or Y)

Advances in Chromosomal Analysis

Identification of individual chromosomes remained difficult until advances in staining techniques such as Q-banding revealed the structural organization of chromosomes. The patterns of bands were found to be specific for individual chromosomes and hence allowed scientists to distinguish the different chromosomes. Also, such banding patterns made it possible to recognize that structural abnormalities or aberrations were associated with specific genetic syndromes. Chromosome disorders, or abnormalities of even a minute segment (or band) are now known to be the basis for a large number of genetic diseases.

Chromosomal disorders and their relationship to health and disease are studied using the methods of **cytogenetics**. Cytogenetic analysis is now an integral diagnostic procedure in prenatal diagnosis. It is also utilized in the evaluation of patients with mental retardation, multiple birth defects, and abnormal sexual development, and in some cases of infertility or multiple miscarriages. Cytogenetic analysis is also useful in the study and treatment of cancer patients and individuals with hematologic disorders. The types of chromosomal abnormalities that can be detected by cytogenetics are numerical aberrations, translocations, duplications, deletions, and inversions.

cytogenetics study of chromosome structure and behavior

Chromosomal Aberrations

Chromosomal abnormalities can result from either a variation in the chromosome number or from structural changes. These events may occur spontaneously or can be induced by environmental agents such as chemicals, radiation, and ultraviolet light. However, mutations are most likely due to mistakes that occur when the genes are copied as the cells are dividing to produce new cells. These abnormalities may involve the autosomes, sex chromosomes, or both. The disruption of the DNA sequence or an excess or deficiency of the genes carried on the affected chromosomes results in

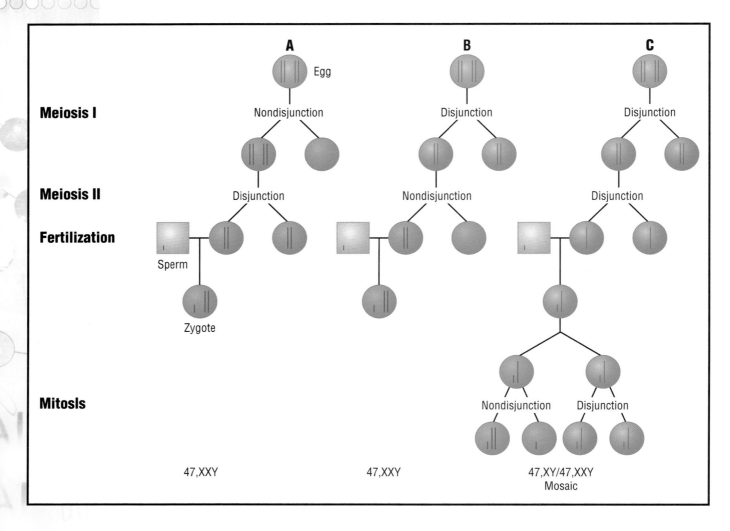

	A	B	C
	Egg		
Meiosis I	Nondisjunction	Disjunction	Disjunction
Meiosis II	Disjunction	Nondisjunction	Disjunction
Fertilization	Sperm		
	Zygote		
Mitosis			Nondisjunction Disjunction
	47,XXY	47,XXY	47,XY/47,XXY Mosaic

a mutation. Such a change may or may not alter the protein coded by a gene. Often, however, a mutation results in the disruption of gene functionality. The resulting altered or missing protein can disrupt the way a gene is meant to function and can lead to clinical disease. Only mutations occurring to the DNA in the **gametes** will potentially pass on to the offspring.

Mutations appear in gametes in one of two ways. A mutation may be inherited from one of an individual's parents. However, a mutation may also occur for the first time in a single gamete, or during the process of fertilization between an egg cell and a sperm cell. In this case the mutation or change is often called a de novo mutation. The parents are not affected by the condition and are not "carriers" of the mutation. The affected individual will have this mutation in all of his or her cells and may be able to pass the mutation on to any offspring. Some common abnormalities and their resulting phenotypes are discussed below.

gametes reproductive cells, such as sperm or eggs

Aneuploidy

Aneuploidy is the gain or loss of individual chromosomes from the normal diploid set of forty-six chromosomes. As in structural anomalies, the error may be present in all cells of a person or in a percentage of cells. Changes in chromosome number generally have an even greater effect upon survival than changes in chromosome structure. Considered the most common type of clinically significant chromosome abnormality, it is always associated with physical and/or mental developmental problems. Most aneuploid patients have a **trisomy** of a particular chromosome. **Monosomy**, or the loss of a chromosome, is rarely seen in live births. The vast majority of monosomic embryos and fetuses are probably lost to spontaneous abortion during the very early stages of pregnancy. An exception is the loss of an X chromosome, which produces Turner's syndrome. Trisomy may exist for any chromosome, but is rarely compatible with life.

Aneuploidy is believed to arise from a process called nondisjunction. Nondisjunction occurs when chromosomes do not separate correctly during meiosis. The direct result is that one gamete will have an extra chromosome and the other will be lacking a chromosome. When these gametes are fertilized by a normal gamete, they have either an extra chromosome (trisomy) or are missing a chromosome (monosomy).

Disorders Associated with Aneuploidy

Three well-known autosomal chromosome disorders associated with trisomies of entire autosomes are sometimes found in live births. These are trisomy 21 (Down syndrome), trisomy 13, and trisomy 18. Growth retardation, mental retardation, and multiple congenital anomalies are associated with all three trisomies. However, each has distinctive morphological characteristics, which are presumably determined by the extra dosage of the specific genes on the additional chromosome.

Down syndrome (chromosome 21) is the most frequent trisomy found in humans, and one of the most common conditions encountered in genetic counseling. General characteristics are mental retardation, distinctive palm prints, and a common facial appearance. The average life expectancy is now much greater thanks to improvements in medical care. Generally, individuals with Down syndrome have affable personalities and are able to be partially independent. The incidence of Down syndrome is about 1 in 800 children and is often associated with later maternal age (as may also be the case with other aneuploids).

Down syndrome appears to be related to the difference in gamete formation (gametogenesis) between males and females. In females, **oocytes** are formed before birth and held in a static state until ovulation. In the case of older mothers, an oocyte may be in this stage for more than forty years, during which time environmental factors may affect the genetic material. In trisomy 13 and trisomy 18 patients, congenital abnormalities are much more severe. These individuals generally do not live much beyond birth. Both trisomy 13 and trisomy 18 result in syndromes characterized by specific dysmorphic features and severe organ malformations.

In addition to trisomies involving the autosomal chromosomes, aneuploidy may also involve the sex chromosomes. Two examples are Turner's syndrome and Klinefelter's syndrome. As mentioned previously, Turner's syndrome is a monosomy involving the X chromosomes. Turner's syndrome females possess forty-five chromosomes (45, X) as compared to clinically normal forty-six (46, XX). They are usually sterile and short in stature with some neck webbing. Klinefelter's syndrome patients have a trisomy involving the

trisomy presence of three, instead of two, copies of a particular chromosome

monosomy gamete that is missing a chromosome

oocytes egg cells

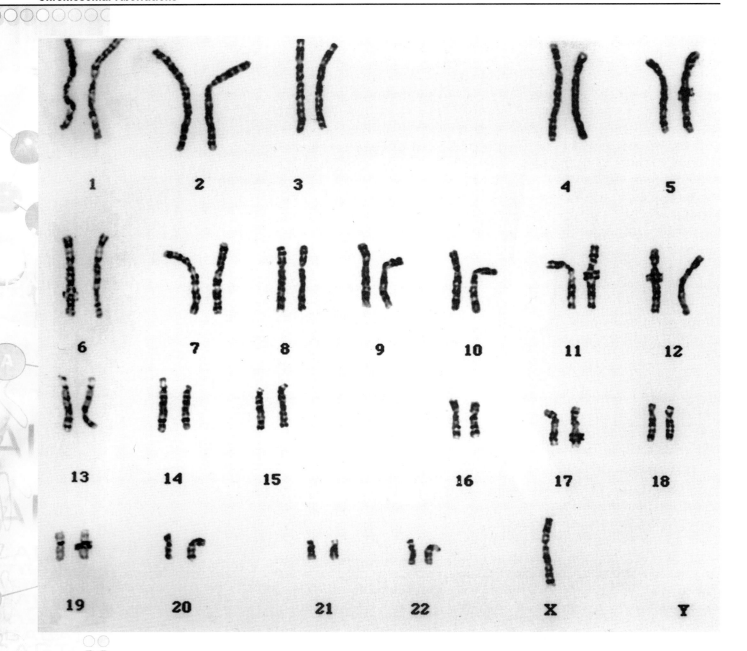

1 2 3 4 5

6 7 8 9 10 11 12

13 14 15 16 17 18

19 20 21 22 X Y

This karyotype of Turner's syndrome is characterized by a missing X chromosome.

sex chromosomes and thus have forty-seven chromosomes (47, XXY). Klinefelter's syndrome individuals are sterile males possessing some female characteristics. These chromosome abnormalities are of interest especially for their implications in infertility and abnormal development.

Abnormalities of Chromosomal Structure

Four types of structural changes may occur in chromosomes: duplications, deletions, translocations, and inversions. All may result when there is breakage of the chromosomes and a rejoining or loss of chromosome fragments. If the same broken ends rejoin, the chromosome becomes intact once again. The resulting effects of such events depend on how large they are and where they occur on the chromosome. Rearrangements may occur in many forms and are less common than abnormalities of chromosome number.

The most common type of rearrangement is called a balanced **translocation** because the amount of genetic information within that cell is normal even though it is repositioned. Therefore the individual with a balanced translocation may appear normal. However, there will be a risk to the children of a carrier of a balanced translocation since that person is likely to produce unbalanced gametes (bearing too little or too much genetic information), and therefore the risk of having abnormal offspring is increased. Rearrangements such as aneuploidy may be found in all cells of an individual, or they may occur only in a percentage of an individual's cells. This latter condition is known as mosaicism. In general, mosaic individuals show a less severe expression of their syndrome than those with chromosome abnormalities in all their cells.

translocation movement of a chromosome segment from one chromosome to another

Unbalanced Chromosome Rearrangements

A rearrangement is considered unbalanced if it results in extra or missing information. Structural rearrangements may be caused by a number of factors including chemicals, some viral infections, and ionizing radiation. Because the complement of DNA or genetic material in the chromosomes is greater or less than the complement of DNA in a normal set of chromosomes, there is likely to be abnormal development.

Deletions

A deletion is the loss of a segment of a chromosome. The amount of deleted material may be any length from a single base to a large piece of the chromosome. The result is a chromosomal imbalance, with the individual being monosomic or possessing half of the required genes present in a normal individual for the segment of DNA missing. Only small deletions are tolerated, and the effect on the individual will depend upon the size of the deleted segment and the number and functionality of the genes that are contained within it. Larger deletions and the deletion of an entire chromosome always result in nonviable embryos. *Cri du Chat's* ("cat's cry") syndrome individuals have a deletion of the short arm of chromosome 5. Although they possess the usual signs of chromosomal anomalies, such as mental retardation and low birth weight, their appearance is not extraordinarily different from normal individuals. One peculiarity is that affected infants make an unusual cry resembling that of a cat, hence the name of the syndrome. Two other interesting diseases are Prader-Willi's syndrome and Angelman's syndrome. In both cases, patients with these diseases possess a deletion in the long arm of chromosome 15. Interestingly, the deletion is in the same location, but the resulting syndrome depends on whether the deletion was in the maternal or paternal chromosome.

Duplications

Duplications also result from the reuniting of broken pieces of **homologous** chromosomes. In some cases the chromosome pieces rejoin in such a way that there is a doubling, or redundancy, of a portion of the chromosome. This changes the number of genes present and may result in a problem with health, development, or growth.

homologous similar in structure

Large insertions and deletions prevent the production of useful proteins. The effect of smaller insertions or deletions depends upon how many bases

Schematic representation of four types of chromosomal aberrations.

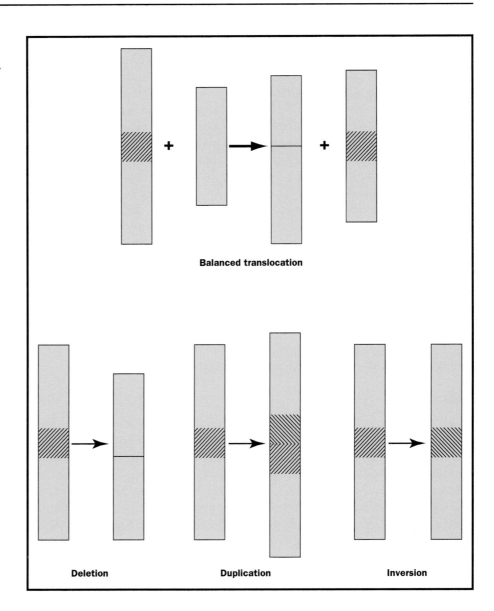

Balanced translocation

Deletion   **Duplication**   **Inversion**

are involved. Sometimes an entire gene can be inserted (in duplications) or deleted. The effect depends upon where in the genome the changes occur and how many base pairs are involved.

Inversions

An inversion is the rotation of a broken chromosome segment in such a way that it rejoins the chromosome in a reversed state, or is flipped, end to end. Inversions are usually characterized by whether the centromere is included in the inverted segment. Inversions containing the centromere are called pericentric. Those not containing the centromere are called paracentric. Although an inversion does not change the overall content of cellular DNA and can be considered a balanced translocation, it can affect a gene at many levels because it alters the normal DNA sequence. The gene may not produce its corresponding protein at all, or a nonfunctioning protein may result. There is a common inversion seen in human chromosomes involving chromosome 9. A small pericentric inversion is present in approximately 1

percent of tested individuals. There appears to be no detrimental effect on the carrier, and it does not appear to cause miscarriage or unbalanced offspring.

Recurrence Risk

Chromosomal aberrations may be inherited from a parent, and because of this many families seek genetic counseling in order to determine if a genetic disorder will recur in another member of the same generation or in generations that will follow. The family needs to know the genetic risk, also known as the recurrence risk, and any means by which transmission may be prevented. A recurrence risk will be calculated based on the accuracy of the diagnosis, the pedigree of the family, and the known genetic mechanisms of the disorder in question. SEE ALSO BIRTH DEFECTS; CHROMOSOME BANDING; CROSSING OVER; DOWN SYNDROME; MEIOSIS; MUTATION; NONDISJUNCTION; PRENATAL DIAGNOSIS.

Jacqueline Bebout Rimmler

Bibliography

Cohen, Jon. "Sorting Out Chromosome Errors." *Science* (Jun. 21, 2002): 2164–2166.

Haines, Jonathan L., and Margaret A. Pericak-Vance. *Approaches to Gene Mapping in Complex Human Diseases.* New York: John Wiley & Sons, 1998.

Klug, William S., and Michael R. Cummings. *Concepts of Genetics.* Upper Saddle River, NJ: Prentice Hall, 1997.

Thompson, Margaret W., Roderick R. McInnes, and Willard F. Huntington. *Genetics in Medicine.* Philadelphia: W. B. Saunders Company, 1991.

Verma, Ram S., and Arvind Babu. *Human Chromosomes: Manual of Basic Techniques.* New York: Pergammon Press, 1989.

Vogel, Friedrich, and Arno G. Motulsky. *Human Genetics.* Berlin, Germany: Springer-Verlag, 1986.

Weaver, Robert F., and Philip W. Hedrick. *Genetics.* Dubuque, IA: W. M. C. Brown Publishers, 1989.

Young, Ian D. *Introduction to Risk Calculation in Genetic Counseling.* Oxford, U.K.: Oxford University Press, 1999.

Chromosomal Banding

A chromosome banding pattern is comprised of alternating light and dark stripes, or bands, that appear along its length after being stained with a dye. A unique banding pattern is used to identify each chromosome and to diagnose chromosomal aberrations, including chromosome breakage, loss, duplication or inverted segments. In the 1950s, chromosomes from the cell's nucleus were identified with a uniform (unbanded) stain that allowed for the observation of the overall length and primary constriction (centromere) of each chromosome, as well as a secondary constriction in chromosomes 1, 9, 16 and the acrocentrics (chromosomes whose centromeres are near the tips). The staining techniques used to make the bands visible were developed in the late 1960s and early 1970s.

Chromosome Structure

To understand what chromosomal bands represent, it is helpful to understand the structure of chromosomes. Eukaryotic chromosomes are composed

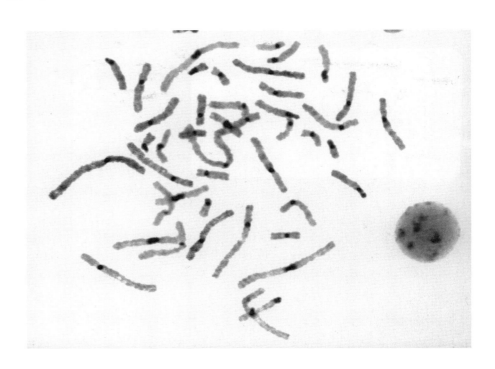

C-banded metaphase cell.

mitosis seperation of replicated chromosomes

metaphase stage in mitosis at which chromosomes are aligned along the cell equator

of chromatin, a combination of nuclear DNA and proteins. At metaphase, which is a phase in the cell cycle after the DNA in the nucleus has been replicated, each chromosome contains two identical strands of DNA. (Each strand contains two complementary strands of nucleotides.) The two strands of DNA, or chromatids, are arranged in a double-helix and are held together at a single point, the centromere, or primary constriction point.

During **mitosis**, each chromatid becomes condensed approximately ten-thousand fold reaching maximal condensation at **metaphase**. DNA that was roughly 5 centimeters (2 inches) long is compacted to 5 micrometers. The DNA wraps around proteins called histones, forming complexes called nucleosomes. The nucleosomes twist around each other and assume a loop formation projecting out from the chromosome's protein backbone, or scaffold. The loops weave and condense further to package the DNA into a chromosome. Some of the looped segments of DNA remain close together and condense more than others, forming regions known as domains. These domains are the darkly-stained chromosomal bands that appear when specific stains are applied (such as Giemsa staining; see below).

Looped domains are also seen in polytene chromosomes, which are found mainly in insects of the order Diptera, including *Drosophila*, which are fruit flies. Polytene chromosomes are large chromosomes that are formed after DNA undergoes repeated rounds of replication without cell division. A polytene chromosome in a *Drosophila* salivary gland cell can contain as many as five thousand alternating dark and light bands. The dark bands correspond to the folded and looped DNA, and the lighter bands are composed of less condensed DNA. The DNA in polytene chromosomes becomes less condensed when genes become active, permitting DNA to be transcribed into messenger RNA. This unraveling is observed as "puffing" of the polytene chromosome. The puffing resolves (the DNA condenses again) as the genes become inactive.

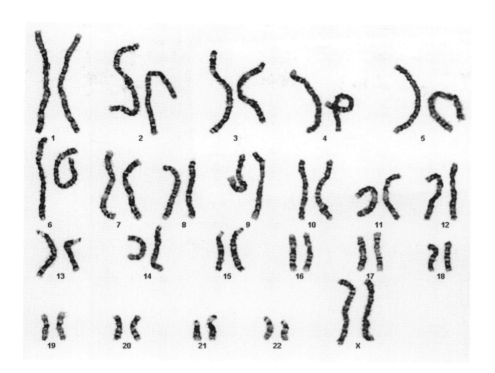

G-banded karyotype from a normal female.

Chromosome Banding Techniques

Quinacrine mustard, an alkylating agent, was the first chemical to be used for chromosome banding. T. Caspersson and his colleagues, who developed the technique, noticed that bright and dull fluorescent bands appeared after chromosomes stained with quinacrine mustard were viewed under a fluorescence microscope. Quinacrine dihydrochloride was subsequently substituted for quinacrine mustard. The alternating bands of bright and dull fluorescence were called Q bands. Quinacrine-bright bands were composed primarily of DNA that was rich in the bases adenine and thymine, and quinacrine-dull bands were composed of DNA that was rich in the bases guanine and cytosine.

Other fluorescent dyes have been used to generate chromosomal banding patterns. The combination of the fluorescent dye, DAPI (4,6-Diamidino-2-Phenylindole) with a non-fluorescent counterstain, such as Distamycin A, will also stain DNA that is rich in adenine and thymine. It will particularly highlight regions that are on the Y chromosome, on chromosomes 9 and 16, and on the proximal short arms of the chromosome 15 **homologues**, or pair.

homologues chromosomes with corresponding genes that pair and exchange segments in meiosis

Giemsa has become the most commonly used stain in cytogenetic analysis. Staining a metaphase chromosome with a Giemsa stain is referred to as G-banding. Unlike Q-banding, most G-banding techniques require pretreating the chromosomes with either salt or a proteolytic (protein-digesting) enzyme. "GTG banding" refers to the process in which G-banding is preceded by treating chromosomes with trypsin. G-banding preferentially stains the regions of DNA that are rich in adenine and thymine. In general, the bands produced correspond with Q-bright bands. The regions of the chromosome that are rich in guanine and cytosine have little affinity for the dye and remain light.

G-banded metaphase from a normal female.

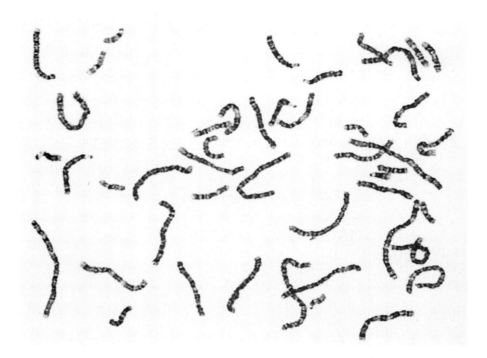

Standard G-band staining techniques allow between 400 and 600 bands to be seen on metaphase chromosomes. With high-resolution G-banding techniques, as many as two thousand different bands have been catalogued on the twenty-four human chromosomes. Jorge Yunis introduced a technique to synchronize cells so they are held at the same stage in the cell cycle. Cells are synchronized by making them deficient in folate, thereby inhibiting DNA synthesis. By rescuing the cells with thymidine, DNA synthesis is initiated and the timing of the prophase and prometaphase stages of the cell cycle can be predicted. Yunis's technique allows more bands to be resolved, as chromosomes produced from either prophase or prometaphase are less condensed and are thus longer than metaphase chromosomes.

Other Banding

R-banding is the reverse pattern of G bands so that G-positive bands are light with R-banding methods, and vice versa. R-banding involves pretreating cells with a hot salt solution that **denatures** DNA that is rich in adenine and thymine. The chromosomes are then stained with Giemsa. R-banding is helpful for analyzing the structure of chromosome ends, since these areas usually stain light with G-banding.

denatures destroys the structure of

C-banding stains areas of heterochromatin, which is tightly packed and repetitive DNA. NOR-staining, where NOR is an abbreviation for "nucleolar organizing region," refers to a silver staining method that identifies genes for ribosomal RNA that were active in a previous cell cycle.

Fluorescence *In Situ* Hybridization

Fluorescence *in situ* hybridization (FISH) is a molecular cytogenetic technique that allows cytogeneticists to analyze chromosome resolution at the DNA or gene level. FISH can be performed on dividing (metaphase) and non-dividing (interphase) cells to identify numerical and structural abnormalities resulting from genetic disorders.

In FISH, cytogeneticists utilize one or more FISH probes that typically fall into one of the following three categories:

1. Repetitive sequences, including alpha satellite DNA, that bind to the centromere of a chromosome;

2. DNA segments, representative of the entire chromosome, that will bind to and cover the entire length of a particular chromosome; and

3. DNA segments from specific genes or regions on a chromosome that have been previously mapped or identified.

A probe is "tagged" either directly, by incorporating fluorescent nucleotides, or indirectly, by incorporating nucleotides with attached small molecules, such as biotin, digoxygenin, or dinitrophenyl, to which fluorescent antibodies can later be bound. The probe and the chromosomes (from either the metaphase or interphase cells) that are being analyzed are denatured and allowed to bind or hybridize to one another. If necessary, **antibodies** with a fluorescent tag are applied to the cells. The cells are then viewed with a fluorescence microscope. The fluorescent signals represent the probe(s) that is bound to the chromosomes. SEE ALSO CELL CYCLE; CHROMOSOMAL ABERRATIONS; CHROMOSOME, EUKARYOTIC; FRUIT FLY: *DROSOPHILA*; *IN SITU* HYBRIDIZATION.

Gail H. Vance

antibodies immune-system proteins that bind to foreign molecules

Bibliography

Alberts, Bruce, et al. *Molecular Biology of the Cell*, 4th ed. New York: Garland Science, 2002.

Craig, J. M., and W. A. Bickmore. "Chromosome Bands—Flavours to Savour." *Bioassays* 15, no. 5 (1993): 349–353.

Earnshaw, W. C. "Meiotic Chromosome Structure." *Bioassays* 9, no. 5 (1988): 47–150.

Sumner, A. T. "The Nature and Mechanisms of Chromosome Banding." *Cancer Genetics and Cytogenetics* 6 (1982): 59–87.

Therman E., and M. Susman. *Human Chromosomes: Structure, Behavior and Effects*, 3rd ed. New York: Springer-Verlag, 1993.

Yunis, J. J., and R. C. Lewandowski. "High-Resolution Cytogenetics." *Birth Defects: Original Article Series* 19, no. 5 (1983): 11–37.

Chromosomal Theory of Inheritance

The chromosomal theory of inheritance is the idea that genes, the units of heredity, are physical in nature and are found in the chromosomes. The theory arose at the turn of the twentieth century, and became one of the cornerstones of the modern understanding of genetics.

The Birth of a Science

Charles Darwin first conceived the idea of hereditary units when he published his theory of pangenesis in 1868. In this model, circulating units called gemmules are accumulated in the gonads and transmitted to the offspring. This theory was discredited by experimental tests performed by Francis Galton in the 1870s. Galton used blood transfusions in rabbits to

show that the alleged gemmules in one rabbit's blood did not alter the heredity of the recipient rabbit's blood. In the 1890s Hugo de Vries took the term "pangenesis" and trimmed it to "pangene" for the assumed units of inheritance. He argued that pangenes remained inside the cell and did not migrate. It was this theory of intracellular pangenesis that led de Vries to independently find what Gregor Mendel had discovered thirty years earlier in his work with contrasting traits in garden peas—there are units of inheritance that are transmitted by reproduction. Wilhelm Johansson introduced the term "gene" to replace several contending and misleading terms for the basic unit of heredity in 1909. The term "genetics" came earlier, when William Bateson coined the word in 1906 to represent the new field that studied heredity, variation, and evolution. The terms "gene," "genetics," and the biblical term "genesis" all share a common Latin root, *gen*, meaning origin.

Mendel identified what he called "factors" (later called genes) as the underlying cause for the appearance of certain traits in peas. He described them as stable units that seemed to disappear in a hybrid (a plant grown from a cross between two parent plants that show differing traits) but would reappear among some of the progeny of such hybrids. Mendel identified two laws—the law of segregation and the law of independent assortment—which together governed the movement of factors from parent to **progeny**. This strongly suggested that the factors of inheritance were discrete physical objects. Shortly before Mendel's work was being rediscovered, advancements in the construction of microscopes had allowed scientists to make careful observations of cell division. This led to the discovery of colored bodies in the cell nucleus that appeared to double and divide just before each division. These were called chromosomes ("colored bodies").

By 1902 the chromosome movements during meiosis had been worked out, and Walter Sutton used them to explain Mendel's laws. He argued that the pairing and separation of **homologues** would lead to the segregation of a pair of factors they carried. Thus, to use one of Mendel's own experiments, hybridizing yellow and green pea plants would yield one yellow and one green gamete apiece. (A gamete is germ cell, sperm or egg, that contains half of a full complement of chromosomes, originating from one of the parent plants.) The result is a gametic ratio of 1:1. The union of **pollen** and **ovules** would result in the 3 yellow to 1 green Mendelian ratio. Similarly, two different pairs of homologs would yield the 9:3:3:1 ratio associated with Mendel's law of independent assortment. Sutton called his union of **cytology** with Mendelian breeding analysis the "chromosome theory of heredity."

progeny offspring

homologues chromosomes with corresponding genes that pair and exchange segments in meiosis

pollen male plant sexual organ

ovules eggs

cytology the study of cells

X-Linked Inheritance in Hybrids

Beginning in 1907, Thomas Hunt Morgan extended Sutton's insights by conducting laboratory studies of the fruit fly, *Drosophila melanogaster*. With his students Alfred Henry Sturtevant, Calvin Blackman Bridges, and Hermann Joseph Muller, he established what is now called classical genetics. Morgan and his students found new phenomena that added to Sutton's chromosome theory of heredity. The first finding, achieved in 1920, was X-linked inheritance, in which white-eyed flies showed a sex-linked inheritance of the trait in a modified 3:1 ratio. In other words, cross-breeding hybrid

red-eyed flies resulted in all the female offspring having red eyes, whereas half the male offspring had white eyes.

Morgan's team explained this modified ratio by proposing that the eye-color genes are carried on the X chromosome, of which females have two but males have only one. The female's two X chromosomes can be **homozygous** (the genes carried are AA or aa) or **heterozygous** (the genes carried are Aa) for an X-linked gene. Males are always homozygous, because the small Y chromosome lacks almost all the genes found on the X chromosome. Thus, they can carry only one gene for the trait: AY or aY. Since males can carry no second copy of the gene for the trait, they will express the white-eyed trait if they inherit the gene for it from a hybrid red-eyed parent. This discovery further strengthened the case in favor of the chromosomal theory of inheritance.

Several additional X-linked mutations arose by 1913. Morgan reported that these mutations produced unusual ratios when subjected to breeding analysis. He explained these findings by proposing that genes could trade places between two homologous chromosomes. Morgan's finding, called crossing over, was used to map the genes along the length of the X chromosome, and Sturtevant used Morgan's data to construct the first linkage map. Working independently, maize geneticists, especially Barbara McClintock, later demonstrated the same phenomenon.

Further Advances in Theory

Also from 1913 to 1916, Bridges found some exceptions to the expected modified 3:1 ratio for white-eyed flies. He inferred, and confirmed by microscopic examination of cells, that these unexpected departures arose from the failure of homologous chromosomes to separate during meiosis. Bridges called this phenomenon nondisjunction and used it as a proof of the chromosome theory of heredity.

While both Mendel and Morgan's group worked with simple, single-gene traits, the relation of genes to most character traits turned out to be more complex. A successful analysis of this was presented by Muller. He argued that the variable wing shapes and lengths of beaded and truncated wings in fruit flies involved several factors. A chief gene was essential, but it required modifier genes that could intensify or diminish its expression. By combining different modifier genes and the chief gene in two parents, Muller could predict the percentages of wing shape and length among the progeny. Muller used this analysis to support a Darwinian model of natural selection of character traits whose variations owe their origins to the highly heterozygous state of natural populations and to new mutations that arise in each generation. Muller's analysis added evolution to cytology and breeding analysis as the three tributaries of classical genetics. SEE ALSO EPISTASIS; FRUIT FLY: *DROSOPHILA*; LINKAGE AND RECOMBINATION; MCCLINTOCK, BARBARA; MENDEL, GREGOR; MORGAN, THOMAS HUNT; NATURE OF THE GENE, HISTORY.

Elof Carlson

homozygous containing two identical copies of a particular gene

heterozygous characterized by possession of two different forms (alleles) of a particular gene

Bibliography

Allen, G. E. *Thomas Hunt Morgan: The Man and His Science.* Princeton, NJ: Princeton University Press, 1978.

Judson, H. P. *The Eighth Day of Creation: The Makers of the Revolution in Biology*. New York: Simon & Schuster, 1979.

Morgan, Thomas Hunt, et al. *The Mechanism of Mendelian Heredity*. New York: Holt Reinhart & Winston, 1915. Reprinted by Johnson Reprint Corporation with an Introduction by Garland E. Allen, 1978.

Sturtevant, Alfred Henry. *A History of Genetics*. New York: Harper & Row, 1965.

Chromosome, Eukaryotic

Living organisms are divided into two broad categories based upon certain attributes of cell structure. The first category, the prokaryotes, includes bacteria and blue-green algae. Eukaryotes include most other living organisms. One of the most important features distinguishing eukaryotes from prokaryotes is the chromosomal arrangement of genetic information in the cells. Eukaryotes enclose their genetic material in a specialized compartment called the nucleus. Prokaryotes lack nuclei.

Basic Organization

In 1883, Wilhelm Roux proposed that the filaments observed when cell nuclei were stained with basic dyes were the bearers of the hereditary factors. Heinrich Wilhelm Waldeyer later coined the word chromosome ("colored body") for these filaments. The eukaryotic chromosome now is defined as a discrete unit of the genome, visible only during cell division, that contains genes arranged in a linear sequence. Eukaryotic organisms contain much more genetic information than prokaryotes. For example, the eukaryotic organism *Saccharomyces cerevisiae* (baker's yeast) contains 3.5 times more DNA in its **haploid** state than the prokaryotic *Escherichia coli*, while higher vertebrate cells contain more than 1,000 times the DNA.

haploid possessing only one copy of each chromosome

The basic component of the eukaryotic chromosome is its DNA, which contains all of the genetic material responsible for encoding a particular organism. Genes are arranged in a linear array on the chromosome. A major distinction between eukaryotic and prokaryotic chromosomes is that eukaryotic chromosomes contain vast amounts of DNA between the genes. The function of most of this "extra" DNA is unknown. It contains repetitive sequences, functionless gene copies called pseudogenes, transposible elements, and other types of DNA.

Eukaryotic genes may be dispersed randomly throughout the chromosome or they may be specifically organized. A gene family is a set of genes that originated from the duplication and subsequent variation of a common, ancestral gene. Members of a gene family may be clustered on the same chromosome, as in the case of the globin genes. Gene duplication events also have resulted in gene clusters in which related or identical genes are arranged in tandem. Examples of gene clusters include the genes for rRNA and histone proteins.

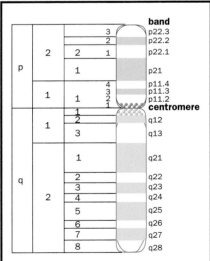

Chromosome staining reveals distinct bands that can be numbered for reference. Adapted from Lewin, 1997.

Higher-Order Organization

The DNA of a eukaryotic cell must be constrained within the confines of the nucleus. In human cells, six billion base pairs are contained on the forty-six chromosomes of double-stranded DNA. This DNA has a total length of

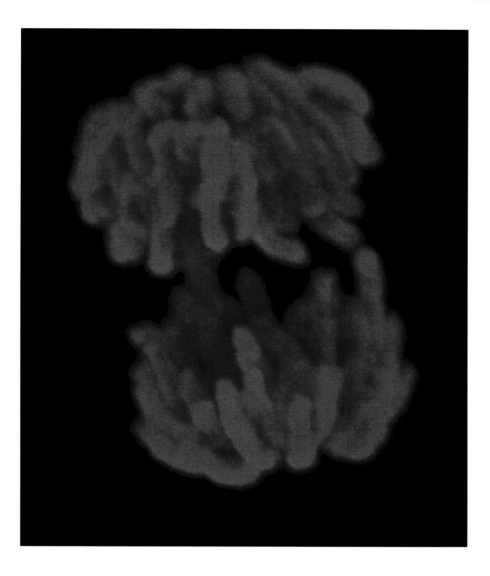

Eukaryotic chromosomes separating during the later stages of cell division.

1.8 meters, yet it must fit into a nucleus with an average diameter of 6 **micrometers**. This feat is accomplished in part by the packaging of DNA into chromatin, a condensed complex of DNA, histones, and nonhistone proteins.

micrometer 1/1000 meter

The basic unit of chromatin is the nucleosome. The nucleosome is composed of approximately 146 base pairs of DNA wrapped in 1.8 helical turns around an eight-unit structure called a histone protein octamer. This histone octamer consists of two copies each of the histones H2A, H2B, H3, and H4. Nucleosomes form arrays along the DNA. The space in between individual nucleosomes is referred to as linker DNA, and can range in length from 8 to 114 base pairs, with 55 base pairs being the average. Linker DNA interacts with the linker histone, called H1, and there are equal numbers of nucleosomes and H1 histone molecules in the chromatin. A nucleosome particle bound to a single molecule of H1 is termed a chromatosome.

Higher-order chromatin structure can be visualized microscopically as fibers 10 and 30 **nanometers** in diameter. The 10-nanometer fiber can be observed under conditions of low ionic strength. This fiber resembles beads

nanometer 10^9 meters; one billionth of a meter

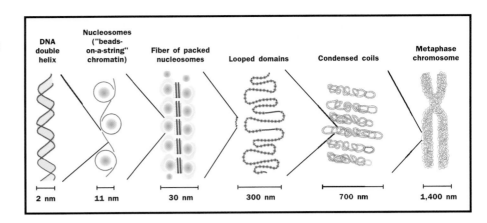

Levels of organization of DNA in the chromosome. The double helix is wound around histone proteins to form nucleosomes. Condensed coils characterize regions that are not expressed. The metaphase chromosomes only appear during cell division. Adapted from Curtis, 1989.

in vivo "in life"; in a living organism, rather than in a laboratory apparatus

in vitro "in glass"; in lab apparatus, rather than within a living organism

interphase the time period between cell divisions

on a string, and is in fact a string of nucleosomes. The structure does not require the presence of histone H1. It is unclear if the 10-nanometer fiber exists **in vivo** or if it is just an artifact of chromatin unfolding during extraction **in vitro**.

Cellular chromatin usually exists as a 30-nanometer fiber. It can be seen under conditions of higher ionic strength. The presence of the linker histone H1 is required for the formation of this fiber, as it helps promote compaction and condensation. The 30-nanometer fiber is formed into a coil, but its exact structure has not been determined.

During cell division, chromosomes condense to an even greater extent. The mechanism by which the 30-nanometer fibers are packed into the highly condensed, organized structure of the mitotic chromosome is unclear. The compaction of chromosomes during cell division is accomplished in part by the organization of chromatin into large, looped structures that are attached at their bases to a protein scaffold. This scaffold remains intact even if the DNA is experimentally removed. It is possible that this scaffold is responsible for maintaining the shape of the chromosomes.

Heterochromatin versus Euchromatin

Chromatin can be divided into two regions, euchromatin and heterochromatin, based on its state of condensation, that is, based on how tightly its constituent elements are packed together. Most of the cellular chromatin is euchromatin, which has a relatively dispersed appearance in the nucleus. It condenses significantly only during mitosis. Genes within euchromatin can be transcriptionally active or repressed at a given point in time.

Heterochromatin, on the other hand, is condensed in **interphase**, usually does not contain genes that are being expressed, and is among the last portions of the genome to be replicated prior to cell division. Heterochromatin frequently is localized at the periphery of the nucleus. It can be subdivided into constitutive and facultative heterochromatin. Constitutive heterochromatin is always inactive. It is often found adjacent to centromeres and telomeres. Facultative heterochromatin refers to DNA sequences that are specifically inactivated as the result of development or a regulatory event. One example of facultative heterochromatin is the mammalian X chromosome. The single X chromosome present in male cells is active. However, in female cells, one of the two copies present is directly and specifically inactivated.

Cytological Features

While the interphase chromatin appears to be a tangled mass within the nucleus, the mitotic chromosome appears as an organized structure with many prominent features. These features include structures known as the **centromere** and the **telomere**.

The centromere is the region of the chromosome to which the spindle apparatus attaches during mitosis and meiosis. The spindle apparatus is the network of fibers along which the chromosomes move during cell division. It also contains the site at which sister chromatids are attached prior to their separation during the stage of cell division known as anaphase. The centromere is responsible for the movement of the chromosome. During mitosis and meiosis, the centromere is pulled by the spindle fibers toward the opposite ends of the dividing cell (poles), as the attached chromosome is dragged behind. The centromere is essential for segregation.

The telomere is a structure that occurs at the end of linear eukaryotic chromosomes and that confers stability. The first telomere to be sequenced was from the organism *Tetrahymena thermophila*, a type of single-cell eukaryotic organism, in 1978 by Elizabeth Blackburn and Joseph Gall. This telomere contains an AACCCC nucleotide sequence that is repeated thirty to seventy times. The sequences of telomeres from other species show the same pattern: a tandem array of a short nucleotide sequence, one DNA strand G-rich and the other DNA strand C-rich. Telomeres are synthesized by an enzyme called telomerase, which adds telomeric sequences back onto chromosome ends, one base at a time.

Banding techniques allow every mitotic chromosome, as well as regions within individual chromosomes, to be distinguished. The first method used, known as Q-banding, uses flourescent derivatives of quinacrine, which for unknown reasons bind preferentially to some regions of chromosomes. When viewed under ultraviolet light, chromosomes appear with bright bands, corresponding to euchromatin, and dark bands, corresponding to heterochromatin. This banding method is useful in identifying some polymorphisms, which are gene variations (**alleles**) within the population of genomes. It is also useful for identifying the Y chromosome.

Later, another banding technique was developed, called G-banding. This technique employs a modified Giemsa stain, which is a dye that specifically binds to DNA. As in Q-banding, a series of identifiable light and dark bands is generated on the chromosome. Euchromatin stains lightly, while most heterochromatin stains darkly.

These banding methods have made it possible to diagrammatically represent each human chromosome, using designated nomenclature for specific chromosomal regions. These techniques have shown that the mitotic chromosome can be divided into a short arm, designated p, and a long arm, q. Each arm is then divided into one to three regions by landmarks, such as the ends of the arm, the centromere, and certain prominent bands. Regions are spaces between adjacent landmarks and are numbered consecutively within each region. These features allow genes to be designated to specific

centromere the region of the chromosome linking chromatids

telomere chromosome tip

alleles particular forms of genes

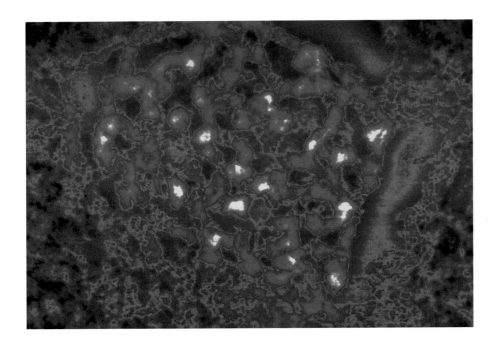

Giant chromosomes from a fruit fly larva's salivary glands enable study of RNA transcription. The faint, "puffed" areas are the sites of active transcription.

regions on the mitotic chromosome. For example, a gene at band 4q14 is localized to chromosome 4, the long arm, region 1, band 4.

Polytene Chromosomes

Due to the compact nature of chromatin in eukaryotic cells, it is virtually impossible to visualize gene expression in vivo. However, there are certain unusual situations in which gene expression can be seen. Such is the case with the polytene chromosomes, which are exceptionally large in comparison to other types of chromosomes.

The salivary glands of *Drosophila melanogaster* (fruit fly) larvae contain greatly enlarged chromosomes. These are polytene chromosomes, and they result from multiple rounds of replication of a diploid pair of chromosomes joined in parallel. The replicated chromosomes remain attached to one another. Each pair of chromosomes can replicate up to nine times; thus, the resultant polytene chromosome can contain up to 1,024 (2^9) strands of DNA.

The vast majority (95%) of DNA in the polytene chromosomes is concentrated in chromosomal bands, called chromomeres, which are microscopically visualized through staining. These chromomeres form a pattern that is characteristic for each *Drosophila* strain. *Drosophila* polytene chromosomes display roughly 5,000 bands. Since the total number of genes in *Drosophila* appears to be greater than the number of bands that can be visualized, it is likely that there are multiple genes located within a given band.

The banding pattern of the polytene chromosomes provides a cytological map, or diagrammatic representation, of the physical location of genes at specific sites in the cell. The positions of individual genes can be determined using a technique called *in situ* hybridization. First, the DNA of an immobilized chromosome preparation is made single-stranded (denatured). A radioactively labeled probe, generally a small piece of DNA corresponding to the gene of interest, is then mixed with the denatured DNA under conditions that permit the radiolabeled DNA to bind to its complementary

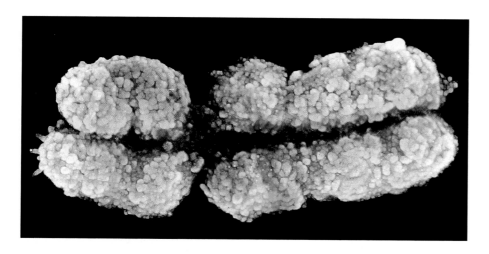

Scanning electron microscope magnification of one human X chromosome. Genetic information is packed within the "bumps" and loops of material pictured here.

DNA strand on the immobilized chromosome preparation. Finally, the chromosomal binding site is determined by way of a procedure called autoradiography, which is used to make the radioactive probe visible on photographic film.

With autoradiography, the sites of gene expression can be visualized along the polytene chromosomes. As DNA decondenses into a more open state, it forms a distinctive swelling known as a chromosomal puff. These puffs are active sites of transcription, or RNA synthesis. Throughout development, they alternately expand and contract, as specific genes are activated or repressed.

Chromosome Organization, Replication, and Transcription

The compact nature of chromatin structure presents a barrier to processes that require access to DNA, such as replication and transcription. It seems likely that the separation of parental DNA strands during replication must disrupt higher-order chromatin structure to at least some degree. In the *Drosophila* polytene chromosomes, this disruption can be seen, at least in part, as puffs at sites of DNA replication. It is unclear, however, what happens to the nucleosomes during this process. If the nucleosomes are removed during replication, they are quickly reassembled, for there is no time period during which microscopic analysis shows the DNA to be free of nucleosomes.

Transcriptional activation of eukaryotic genes requires that the machinery responsible for synthesizing the RNA gain access to the regulatory regions of the DNA that control gene expression. This again necessitates some decondensation of the chromatin structure. Decondensation can be facilitated by protein complexes known as chromatin-remodeling **enzymes**. Chromatin-remodeling enzymes alter the structure of chromatin in such a way that regulatory factors can gain access to the DNA. These enzymes are divided into two groups, those that chemically modify histones and those that use the energy derived from ATP hydrolysis to alter histone-DNA linkages.

enzymes proteins that control a reaction in a cell

Chemical Modification of Chromatin Structure

The best-characterized of the enzymes capable of chemically modifying histones to open chromatin structure are the histone acetyltransferases, or

HATs. These enzymes add acetyl groups to lysine residues within the amino termini (also known as the tails) of H3 and H4 histones. This adds a negative charge to the histone tails. The negative charge is believed to cause them to push away from the DNA backbone, resulting in a somewhat less condensed chromatin structure. Hyperacetylation of histones in the promoter regions of genes is associated with active, ongoing gene expression, while histone hypoacetylation is associated with genes that are transcriptionally silent.

The association between histone acetyltransferases and gene activation was first suggested when it was found that the *Tetrahymena thermophila* HAT p55 was structurally similar to the yeast protein GCN5. GCN5 previously had been shown to be involved in gene activation. Later it was discovered that GCN5-regulated genes are hyperacetylated when active, and that certain mutations in GCN5 that affect its ability to activate target genes result in diminished levels of acetylation of these regions. Histone acetyltransferases also are found in mammalian cells. It is believed that the HATs are recruited to specific genes by specific transcriptional activators.

The activity of the histone acetyltransferases is opposed by histone deacetylases. These enzymes remove acetyl groups from the histone tails, resulting in the repression of gene activation. As with the HATs, these enzymes are believed to be recruited to chromatin by repressor proteins to aid in the inactivation of specific genes.

Transcriptional activation also can be repressed by the methylation of specific cytosine residues found in some genes. It is unclear how methylation results in transcriptional repression, but it is known to cause the inhibition of specific transcriptional activators and to initiate the recruitment of specific repressors that bind methylated DNA. At least one methylated DNA-binding repressor can be isolated from cell extracts together with (and therefore appears to be associated with) histone deactylases, thereby providing a link between methylation and deacetylation and gene inactivation. Other enzymes chemically modify histones by adding or removing phosphate, ubiquitin, and other chemical groups.

ATP-Dependent Chromatin-Remodeling Complexes

The presence of enzymes that can alter the structure of chromatin was suggested by yeast genetic studies that identified a number of genes, called SWI and SNF genes. These genes are required for multiple transcriptional activation events. A key breakthrough came when it was discovered that yeast cells could compensate for a deficiency in these SWI and SNF gene products by altering their chromatin structure. This led to the hypothesis that SWI and SNF genes are involved in the regulation of chromatin structure. It is now known that SWI and SNF proteins form a large, multisubunit complex, termed SWI/SNF, that can hydrolyze ATP and use the energy thus generated to alter chromatin structure.

Similar proteins that can hydrolyze ATP are present throughout the eukaryotic kingdom, and these form related multiprotein enzymes that also possess chromatin-remodeling properties. The mechanism by which these complexes alter the chromatin structure is unclear, but it is likely that the enzymes break or loosen the linkage between histone and DNA in a manner that increases the mobility and flexibility of the DNA wrapped around the histone core. It is important to keep in mind that these enzymes can be involved both in gene-activation events, by facilitating the binding of transcriptional activators, and in gene-repression events, perhaps by facilitating the binding of a transcriptional repressor or by directly promoting compaction of the chromatin structure. SEE ALSO CELL CYCLE; CELL, EUKARYOTIC; CHROMOSOMAL BANDING; CHROMOSOME, PROKARYOTIC; DNA; EVOLUTION OF GENES; GENE; GENE EXPRESSION: OVERVIEW OF CONTROL; *IN SITU* HYBRIDIZATION; MEIOSIS; MITOSIS; MOSAICISM; REPETITIVE DNA ELEMENTS; REPLICATION; TELOMERE; X CHROMOSOME; Y CHROMOSOME.

Cynthia Guidi and Anthony N. Imbalzano

Bibliography

Alberts, Bruce, et al. *Molecular Biology of the Cell*, 3rd ed. New York: Garland Science, 1994.

Curtis, Helena, and Sue Barnes. *Biology*, 5th ed. New York: Worth, 1989.

King, Robert C., and William D. Stansfield. *A Dictionary of Genetics*, 4th ed. New York: Oxford University Press, 1990.

Lewin, Benjamin. *Genes VI*. Oxford U.K.: Oxford University Press, 1997.

Robinson, Richard, ed., *Biology*. New York: Gale Group, 2001.

Strahl, B. D., and C. D. Allis. "The Language of Covalent Histone Modifications." *Nature* 403 (2000): 41–45.

Voet, Donald, and Judith G. Voet. *Biochemistry*, 2nd ed. New York: John Wiley & Sons, 1995.

Watson, James D., et al. *Molecular Biology of the Gene*, 4th ed. Menlo Park, CA: Benjamin Cummings, 1987.

Chromosome, Prokaryotic

The bacterial or prokaryotic chromosome differs in many ways from that of the eukaryote. The term "eukaryote" comes from the Greek and means "true nucleus." Eukaryotic cells have a double membrane (the **nuclear membrane**) surrounding the nucleus, the organelle that contains several chromosomes. In contrast, the term "prokaryote" means "primitive nucleus," and, indeed, cells in prokaryotes have no nucleus. Instead, the prokaryotic chromosome is dispersed within the cell and is not enclosed by a separate membrane.

nuclear membrane membrane surrounding the nucleus

This dispersed chromosome is called the bacterial "nucleoid," which can be seen in electron micrographs of thin sections, as shown in Figure 2. Although bacteria (now called eubacteria) are highly diverse, the prototypical bacterial species is *Escherichia coli*, which has served as a model organism for genetic, biochemical, and biotechnological research for many decades.

The *E. coli* chromosome is a single circle. Because the single DNA molecule forming the chromosome is so long (about 4.6 million base pairs), it is easily broken when researchers try to isolate it. However, in the early 1960s, the Australian biochemist John Cairns was able to gently **lyse** *E. coli* cells without breaking the chromosome. He was interested in chromosomal replication and had labeled the DNA with tritium (^{3}H), a radioactive form of hydrogen. Autoradiograms of the DNA demonstrated that the bacterial chromosome is a circular molecule. While the vast majority of bacterial species possess a single unique chromosome, there are a few rare species,

lyse break apart

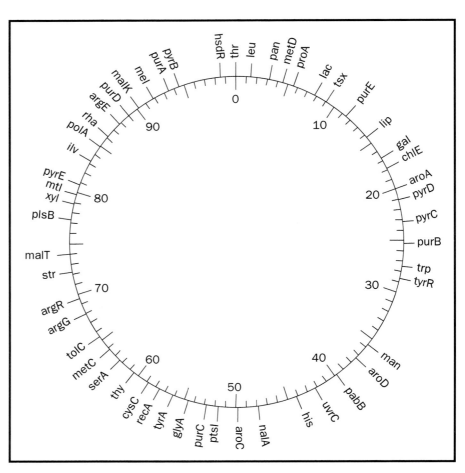

Figure 1. The circular genetic map of the *Escherichia coli* chromosome. The map is divided into 100 map units, and representative genes are shown. The lac operon is at approximately 8 minutes. The origin of replication (not shown) is at 83.5 minutes. Based on Ingraham, 2000.

such as *Vibrio cholerae* (the agent that causes the disease cholera) and *Deinococcus radiodurans*, that have two different chromosomes.

It is also quite common for bacterial species to possess extrachromosomal genetic elements called plasmids. These are small, circular DNA molecules which, when present, vary in number from one to about thirty identical copies per cell. Plasmids include the fertility factor (F$^+$ plasmid), described below, as well as plasmids that carry drug-resistance genes. Indeed, these drug-resistance plasmids may be passed from species to species and are a major problem in the spread of antibiotic resistance. Whereas most bacteria that contain plasmids have just a single kind of plasmid, some bacterial species simultaneously possess a number of different plasmids, each of which, in turn, is present in varying numbers within the bacterial cell.

The bacterial chromosome is condensed into chromosomal domains. The bacterial chromosome must be tightly packed to fit into the small volume of the bacterial cell. Figure 3 shows the relative sizes of the unfolded chromosome and the *E. coli* cell. During the 1980s, techniques were developed to isolate intact bacterial nucleoids by gentle lysis, under conditions that prevented the DNA of the chromosome from uncoiling. These isolated

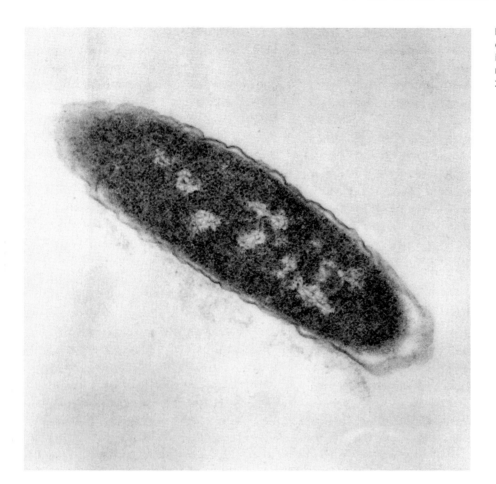

Figure 2. A transmission electron microscope image of *Escherichia coli* magnified approximately 20,000 times.

nucleoids were highly condensed into a very compact structure, as shown in Figure 4.

Compacting the DNA involves supercoiling, or further twisting the twisted chromosome. The chromosome's fifty or so DNA domains are held together by a scaffold of RNA and protein, and the entire nucleoid is attached to the cell membrane. This membrane attachment aids in the segregation of the chromosomes after they replicate in preparation for cell division. Bacteria lack the histone proteins that are found bound to the DNA and that form the nucleosomes of eukaryotic chromosomes. However, it is believed that polyamines (organic molecules with multiple $-NH_2$ amine groups) such as spermidine, as well as some basic proteins, aid in compacting the bacterial chromosome. These basic proteins have a net positive charge that bind them to the negative charge of the phosphates in the DNA backbone.

Replication of the circular chromosome begins at a single point, called OriC, and proceeds in both directions around the circle, until the two replication forks meet up. The result is two identical loops. Replication takes approximately forty minutes.

The *E. coli* genetic chromosome. The field of bacterial genetics began in 1946, even before the structure of DNA was determined, with the discovery by the geneticists Joshua Lederberg and Edward Tatum at the University of Wisconsin of sex in bacteria, in the form of conjugational genetic exchange between *E. coli* bacteria. In the conjugation process, a fertility

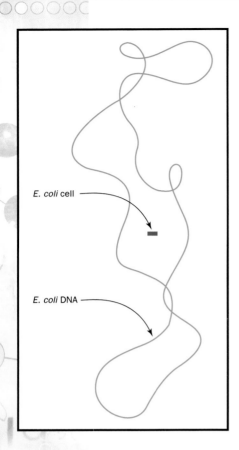

E. coli cell

E. coli DNA

Figure 3. Relative size of an *E. coli* cell and its chromosome.

pathogens disease-causing organisms

factor (F plasmid) recombines with (splices itself into) the *E. coli* chromosome at a specific site. It then acts as a "molecular motor" to drive the transfer of the entire *E. coli* chromosome to a recipient (F⁻) cell. The transferred molecule can then recombine with the host chromosome, increasing the genetic diversity of the host. Transferring the entire chromosome takes approximately one hundred minutes, and thus the genetic map is divided into one hundred minutes (which were later defined as one hundred map units). As more and more genetic markers were found and mapped, it became apparent that the genetic chromosome map formed a circle, as shown in Figure 1.

The DNA sequence of the *E. coli* chromosome. *E. coli* was chosen as one of the genetic model organisms whose chromosome was to be sequenced as part of the Human Genome Project. Although it was not the first bacterial species to be completely sequenced, it was one of the most important ones. In 1997, Fredrick Blattner of the University of Wisconsin and colleagues published the sequence of 4,639,221 base pairs of the K-12 laboratory strain. *E. coli* is estimated to have 4,279 genes.

Many sets of genes on the *E. coli* chromosome are organized into operons. An operon is a set of functionally related genes that are controlled by a single promoter and that are all transcribed at the same time.

Comparative bacterial genomes. As of June 2002, the genomes of sixty-five different bacterial species had been completely sequenced. Several of these are listed in Table 1, along with the genomes' size and number of genes. Many of the species sequenced are human **pathogens**. Having the DNA sequence will prove useful in designing drugs and antibiotics to combat infections and bacterial toxins. DNA sequences may be found on the Internet, at the Genome Web site of the National Center for Biotechnology Information.

CHROMOSOME SIZE AND NUMBER OF GENES FOR SEVERAL BACTERIAL SPECIES SEQUENCED 1995–2000

Bacterial species	Chromosome size (base pairs)	Number of genes	Year sequence completed
Haemophilus influenzae	1,830,138	1714	1995
Mycoplasma genitalium	580,074	480	1995
Synechocystis sp.	3,573,470	3167	1996
Mycoplasma pneumoniae	816,394	1054	1996
Helicobacter pylori	1,667,867	1576	1997
Escherichia coli	4,639,211	4279	1997
Bacillus subtilis	4,214,814	4112	1997
Borrelia burgdorferi	910,724	851	1997
Aquifex aeolicus	1,551,335	1529	1998
Mycobacterium tuberculosis	4,411,529	3927	1998
Treponema pallidum	1,138,011	1036	1998
Rickettsia prowazekii	1,111,523	835	1998
Chlamydia trachomatis	1,042,519	895	1998
Chlamydiophila pneumonia	1,230,230	1054	1999
Thermotoga maritima	1,860,725	1858	1999
Campylobacter jejuni	1,641,481	1654	2000
Neisseria meningitidis	2,272,351	2079	2000
Buchnera sp.	640,681	564	2000
Bacillus halodurans	4,202,353	4066	2000

Table 1.

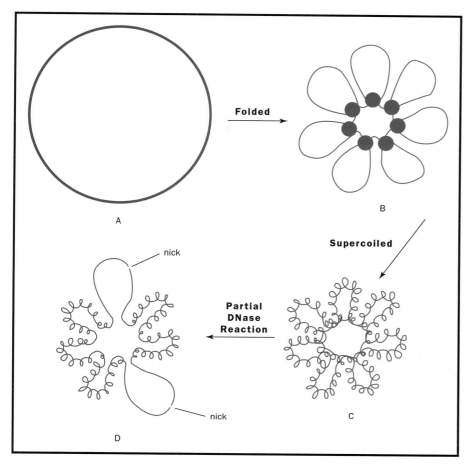

Figure 4. Model of the condensed bacterial chromosome. The circular DNA molecule (a) is folded into several domains (b). Seven domains are shown, but in the cell there are approximately fifty. These domains are held together by RNA and proteins. The DNA is further condensed by supercoiling (c). If the DNA backbone is nicked by a nuclease, the DNA domain opens (d). Adapted from Pettijohn, David E., 1996.

Minimal-gene-set concept. One of the interesting features of studying bacterial chromosomes has been the concept of the minimal number of genes a cellular life form would need to survive. (This excludes viruses and viroids, which need living cells of a host in which to carry out their life cycle.) We know from the sequence of the *Mycoplasma genitalium* chromosome, the smallest genome sequenced so far, that the upper limit of the minimal gene set is 480, as shown in Table 1. After the sequence of the *Haemophilus influenzae* chromosome was completed, a comparison of the genes that were identical (or highly conserved) in the two species led to an estimate of 256 as the minimal gene set. The National Institutes of Health scientist Eugene Koonin, with the availability of many more sequenced species, has also estimated a minimal size of about 250 genes. It may be possible in the future for scientists to construct a minimal life-form by removing nonessential genes from an organism such as *M. genitalium*. SEE ALSO ANTIBIOTIC RESISTANCE; ARCHAEA; CHROMOSOME, EUKARYOTIC; CONJUGATION; *ESCHERICHIA COLI* (*E. COLI* BACTERIUM); EUBACTERIA; HUMAN GENOME PROJECT; OPERON; REPLICATION.

Ralph R. Meyer

Bibliography

Berlyn, Mary K. B., K. Brooks Low, and Kenneth E. Rudd. "Linkage Map of *Escherichia coli* K-12." In *Escherichia coli and Salmonella: Cellular and Molecular Biology*, 2nd ed., Frederick C. Neidhardt, et al., eds. Washington, DC: ASM Press, 1996.

Ingraham, John L., and Catherine A. Ingraham. *Introduction to Microbiology*, 2nd ed. New York: Brooks/Cole, 2000.

Koonin, Eugene V. "How Many Genes Can Make a Cell: The Minimal-Gene-Set Concept." *Annual Review of Genomics and Human Genetics* 1 (2000): 99–116.

Pettijohn, David E. "The Nucleoid." In Escherichia coli *and Salmonella: Cellular and Molecular Biology*, 2nd ed., Frederick C. Neidhardt et. al. eds. Washington, DC: ASM Press, 1996.

Internet Resource

Entrez-Genome. National Center for Biotechnology Information. <http://www.ncbi.nlm.nih.gov/Entrez/Genome/org.html>.

Chromosomes, Artificial

Artificial chromosomes are laboratory constructs that contain DNA sequences and that perform the critical functions of natural chromosomes. They are used to introduce and control new DNA in a cell, to study how chromosomes function, and to map genes in genomes.

Natural Chromosome Function

DNA, which constitutes the genome of a cell, is always packaged with a variety of proteins, and together these make up the chromosomes. A chromosome serves to compact the DNA and protect it from the damage, while at the same time allowing the genes it contains to be available for transcription into RNA. In addition to these functions, extra ones are necessary when the cell divides. Prior to cell division the DNA must be copied and these copies separated (segregated) and delivered to different parts of the cell, ensuring that each of the new cells receives only a single copy.

To ensure correct segregation, chromosomes have to have distinct components that are composed of specific DNA sequences and associated proteins. Bacterial chromosomes, (**plasmids**) which are circular, have a single site at which DNA replication originates, and attachment to the cell membrane results in segregation. Artificial bacterial chromosomes (BACs) mimic this using appropriate origin sequences.

In organisms with multiple linear chromosomes (eukaryotic organisms) the process is more complicated. The ends of the chromosomes must be protected from degradation and from the mechanisms that the cell uses to protect itself against broken DNA. **Telomeres**, which provide these functions, are arrays of short, repeated sequences with complexes of specific proteins attached. To ensure segregation complexes of other proteins, DNA sequences known as kinetochores form at sites known as **centromeres**. These contain molecular motors, systems to monitor correct segregation, and sites for attachment of **microtubules**. Chromosomes will contain one or more origins of replication.

Yeast Artificial Chromosomes

Artificial chromosomes for use in yeast and mammalian cells aim to replicate these components on a single DNA molecule. In bakers yeast (*Saccharomyces*

plasmids small rings of DNA found in many bacteria

telomeres chromosome tips

centromeres regions of the chromosome linking chromatids

microtubules protein strands within the cell, parts of the cytoskeleton

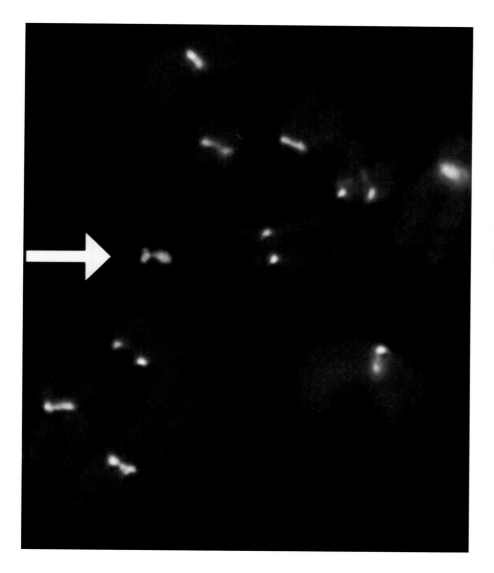

Fluorescence *in situ* hybridization (FISH) visualization of one of the artificial human chromosomes successfully created by investigators at the Case Western Reserve University School of Medicine and Athersys, Inc. The arrow highlights the synthetic microchromosome, a miniature version of the surrounding native human chromosomes. The chromosome material appears dark blue, and the centromeres white.

cerevisiae), telomeres, centromeres, and origins of replication have all been defined using genetics and have been cloned. When assembled they can be grown as a small chromosome in bacteria and form a **vector** capable of incorporating up to a million bases of other DNA as a chromosome in yeast (a YAC).

vector carrier

This technology has been used to investigate the properties of yeast chromosomes but has been most extensively used in the early phases of genome mapping projects. By cloning complete representations of the human genome into large YACs, the order of these YACs could be deduced by a number of methods and overlapping ones assembled conceptually into a representation of regions of the human genome. These are useful for finding genes from information about the inheritance of genetic diseases. They have also been useful for testing the function of genes in mice. Because of the size of YACs, they frequently contain all of the DNA needed to control the expression of genes with the correct developmental and tissue specificity. When injected into developing mouse eggs, they can fully correct mutations. More recently YACs have been largely supplanted by BACs, because the latter are easier to manipulate and prepare in the laboratory.

Mammalian Artificial Chromosomes

Mammalian artificial chromosomes (MACs) are conceptually similar to YACs, but instead of yeast sequences they contain mammalian or human ones. In this case the telomeric sequences are multimers (multiple copies) of the sequence TTAGGG, and the commonly used centromeric sequence is composed of another repeated DNA sequence found at the natural centromeres of human chromosomes and called alphoid DNA.

Because the alphoid DNA is needed in units of many kilobases, these MAC DNAs are grown as YACs or, more recently, as BACs. When added to suitable cell lines, these MAC DNAs form chromosomes that mimic those in the cell, with accurate segregation and the normal complement of proteins at telomeres and centromeres. Their primary use is not in genome mapping but as vectors for delivery of large fragments of DNA to mammalian cells and to whole animals for expression of large genes or sets of genes. They are still in development, and although gene expression has been demonstrated they have not been used in a practical application. SEE ALSO CHROMOSOME, EUKARYOTIC; HUMAN GENOME PROJECT; MAPPING; TELOMERE.

Howard Cooke

Bibliography

Grimes, B., and H. Cooke. "Engineering Mammalian Chromosomes." *Human Molecular Genetics* 7, no. 10 (1998): 1635–1640.

Willard, H. F. "Genomics and Gene Therapy: Artificial Chromosomes Coming to Life." *Science* 290 (2000): 1308–1309.

Classical Hybrid Genetics

Common garden peas (*Pisum sativum*) are wonderful when eaten raw. Gregor Mendel (1822–1884) no doubt ate his share. Today he is recognized for using peas to establish the science of genetics.

Mendel investigated hereditary patterns of hybrids. Hybrids are offspring from two organisms that are of different breeds, varieties, or species. Hybrids create new **cultivars**, from new apple varieties to tangelos to hybrid corn. Some mammals produce hybrids; a mule is the progeny of a horse and a donkey.

Mendel was interested in new flower varieties and absorbed by what hybrids reveal about inheritance. Nineteenth-century scientists wanted to know how organisms created a vast diversity of forms while faithfully maintaining distinct sets of characteristics. Constancy was a known quality of life. Then, as now, people had no trouble recognizing the difference between a housefly and a bee. But what prevented people from suddenly sprouting flowers or losing their human features? Genes, DNA, meiosis, and chromosomes were all unknown. Hybrid genetics was a means to discover answers to fundamental biological questions.

Mendel's aim was to discover the mathematical rules behind the reappearing patterns he saw in hybrids. After testing many varieties of peas, he decided to study seven specific traits: shape of the ripe seeds, seed color,

cultivars plant varieties resulting from selective breeding seed coat color, form of the ripe pods, color of the unripe pods, position of the flowers, and length of the stems.

Initially, Mendel determined the inheritance patterns for one trait at a time, ignoring other traits. He crossed two varieties that differed sharply in a specific trait. For instance, for the trait of seed shape, he used one variety whose seeds were wrinkled and another whose seeds were smooth. The seed-shape trait was represented by two specific inherited features, smooth and wrinkled, that were **alleles**.

Mendel also crossed plant varieties , each with specific combinations of alleles, in order to follow the fates of two or even three traits at the same time. For instance, he crossed a <u>tall</u> plant with <u>wrinkled,</u> <u>green</u> seeds and a <u>short</u> plant with <u>round,</u> <u>yellow</u> seeds. (The underlined words represent alleles for three traits: height, seed shape, and seed color.) To cross two varieties, Mendel first had to make sure the flowers did not pollinate themselves. To do that, he cut off their anthers, their pollen-bearing parts. Then he used the anthers from one variety to pollinate another.

Mendel was the first person to follow specific alleles (which he called "factors") as they were passed from parental varieties through several generations of offspring. He selected plant varieties that were true-breeding: Each generation of plants looked like the antecedent generations, with regard to the studied traits. Mendel used artificial pollination (described above) to perform an initial outcross, a mating between individuals differing in their alleles for at least one trait. The outcross created hybrids, or, more precisely, **heterozygotes** for the chosen traits. Next, Mendel let the heterozygotes self-pollinate, or intercross (i.e., the mating of inbred animals or self-pollinating plants), that are heterozygous for one or more traits. He then let two subsequent generations of **progeny** self-pollinate.

Mendel discovered that hybrids (heterozygotes) resembled only one parental variety, despite clearly identified input from both parents. Regarding the stem-length trait for example, one parental variety had tall stems, whereas the other variety had very short (dwarf) stems, yet all their hybrid offspring had only tall stems. Mendel named the displayed feature "dominant," because the hybrid's appearance of a tall stem (**phenotype**) reflected input from only one parent. He identified the dominant factor or allele for each of the seven traits (seed shape, seed color, length of stem, etc.). The other hidden feature (named "recessive" by Mendel) resurfaced in the next generation, when both hybrid (heterozygous) parents donated their hidden (recessive) allele to a descendent. For example, dwarf stems appeared again in progeny, instead of the conspicuous tall stems seen in the heterozygous parents. A descendent with dwarf stems had a "homozygous recessive" **genotype** for this trait. An individual's genotype described the two alleles received for a trait, whether hidden or displayed. If a plant had a dominant phenotype for one or several traits, but the genotype was unknown, a method called a test cross was used to reveal the presence of a recessive allele for each of the traits in question. The dominant phenotype to be tested might carry one dominant and one recessive allele, or two dominant alleles (see Mendel's laws, below), depending on what its parents had donated. The dominant phenotype was crossed with a plant known to be homozygous recessive for a particular trait. If half the offspring showed the dominant phenotype and the other half the recessive

alleles particular forms of genes

heterozygotes individuals whose genetic information contains two different forms (alleles) of a particular gene

progeny offspring

phenotype observable characteristics of an organism

genotype set of genes present

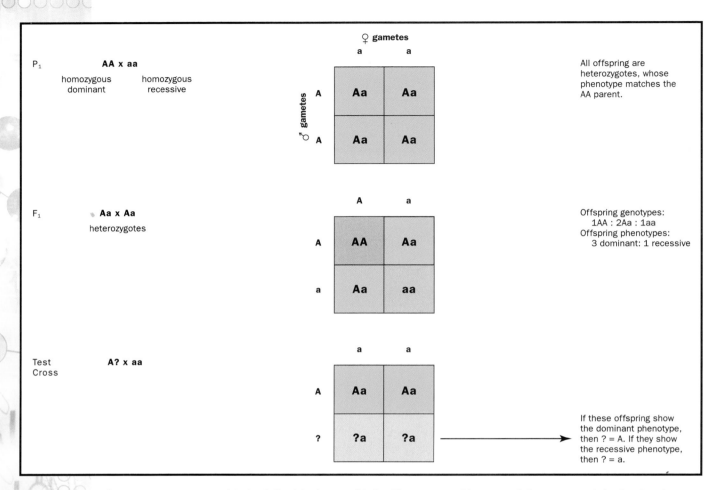

Punnett squares are used to track the inheritance of traits. The upper- and lowercase letters represent dominant and recessive alleles, respectively. P₁ is the parental generation, and F₁ is the first filial, or offspring, generation. A test cross is performed to determine the genotype of an organism showing the dominant trait, but whose genotype is unknown.

form, then the parent with the dominant phenotype had to carry one recessive and one dominant allele for that trait. In contrast, if all its offspring showed only the dominant form of the trait, the plant with the dominant phenotype must have contained only dominant alleles.

By categorizing and counting offspring of several generations of plants, Mendel discovered two laws of inheritance, described as segregation and independent assortment. Segregation meant that a gamete or reproductive cell received only one allele out of a choice of two alleles carried for each trait (gene) within a parent cell. The gamete had an equal chance of receiving either allele. Independent assortment indicated that traits entered into a gamete independently of each other. This is only true for gene or trait collections located on long strings of hereditary material (chromosomes) that are now termed "nonhomologous" (not alike). Nonhomologous chromosomes carry unique collections of specific genes or traits, whereas two homologous chromosomes carry the same collection of genes, but may carry two distinct alleles for a specific trait or gene.

Armed with his laws, Mendel was able to predict the frequency at which various alleles for several traits would co-appear in descendents of hybrids of the common garden pea. Mendel's principles were not appreciated for about thirty years, until Walter Sutton (1877–1916), an American physician and geneticist, described chromosome movements during meiosis and identified chromosomes as the carriers of genes and heredity.

Back crosses, which are based on Mendel's principles, are used to develop commercially useful plant or animal varieties. In a back cross, a heterozygous plant and its offspring are crossed repeatedly to one of their parents to develop a line of plants that mostly resemble one parent but that have an allele of interest from the other. For instance, a rare flower color is integrated into a line of nematode-resistant stock of roses, joining beauty and disease-resistance. SEE ALSO GENETICS; INHERITANCE PATTERNS; MENDEL, GREGOR; MENDELIAN GENETICS; PROBABILITY.

Susanne D. Dyby

Bibliography

Sutton, Walter S. "The Chromosomes in Heredity." *Biological Bulletin* 4 (1903): 231–251.

———. "On the Morphology of the Chromosome Group in Brachystola Magna." *Biological Bulletin* 4 (1902): 24–39.

Internet Resource

Mendel, Gregor. "Experiments in Plant Hybridization." (1866). C. T. Druery and William Bateson, trans. MendelWeb. <http://www.netspace.org/MendelWeb>.

Clinical Geneticist

Medical genetics is the application of genetics to the study of human health and diseases. As a profession, medical genetics is usually a mixture of both clinical services and research. Worldwide, services can include diagnosis, counseling, and management of birth defects and genetic disorders. How medical genetics is actually practiced depends on several factors, including the expertise and training of the professionals involved, the expertise available within any given medical facility, and the structure of the practice of medicine within a given society.

In the United States, the practice of medical genetics includes two different career tracks, both requiring certification by the American Board of Medical Genetics (ABMG): the medical geneticist and the clinical geneticist. A medical geneticist holds a Ph.D. and is certified in medical genetics. Typically a medical geneticist is a highly trained research laboratory professional who can additionally take on the role of consultant to physicians. A clinical geneticist is a physician (either a doctor of osteopathy or a medical doctor) involved in all parts of clinical practice related to genetic disorders. Working closely with patients, clinical geneticists identify, diagnose, determine the prognosis of, develop predictive tests for, treat, and manage genetic diseases. They can also be active in conducting research on genetic disorders and studying theoretical genetics, and they usually help to administer and set policies for the clinical genetics profession and for medical centers in general.

Clinical geneticists also can be involved in the bioethical debates and policy-making issues concerning how genetic information is gathered, who has access to it, and how that access should be regulated. This role is becoming increasingly important as society struggles to deal with the tremendous explosion of genetic information arising as a consequence of the Human

Genome Project. Administrative roles for clinical geneticists can include formulating plans and procedures for clinical genetic services, scheduling the use of medical genetics facilities, and teaching interns and residents the methods and procedures involved in the diagnosis and management of genetic disorders.

Very few clinical geneticists develop a private practice. Instead, they typically work in a team environment within regional medical centers alongside scientists, medical geneticists, genetic counselors, and other academics. Most hospitals specializing in pediatric care will also have clinical geneticists on their staff. Some large national clinics have entire departments devoted to the practice of medical and clinical genetics.

Physicians are attracted to the practice of clinical genetics for a variety of reasons. Many enjoy understanding the evolving human gene map, the rapid technological advances in the field, and the opportunity to perform laboratory research as well as practice medicine. They enjoy the challenge of applying the advances in the molecular basis of disease to the care of patients. As a group, clinical geneticists derive satisfaction by remaining close to the "cutting edge" of new discoveries in genetic diseases, which challenges them to remain current while constantly using their knowledge and skills to provide innovative and effective medical services.

Many are attracted to the profession because it allows them to develop long-term relationships with patients and their families. Others find that the narrow focus of clinical genetics is more to their liking than the broader disciplines of internal medicine or pediatrics. Within clinical genetics, physicians can develop their own disease specialty if they choose, which for some provides a more rewarding work environment, giving them the opportunity to make an impact on both research and the lives of patients whose diseases may be rare and often poorly understood by other medical practitioners. Clinical geneticists enjoy complex problem-solving, taking care of people, and paying attention to details that others may miss. They are good listeners.

Students interested in clinical genetics as a profession should become familiar with mathematics, chemistry, biology, and some physics, while still in high school. Courses aimed at developing communication and writing skills are also valuable for students preparing for a career in clinical genetics. Because the practice of clinical genetics requires a medical degree, students must first receive a bachelor's degree, enrolling in courses that meet medical school admission requirements.

After obtaining a medical degree, clinical geneticists typically complete three to five years of residency in medical disciplines approved by the Accreditation Council for Graduate Medical Education (ACGME), followed by a two- to three-year fellowship that is approved by the ABMG, in clinical genetics itself. Certification can be in clinical genetics or in more focused subspecialties, including clinical cytogenetics, clinical biochemical genetics, clinical molecular genetics, and molecular genetic pathology. Certification requires the successful passage of a national examination that is given at regularly scheduled intervals. To maintain certification, clinical geneticists must fulfill continuing education requirements throughout the duration of their career.

Both the ACGME and ABMG maintain a list of approved programs that lead to certification as a clinical geneticist. Approved residency programs include clinical and academic components. Approved programs expose the resident to a patient population large enough to develop an understanding of the wide variety of medical genetic problems. They enable direct involvement in genetic research laboratories, where students learn to critically interpret laboratory data. They include graduate-level course work in basic, human, and medical genetics, as well as clinical teaching conferences, and they foster the development of the communication skills necessary to interact and sustain a long-term therapeutic relationship with patients and their families.

One need not necessarily decide upon the profession of clinical genetics prior to entering medical school or even upon receiving a medical degree or completing a full residency. Traditionally, the ABMG has accepted physicians into clinical genetics programs who come from approved residency programs in pediatrics, obstetrics-gynecology, and internal medicine. This is because the profession of clinical genetics originated from the treatment of inherited diseases that are initially observed in newborn infants and children. It is still within these disciplines that many physicians develop a secondary interest in clinical genetics, during residency or even later in their medical career. However, the field of medical genetics is rapidly changing, due to the recognition that many genetic disorders result in symptoms that are delayed until adulthood. As a consequence, clinical geneticists are now beginning to provide their services to adults as well.

The Human Genome Project has also brought the realization that clinical genetics involves more than single-gene and single-chromosome conditions. Genetic medicine is becoming applicable to many kinds of complex diseases and disorders such as cancer, heart disease, and asthma, to name a few well-known conditions.

The 1,100 certified clinical geneticists registered in the United States in the year 2000 were too few to keep up with an ever-increasing demand for their services. Students interested in pursuing clinical genetic careers can expect that the number of subspecialties will continue to grow, that both the ACGME and ABMG will continue to strive to provide new certification programs and career tracks, and that the concept of clinical genetics may become an integral component of "well" medical health care. SEE ALSO DISEASE, GENETICS OF; GENETIC COUNSELOR; GENETIC TESTING; GENETIC TESTING: ETHICAL ISSUES; PRENATAL DIAGNOSIS.

Diane C. Rein

Bibliography

Internet Resources

Accreditation Council for Graduate Medical Education. <http://www.acgme.org>.

American Board of Medical Genetics. <http://www.abmg.org>.

Careers in Human Genetics. University of Kansas Medical Center's Genetics Education Center. <http://www.kumc.edu/gec/prof/career.html>.

Guide to North American Graduate and Postgraduate Training Programs in Human Genetics. American Board of Medical Genetics. <http://www.abmg.org/genetics/ashg/tpguide/intro.htm>.

National Coalition for Health Professional Education in Genetics. <http://www
.nchpeg.org>.

Webliography for Clinical Geneticists. The Federation of American Societies for Experimental Biology. <http://www.faseb.org/genetics/webliog.htm>.

Cloning Genes

Gene cloning, or molecular cloning, has several different meanings to a molecular biologist. A clone is an exact copy, or replica, of something. In the
literal sense, cloning a gene means to make many exact copies of a segment
of a DNA molecule that encodes a gene. This is in marked contrast to
cloning an entire organism—regenerating a genetically identical copy of the
organism—which is technically much more difficult (with animals) and can
involve ethical ramifications not associated with gene cloning. Molecular
biologists exploit the replicative ability of cultured cells to clone genes.

Purposes of Gene Cloning

To study genes in the laboratory, it is necessary to have many copies on
hand to use as samples for different experiments. Such experiments include
Southern or Northern blots, in which genes labeled with radioactive or fluorescent chemicals are used as probes for detecting specific genes that may
be present in complex mixtures of DNA.

Cloned genes also make it easier to study the proteins they encode.
Because the genetic code of bacteria is identical to that of **eukaryotes**, a
cloned animal or plant gene that has been introduced into a bacterium can
often direct the bacterium to produce its protein product, which can then
be purified and used for biochemical experimentation. Cloned genes can
also be used for DNA sequencing, which is the determination of the precise order of all the base pairs in the gene. All of these applications require
many copies of the DNA molecule that is being studied.

Gene cloning also enables scientists to manipulate and study genes in
isolation from the organism they came from. This allows researchers to conduct many experiments that would be impossible without cloned genes. For
research on humans, this is clearly a major advantage, as direct experimentation on humans has many technical, financial, and ethical limitations.

Cloning Techniques

Cloning genes is now a technically straightforward process. Usually, cloning
uses recombinant DNA techniques, which were developed in the early
1970s by Paul Berg, of Stanford University, and, independently, by Stanley Cohen and Herbert Boyer, of Stanford and the University of California. These researchers devised methods for excising genes from DNA at
precise positions, using **restriction enzymes** and then using the enzyme
known as DNA ligase to splice the resulting gene-containing fragment into
a plasmid **vector**.

Plasmids are small, circular DNA molecules that occur naturally in many
species of bacteria. The plasmids naturally replicate and are passed on to
future generations of bacterial cells. To replicate, all plasmids must contain
a sequence, called an origin of replication, which directs the bacterial DNA

eukaryotes organisms
with cells possessing a
nucleus

restriction enzymes
enzymes that cut DNA
at a particular sequence

vector carrier

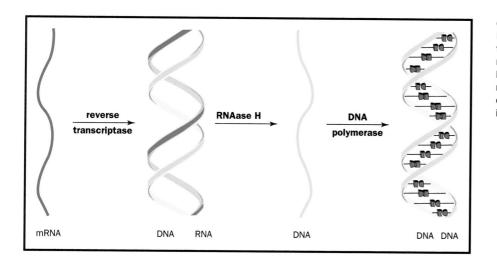

reverse transcriptase → RNAase H → DNA polymerase

mRNA   DNA   RNA   DNA   DNA   DNA

cDNA is double-stranded DNA that is synthesized from single-stranded messenger RNA. cDNA has all the coding and regulatory regions of the original gene, but no introns.

polymerase to replicate the DNA molecule. In addition, recombinant plasmids contain one or more selectable markers. A selectable marker is a gene that confers on the bacterium harboring the plasmid the ability to survive under conditions in which bacteria lacking the plasmid would otherwise die. Usually, such genes encode enzymes that enable the bacterium to live and grow despite the presence of an antibiotic drug.

The recombinant plasmid is then introduced into a host cell, such as an *Escherichia coli* bacterium, by a process called transformation, and the cell is allowed to multiply and form a large population of cells. Each of these cells harbors many identical copies of the recombinant plasmid. The cells are then cultured in growth media containing the antibiotic to which the plasmid confers resistance. This ensures that only cells containing the recombinant plasmid will survive and replicate. A researcher then harvests the cells and can extract and purify many copies of the plasmid.

Another method to produce many copies of a DNA molecule, which is even simpler than traditional recombinant cloning methods, is the polymerase chain reaction (PCR). PCR amplifies the DNA in a reaction tube without the need for a plasmid to be grown in bacteria.

Importance for Medicine and Industry

The ability to clone a gene is not only valuable for conducting biological research. Many important pharmaceutical drugs and industrial enzymes are produced from cloned genes. For example, insulin, clotting factors, human growth hormone, cytokines (cell growth stimulants), and several anticancer drugs in use are produced from cloned genes.

Before the advent of gene cloning, these proteins had to be purified from their natural tissue sources, a difficult, expensive, and inefficient process. Using recombinant methods, biomedical companies can prepare these important proteins more easily and inexpensively than they previously could. In addition, in many cases the product that is produced is more effective and more highly purified. For example, before the hormone insulin, which many diabetes patients must inject, became available as a recombinant human protein, it was purified from pig and cow pancreases. However, pig and cow insulin has a slightly different amino acid sequence than the

polymerase enzyme complex that synthesizes DNA from individual nucleotides

Escherichia coli common bacterium of the human gut, used in research as a model organism

153

human hormone. This sometimes led to immune reactions in patients. The recombinant human version of the hormone is identical to the natural human version, so it causes no immune reaction.

Gene cloning is also used to produce many of the molecular tools used to study genes. Even restriction enzymes, DNA ligase, DNA polymerases, and many of the other enzymes used for recombinant DNA methods are themselves, in most cases, produced from cloned genes, as are enzymes used in many other industrial processes.

Genomic Versus cDNA Clones

A gene can take varying forms, and so can gene clones. The protein-coding regions of most eukaryotic genes are interrupted by noncoding sequences called introns, which are ultimately excluded from the mature messenger RNA (mRNA) after the gene is transcribed. In addition to the protein-coding sequences, all genes contain "upstream" and "downstream" regulatory sequences that control when, in which tissues, and under what circumstances the gene is transcribed. A clone containing the entire region of a gene as it exists on the chromosome, including introns and nontranscribed regulatory sequences, is called a genomic clone because it is derived directly from genomic, or chromosomal, DNA.

It is also possible to clone a gene directly from its messenger RNA transcript, from which all introns have been removed. This type of clone, called a complementary DNA or cDNA clone, includes only the protein-coding sequences and upstream and downstream sequences that do not code for amino acids but that may control how the mRNA transcript gets translated to protein.

To prepare cDNA a researcher starts with mRNA and then makes a complementary single-stranded DNA copy using the enzyme reverse transcriptase. Reverse transcriptase is a DNA polymerase that synthesizes DNA based on an RNA template that is produced by retroviruses. After the mRNA strand is digested away by another enzyme, called RNase H, DNA polymerase can synthesize a second DNA strand by using the newly made first strand cDNA as a template.

Because cDNAs lack introns, the protein-coding region in a cDNA molecule is contained in a single uninterrupted sequence, called an open reading frame, or ORF. This makes cDNA clones extremely useful for predicting the amino acid sequence of the protein that a gene encodes. It also makes it possible to direct protein synthesis from a eukaryotic cDNA clone in a bacterium, which cannot splice introns. With introns still present in a cloned gene, the bacteria will misinterpret the intron sequences as protein-encoding sequences. The resulting incorrect messanger RNA will encode a protein with an incorrect amino acid.

"Gene Cloning" Usually Means "Gene Identification"

When researchers report in a scientific journal that they have "cloned a gene" they are not referring to the rather mundane process of amplifying copies of a DNA molecule. What they are really talking about is the molecular identification of a previously unknown gene, and determination of its precise position on a chromosome. There are many different methods that can be used to identify a gene. Two of the most common approaches are discussed below.

A gene can be defined in several ways. In fact, the concept of the gene is undergoing a re-evaluation as scientists are analyzing the complete genomes of more and more organisms and finding that many sequences encode more than one protein product. Gregor Mendel identified genes—for example, he identified the factor that made peas either yellow or green—long before he or anyone else knew that genes were encoded on segments of the DNA that made up chromosomes. Studying genetics in the fruit fly, *Drosophila melanogaster*, Morgan and Sturtevant demonstrated that genes are entities that reside at measurable locations, or loci, on chromosomes, although they did not yet understand the biochemical nature of genes.

Modern geneticists often use the same methods as Mendel and Morgan to identify genes by physical traits, or phenotypes, that mutations in them can cause in an organism. But today we can go even further. Using a broad range of molecular biology techniques, including gene cloning, researchers can now determine the precise DNA coding sequence that corresponds to a particular **phenotype**. This capability is tremendously powerful, because discovering the gene responsible for a trait can help humankind understand the cellular and biochemical processes underlying the trait. For example, geneticists have learned a great deal about the basis of cancer by identifying genes that, when mutated, contribute to cancer. By studying these genes, researchers now know that many of them control when cells divide (e.g., proto-oncogenes and tumor suppressor genes) or when they die (e.g., the apoptosis genes). Under some circumstances, when such genes are damaged by mutation, cells divide when they shouldn't, or don't die when they should, leading to cancer.

phenotype observable characteristics of an organism

Positional Cloning

Positional cloning starts with the classical methods developed at the turn of the twentieth century by Thomas Hunt Morgan, Alfred Sturtevant, and their colleagues, of genetically mapping a particular phenotype to a region of a chromosome. A detailed discussion of genetic mapping is beyond the scope of this section, but, in general, it is based on conducting genetic crosses between individuals with two different mutant traits and analyzing how often the traits occur together in the progeny of subsequent generations.

Genetic mapping provides a general idea of where a gene is located on a particular chromosome, but it does not identify the precise DNA sequence that encodes the gene. The next step is to locate the gene on what is called the physical map of the chromosome. A physical map is a high-resolution map of all the DNA sequences that make up a chromosome. One type of physical map is a restriction map, which depicts the order of DNA fragments produced when a large DNA molecule is cut with restriction endonucleases (restriction enzymes).

Restriction maps have been made for the complete genomes of several model genetic organisms, such as the fruit fly (*Drosophila melanogaster*), and the roundworm, (*Caenorhabditis elegans*). For these organisms, individual large DNA fragments—on the order of forty to one hundred thousand base pairs from the whole genome—have been cloned in bacterial plasmid vectors to make a "library" of the genome. Each fragment is mapped to a known

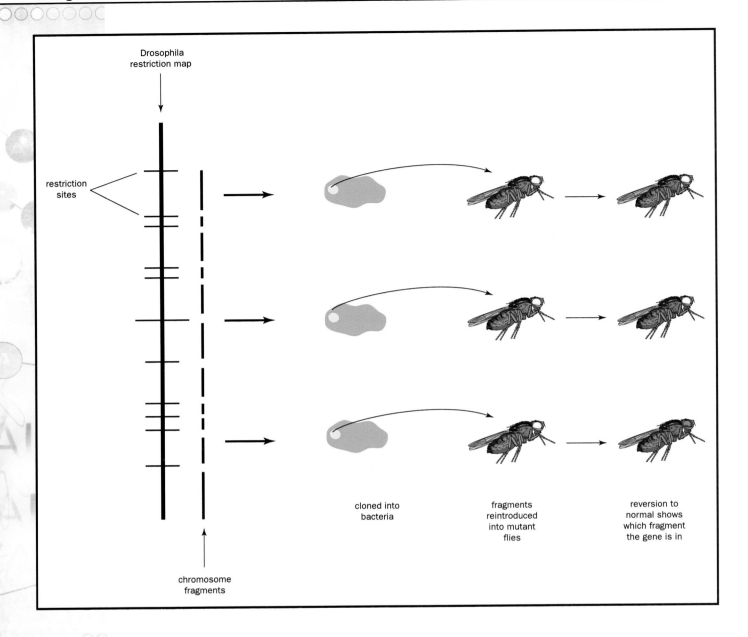

Drosophila
restriction map

restriction
sites

cloned into
bacteria

fragments
reintroduced
into mutant
flies

reversion to
normal shows
which fragment
the gene is in

chromosome
fragments

The location of a gene can be found by cutting the DNA with restriction enzymes, inserting the fragments into bacteria, and then reintroducing them individually into mutant flies.

position, but the identify of the gene or genes it contains is originally unknown. To identify the genes, a cloned fragment is introduced into a mutant fly or roundworm.

To pinpoint the location of a particular gene, a researcher can introduce one or several of the plasmid clones from the physical map that are in the general vicinity of the region on the genetic map where the gene is thought to lie into a mutant that is defective in the gene of interest. If the introduced DNA corrects the mutant's defect, that DNA probably contains a normal copy of the defective gene. But these large clones usually contain several genes. By further "trimming" the DNA into smaller subfragments and testing the ability of each subfragment to rescue mutants, the researcher can eventually home in on the gene. As further confirmation that this gene is the cause of the mutant phenotype, the researcher can isolate the corresponding gene from the mutant and determine its DNA sequence to see if

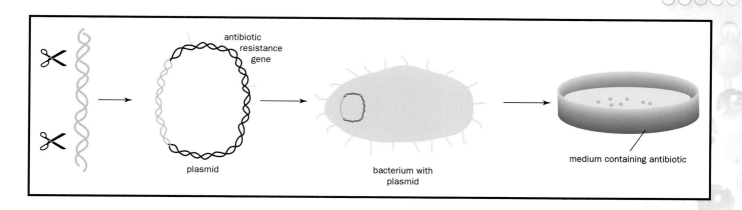

it contains a mutation (a DNA sequence alteration) relative to the normal gene sequence.

Expression Cloning

In some cases, a researcher becomes interested in studying a gene not because mutations in it cause an interesting phenotype but because the protein it encodes has interesting properties. A prominent example is beta-amyloid protein, which accumulates in the brains of Alzheimer's disease patients.

Expression cloning is a method of isolating a gene by looking for the protein it encodes. If the protein of interest is an enzyme, it can be found by testing for its biochemical activity. A very common method for identifying a particular protein is by using antibodies, or immunoglobulins, that bind specifically to that protein. Expression cloning usually uses a cDNA library, in which protein-coding sequences are uninterrupted by introns. Each cDNA is inserted into an "expression vector," which contains all the necessary signals for the DNA to be transcribed into mRNA. The mRNA can then be translated into protein. Thus the host cell harboring the clone will produce the gene's protein product, and the protein can then be detected by biochemical or immunologic methods. Once the cell making the protein is found, the cDNA can be re-isolated and the gene sequenced by standard means.

Gene cloning techniques continue to advance rapidly, aided by the Human Genome Project and bioinformatics. It is likely that positional cloning will take on a secondary role, and that bioinformatics and proteomics methods will begin to contribute more, as more progress in these fields is made. SEE ALSO Bioinformatics; Blotting; Chromosomes, Artificial; Cloning Organisms; Cloning: Ethical Issues; DNA Libraries; Gene; Gene Discovery; Human Genome Project; Linkage and Recombination; Marker Systems; Morgan, Thomas Hunt; Plasmid; Polymerase Chain Reaction; Recombinant DNA; Restriction Enzymes; Reverse Transcriptase; RNA Processing; Sequencing DNA; Transformation.

Paul J. Muhlrad

Restrictive enzymes (scissors) cut a gene out of its normal chromosomal position. Other enzymes insert it into a plasmid, which is then introduced into a bacterium. Only those bacteria that took up the plasmid survive on the growth medium. These bacteria can then be grown in bulk to produce many gene copies.

Bibliography

Alberts, Bruce, et al. *Molecular Biology of the Cell*, 4th ed. New York: Garland Science, 2002.

Lodish, Harvey, et al. *Molecular Cell Biology*, 4th ed. New York: W. H. Freeman and Company, 2000.

Micklos, David A., and Greg A. Freyer. *DNA Science: A First Course in Recombinant DNA Technology.* Cold Spring Harbor, NY: Cold Spring Harbor Laboratory Press, 1990.

Watson, James D., et al. *Recombinant DNA*, 2nd ed. New York: Scientific American Books, 1992.

Cloning: Ethical Issues

Cloning is the creation of an individual that is a genetic replica of another individual. The process transfers a nucleus from a somatic nonreproductive cell into an "enucleated" fertilized egg, one that has had its own nucleus destroyed or removed. The genes in the transferred nucleus then direct the development of a complete organism from the altered fertilized egg. Two individuals who are clones have identical genes in their cell nuclei, but differ in characteristics that are acquired in other ways.

Cloning in Context

Cloning is a natural phenomenon in species as diverse as armadillos, poplar trees, aphids, and bacteria. Identical twins are clones. Biologists have been cloning some organisms, such as carrots, for decades. Attempts to clone animals have been far less successful. They began long before the February 1997 announcement of the birth of Dolly, a sheep cloned from a mammary gland cell nucleus of a six-year-old sheep.

Oxford University developmental biologist John Gurdon cloned frogs in the 1960s, but in a limited way. He showed that a nucleus from a tadpole's intestinal lining cell could be transferred to an enucleated fertilized egg and support development to adulthood, and that a nucleus from an adult cell could support development as far as the tadpole stage. However, he was unable to coax a nucleus from an adult amphibian's cell to support development all the way to adulthood. In the 1980s several companies tried to commercialize cloning of livestock from nuclei taken from embryos or fetuses. The efforts failed because the cloned animals were nearly always very unhealthy newborns and did not survive for long. Currently, livestock cloning is limited to research, although some companies offer tissue preservation services in anticipation of future advances in commercial livestock cloning. There is no reason to believe that human clones would fare any better in terms of health or survivability than most cloned animals do.

The Cloning Ban

Ethical concerns about whether an action is "right" or "wrong" are often clouded by subjectivity, emotion, and perspective. Cloning members of an endangered species, for example, is generally regarded as a positive application of the technology, whereas attempting to clone an extinct woolly mammoth from preserved tissue elicits more negative responses, including that this interferes with nature. A project at Texas A&M University, funded by a dog lover wishing to clone a beloved deceased pet, announced the first successful cloning of a domestic animal, a cat, in February 2002. Cloning pets when strays crowd shelters might be seen as unethical. A different set of ethical issues emerges when considering the cloning of humans, which a few scientists and physicians have proposed doing outside of the United States.

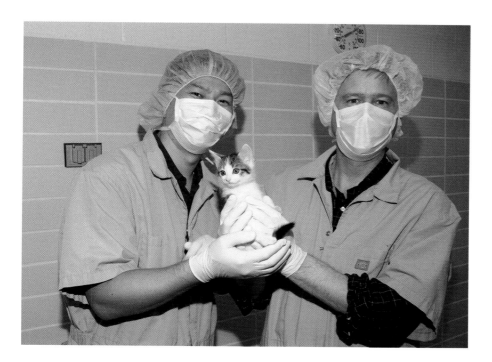

The first cloned cat, named "cc," (short for "copycat") is proudly displayed by doctors Mark Westhusin (right) and Tae Young (left) of the College of Veterinary Medicine at Texas A&M University on Febuary 8, 2002.

Bioethics is concerned with the rights of individuals, such as the right to privacy and the right to make informed medical decisions. It is difficult to see how these issues would apply to cloning, unless someone was forced or paid to provide material for the procedure, or if an individual was cloned and not informed of his or her origin. Ethical objections to cloning seem to focus more on the fact that this is not a normal way to have a baby. Accordingly, the U.S. House of Representatives voted overwhelmingly on July 31, 2001 to pass legislation that would outlaw human cloning for any reason. However, the broadness of this action may impede other types of medical research, thus introducing a different bioethical dilemma.

The legislation seeks to ban all human cloning, both "reproductive cloning" that would be used to create a baby, and "therapeutic cloning." In therapeutic cloning, a nucleus from a somatic cell is transferred to an enucleated donor egg, and an embryo is allowed to develop for a few days. Then, cells from a part of the embryo called the inner cell mass are used to establish cultures of embryonic stem cells that are genetically identical to the individual who donated the somatic cell nucleus.

If this person has a spinal cord injury or a neurodegenerative disease, the embryonic stem cells might specialize into needed neural tissue. To treat muscular dystrophy, the cells might be coaxed to differentiate into muscle-cell precursors. Such tailored embryonic stem cells would have many applications, and a person's immune system would not reject what is essentially its own tissue. Some people argue that therapeutic cloning violates the rights of early-stage embryos; others argue that banning this research violates the rights of people who might benefit from embryonic stem cell therapy.

According to the bill's ban on producing or selling "any embryo produced by human cloning," scientists caught in the act could expect a fine of up to $1 million or ten years in prison. Proposals to exempt therapeutic

cloning were defeated. The criminalization of basic research is unprecedented: Before 2001, bans on using embryonic stem cells applied only to federally funded research, and work using a small number of previously existing stem cell lines was permitted. Since the 2001 ruling, some researchers have moved to nations that permit them to derive new embryonic stem cell lines. Stem cells that are normal parts of adult bodies are being investigated as alternative sources of replacement tissues.

Cloning Misconceptions

The premise that a clone is an exact duplicate of another individual is flawed, and so if the intent of cloning is to create such a copy, it simply will not work. For example, the tips of chromosomes, called **telomeres**, shorten with each cell division. A clone's telomeres are as short as those from the donor nucleus, which means that they are "older" even at the start of the clone's existence. DNA in the donor nucleus has also had time to mutate, that is to say, it has had time to undergo modification from its original sequence, thus distinguishing it genetically from other cells of the donor. A mutation that would have a negligible or delayed effect in one cell of a many-celled organism, such as a cancer-causing mutation, might be devastating if an entire organism develops under the direction of that nucleus. Finally, the clone's **mitochondria**, the cell organelles that house the reactions of metabolism and contain some genes, are those of the recipient cell, not the donor, because they reside in the cytoplasm of the egg. Mitochondrial genes, therefore, are different in the clone than they are in the nucleus donor. The consequences of nuclear and mitochondrial genes from different individuals present in the same cell are not known, but there may be incompatibilities.

Perhaps the most compelling reason why a clone is not really a duplicate is that the environment affects gene expression. Cloned calves have different color patterns, because when the animals were embryos, the cells that were destined to produce pigment moved in different ways in each calf. For humans, consider identical twins. Nutrition, stress, exposure to infectious diseases, and other environmental factors greatly influence our characteristics. For these reasons, cloning a deceased child, the application that most would-be cloners give for pursuing the technology, would likely lead to disappointment.

Bioethical concerns over cloning may be moot, because the procedure is extremely difficult to do. Dolly was one of 277 attempts; Cumulina, the first cloned mouse, was among 15 liveborn mice from 942 tries. Cloning so often fails, researchers think, because it is not a natural way to start the development of an animal. That is, the DNA in a somatic cell nucleus is not in the same state as the DNA in a fertilized **ovum**. The donor DNA in cloning does not pass through an organism's germ line, the normal developmental route to sperm or egg, where gene activities are regulated as a new organism develops.

Ethical objections to human cloning are more philosophical than they are practical. The very idea of cloning assumes that our individuality can be understood so well that we can duplicate it. If human cloning ever became a reality, that this is not true would become evident. After all, we are more than a mere collection of genes. SEE ALSO BIOTECHNOLOGY: ETHICAL ISSUES;

telomeres chromosome tips

mitochondria energy-producing cell organelle

ovum egg

CLONING GENES; CLONING ORGANISMS; MITOCHONDRIAL GENOME; STEM CELLS; TELOMERE.

Ricki Lewis

Bibliography

Annas, George J. "Cloning and the U.S. Congress." *The New England Journal of Medicine* 346 (2002): 1599.

Holden, Constance. "Would Cloning Ban Affect Stem Cells?" *Science* 293 (2001): 1025.

Lewis, Ricki. "The Roots of Cloning." In *Discovery: Windows on the Life Sciences.* Medford, MA: Blackwell Science, 2000.

Mayor, Susan. "Ban on Human Reproductive Cloning Demanded." *British Medical Journal* 322 (Jun., 2001): 1566.

Cloning Organisms

There are two distinct types of cloning: molecular and organismal. Molecular cloning is the removal of a stretch of DNA, usually a gene, from an organism, and its insertion into another piece of DNA, such as a **plasmid**, to form a substance called recombinant DNA. This recombinant DNA may then be expressed in, or simply carried passively by, another organism, such as bacteria. Organismal cloning, the subject of this entry, is the production of genetically identical organisms and, as such, can be used to produce genetically identical copies of livestock or may be used to produce new members of endangered or even extinct species. It may be especially cost-effective to clone animals that produce therapeutic proteins such as blood clotting factors, thus combining both types of cloning. Cloning is controversial, however, because our understanding of the procedures needed to clone mammals may be applied to human cloning, which gives rise to profound ethical issues.

plasmid a small ring of DNA found in many bacteria

The History of Cloning

Cloning has a long history. Animals that reproduce sexually produce clones whenever identical twins are born. These twins are genetically indistinguishable, and are formed when a fertilized egg separates at a very early stage of development. Clones are also the natural product of asexual reproduction, although in this case perfect clones cannot be maintained through an infinite number of generations, because spontaneous mutations can and do occur. Lastly, clones can be produced by regeneration in both plants and animals. For example, plant cuttings will regenerate roots and, ultimately, an entire "new" plant, and some invertebrates, such as planaria, can regenerate two identical animals if the adult is cut in half. In these forms, cloning has been with us for a very long time.

Since the mid-1960s, scientists have been able to culture plant cells, that is, grow cells from plants such as tobacco and carrots in a petri dish, to get thousands of genetically identical cells. From such cultured cells an unlimited quantity of cloned plants can then be grown. These cultured cells can be modified to contain recombinant, or cloned, DNA as well.

Cloning Amphibians

The first cloning of a vertebrate by nuclear transfer was reported by John Gurdon of the University of Cambridge in the 1950s. In nuclear

The process used to clone Dolly. Adapted from Gurdon, J. B., 1999.

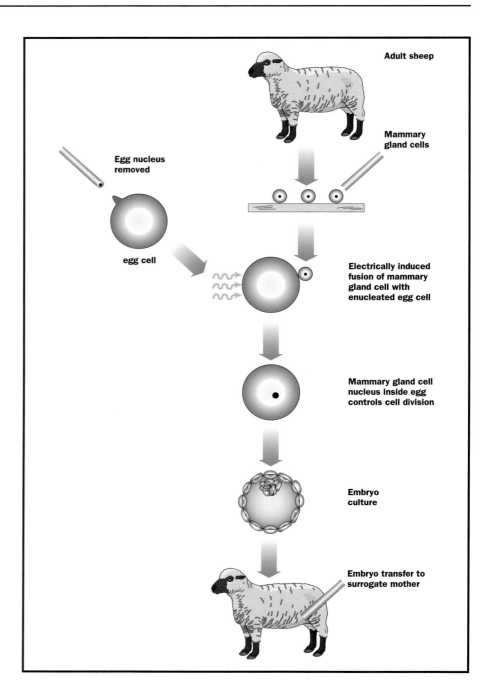

transplantation, the nucleus of an unfertilized donor egg is either mechanically removed or it is destroyed by ultraviolet light in a process called enucleation. The original nucleus is then replaced by a nucleus containing a full set of genes that has been taken from a body cell of an organism. This procedure eliminates the need for the fertilization of an egg by a sperm.

The most successful nuclear transplants have been achieved after serially transferring donor intestinal nuclei, that is, putting an adult nucleus from an intestinal cell into an egg whose nucleus was destroyed, allowing the egg to divide only a certain number of times, removing nuclei from these cells, and repeating this process several times before allowing the embryo to complete development. Eventually, transplantation of nuclei from albino embryonic frog cells into enucleated eggs from a dark green female frog led to the production of adult albino frog clones, demonstrating that a properly treated adult nucleus could support the full development of an egg into an adult clone. Later experiments demonstrated that nuclei from cells of other tissues, even **quiescent** cells such as blood cells, could also be used if properly treated. Despite these successes, no adult frog has been cloned when a nucleus from an adult cell was used without serial transfer. Without serial transfer of the nuclei, the animals would only develop to the tadpole stage, and then they would die.

Cloning of Mammals: Dolly

Nuclear transplantation has also been successful in producing mammalian clones, most notably of sheep, cattle, pigs, and mice. The most famous cloned mammal is a sheep named "Dolly," the first animal to be cloned directly from an adult cell. Experiments leading to the birth of Dolly were done at the Roslin Institute with collaborators at Pharmaceutical Proteins Limited, both in Scotland. This group had earlier produced Megan and Morag, the first mammals to be cloned from cultured cells. These two sheep were produced from embryonic cells, however, not from cells of an adult animal.

Dolly was born in the summer of 1996, the product of a nucleus from the mammary gland of a six-year-old female Finn-Dorsett sheep and an egg from a Scottish Blackface female. Mammary gland cells were grown in a petri dish and were deprived of nutrients so that they would stop dividing, just like an unfertilized egg. Donor eggs were taken from sheep soon after **ovulation**, and nuclei were mechanically removed from them. These enucleated eggs were then fused with the cultured mammary gland cells so that a mammary gland nucleus would be inside an unfertilized egg. Two hundred and seventy-seven such embryos were constructed and temporarily allowed to divide in a petri dish, and then all of them were transferred into the **oviduct** of a temporary surrogate mother. Of the original 247 embryos, only 29 developed further, and these were transferred to 13 hormonally treated surrogate mothers.

Only one surrogate mother became pregnant, and she only had one live lamb, named Dolly. The success rate was very low, but Dolly has been proven to be a true clone: She has all the characteristics of a Finn-Dorsett sheep. Independent scientists used a technique called DNA fingerprinting to show that Dolly's DNA matched the donor mammary cells but did not match that of other sheep in the Finn-Dorsett flock, nor did her DNA match that of her surrogate mother or the egg donor. Similar results have been obtained by Ryuzo Yanagimachi at the University of Hawaii, who worked with several generations of cloned mice.

In 1997 Polly, a sheep created with a combination of both molecular and organismal cloning techniques, was born. Polly was derived from a fetal sheep cell that had been engineered to contain the human gene that makes coagulation factor IX. Factor IX is missing in people with a disease called hemophilia type B. Polly and two other sheep were engineered to produce factor IX in their milk, thus providing people with hemophilia access to a safer and less expensive source of clotting factor than was previously available. Because Polly was made from more easily cultured and, therefore, more easily engineered embryonic cells, it is thought that this type of cloning

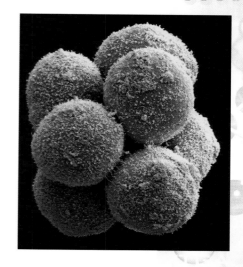

Scanning electron microscope scan of human embryo, day three, at the eight-cell stage.

quiescent nondividing

ovulation release of eggs from the ovaries

oviduct a tube that carries the eggs

technology holds the most promise for the future of pharmaceutical production of proteins that cannot be made in bacteria.

In January 2001, the first cloned member of an endangered species was born. This was a gaur, a wild ox native to India and southeast Asia, which the researchers named Noah. The gaur was chosen by Advanced Cell Technology as a candidate for cloning after the company had successfully cloned domestic cattle, which are related to the gaur species.

The embryo from which Noah developed was created from the nuclei of frozen skin cells that had been taken from an adult male gaur that had died eight years earlier. Skin cell nuclei were fused with enucleated domestic cow eggs to produce forty embryos. One of these forty was carried to full term in a surrogate cow mother. Unfortunately, Noah died of an infection two days after his birth (the infection is thought to be unrelated to his origin as a cloned animal). Despite Noah's death, it is likely that cloning will eventually be used to aid the conservation of endangered species. In the future, scientists may attempt to clone a recently extinct species, should intact DNA for an extinct species be obtained.

Problems with Cloning

In general, the success rate of mammalian cloning is low, with less than 0.1 to 2.0 percent of transplanted nuclei yielding a live birth. The vast majority of transplants fail to divide or to develop normally, indicating there is much we still do not understand about reprogramming an adult nucleus to support embryonic development. One thing that is clear, however, is that having both the donor cell and host egg cell in a nondividing state is essential for success.

telomeres chromosome tips

What might be both the most vexing and most interesting problem with cloning is related to aging. Chromosomes "show their age" by a shortening in their tips, or **telomeres**, a process that occurs every time the cell they are in divides. This telomere shortening occurs in all cells except eggs, sperm, and most cancer cells, and shortened telomeres are correlated with the aging of organisms. Since the nuclear DNA in most cloned animals is taken from an adult, the chromosomes of cloned animals are expected to have shorter telomeres than animals of the same birth age that are produced by sexual reproduction, causing researchers to wonder whether cloned animals will age prematurely. Shorter telomeres have been found in Dolly and other cloned sheep, but telomeres are reported not to be shorter in cloned mice or cattle. Underlying reasons for the different results may include differences between cell types or species used.

The Myth of the Perfect Clone

Cloned animals are not 100 percent identical to their "parents." Whenever nuclear transplantation is used to produce cloned organisms, the offspring display some differences from the organism that donated the nuclei. The egg donor contributes mitochondria, the energy producers of eukaryotic cells, and these mitochondria have their own small amount of DNA-containing genes used for energy metabolism. Since mitochondria are inherited only with egg cytoplasm, they will not match the mitochondria of the animal from which the nucleus was taken. In addition, maternally derived gene products, both mRNA (messenger RNA) and protein, which serve to

begin embryonic development, will differ from that of the nuclear donor, as will the uterine environment and the external environment. Thus, for example, clones produced by nuclear transplantation will be significantly less identical than will clones produced by twinning. SEE ALSO CLONING: ETHICAL ISSUES; CLONING GENES; CONSERVATION BIOLOGY: GENETIC APPROACHES; HEMOPHILIA; MITOCHONDRIAL GENOME; REPRODUCTIVE TECHNOLOGY; TELOMERE; TRANSGENIC ANIMALS; TWINS.

Elizabeth A. De Stasio

Bibliography

Gurdon, J. B., and Alan Colman. "The Future of Cloning." *Nature* 402 (1999): 743.

Lanza, Robert P., Betsy L. Dresser, and Philip Damiani. "Cloning Noah's Ark." *Scientific American* (Nov., 2000): 84–89.

Wilmut, Ian. "Cloning for Medicine." *Scientific American* (Dec., 1998): 58–63.

Wilmut, Ian, Keith Campbell, and Colin Tudge. *The Second Creation: Dolly and the Age of Biological Control*. Cambridge, MA: Harvard University Press, 2000.

College Professor

Professors of genetics at colleges and universities teach, perform research, and handle administrative responsibilities. Professors may teach at the undergraduate or graduate levels, or both. Undergraduate teaching involves class lectures, small group seminars, and hands-on laboratory sessions. Professors evaluate students based on their performances on examinations, essays, and laboratory work, and may also work individually with students to advise them about their college careers or to mentor them in independent laboratory studies. Graduate teaching provides advanced instruction in the field of genetics, usually in smaller classes. Professors also act as mentors to graduate students, providing a supportive environment for conducting research.

Research is an integral component of professorship. Professors apply for grants to fund genetic experiments, perform original research, analyze the results, and submit their findings for publication. They must keep current with the published results of other scientists in the genetics field by reading journals and books, attending the conferences of professional societies, and interacting with other researchers. In addition, professors must fulfill administrative responsibilities, including participation on departmental and faculty committees that consider issues such as courses of study for students (curricula), budgets, hiring decisions, and allocating resources.

Most genetics professors at four-year universities hold a doctorate degree (Ph.D.) in a specialized area of genetics or molecular biology. Earning a Ph.D. first requires completion of an undergraduate bachelor's degree. The student must then complete graduate school, typically consisting of about three years of advanced coursework and two to three years of original, independent laboratory research. The results of this research are written up in an extensive report, called a dissertation.

The career path of a college professor can be described as a rise through four levels: instructor, assistant professor, associate professor, and full professor. Instructors are either completing or have already earned their Ph.D., and are beginning their teaching careers. They usually spend about nine to twelve hours a week teaching. The *Occupational Outlook Handbook*, published

Genetics professors are often specialists in a specific area of molecular biology or genetics and are expected to juggle the roles of educator and researcher. Here, a biology lecture is given by a professor at Morehouse College in Atlanta, Georgia.

by the U.S. Department of Labor, lists the average instructor's salary as $33,400. After instructors have taught for several years, their teaching, research, and publications are reviewed by their academic institution. If found satisfactory, they are eligible for promotion to the position of assistant professor. Assistant professors also teach nine to twelve hours weekly, but are more likely to lecture in large undergraduate courses. They are expected to conduct research projects and publish their work. Assistant professors earn approximately $43,800.

With continuing success in research and teaching responsibilities, a professor can obtain the position of associate professor. These faculty members spend fewer hours on undergraduate teaching (about six to nine hours a week), and are likely to lead graduate classes and advise graduate students on their dissertation projects. They can expect to make about $53,200 yearly. Promotion to full professorship is based on the quality of one's research and reputation with the field. Teaching is less emphasized, usually occupying only three to six hours per week. Professors take an active role in the research projects and dissertations of doctoral candidates. Further advancement opportunities include positions in administration such as department chair, dean of students, or college president.

College professors may work at public or private institutions. They generally teach for nine months of the year, allowing them to work in other environments as well. Instructors may teach additional classes, act as consultants to private, governmental, or nonprofit organizations, or author publications in their field of expertise. The ability to make one's own schedule, conduct original research, teach and mentor students, take paid leaves of absence, and have access to campus facilities makes professorship an attractive and competitive choice for motivated individuals. SEE ALSO EDUCATOR; GENETICIST.

Regina M. Carney

Bibliography

Bureau of Labor Statistics, U.S. Department of Labor. *Occupational Outlook Handbook: 2000–2001 Edition.* Bulletin 2520. Washington, DC: Superintendent of Documents, U.S. Government Printing Office, 2000.

Wright, John W., and Edward J. Dwyer. *The American Almanac of Jobs and Salaries.* New York: Avon Books, 1990.

Colon Cancer

Colon cancer is the second leading cause of cancer death in the United States, occurring in approximately 5 percent of the population and resulting in roughly 55,000 deaths annually. New cases of colorectal cancer are diagnosed in approximately 90 per 100,000 people annually. The majority of cases occur in individuals older than age fifty. Of those who suffer from colorectal **malignancy**, an estimated 40 percent will die from the disease.

malignancy cancerous tissue

Colon cancer–related health-care costs, consisting of outpatient visits, hospitalizations, hospice and home health care, medications, and physician services, exceed $5 billion per year. This figure does not include the indirect costs of wages lost and reduced productivity.

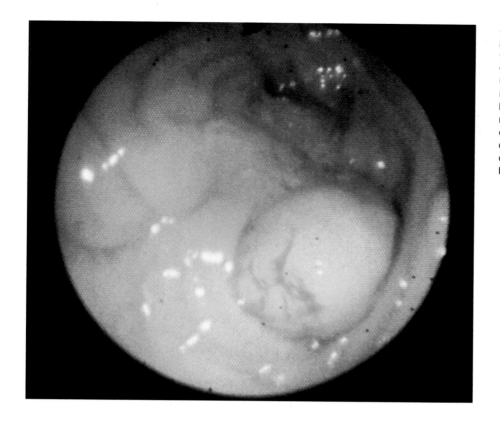

A colonic polyp, or visible growth, is created by the overgrowth of colonic epithelial cells. Such uncontrolled cell division is not always cancerous, but may become malignant. DNA mutations of the colonic epithelial cells—the cause of the overgrowth—may be hereditary.

Developing Cancer

The colon, also known as the large intestine, is the final portion of the digestive tract and consists of a tube (called the lumen) lined by specialized cells called colonic epithelial cells. These cells are constantly reproducing in a regulated manner, but when the growth and division of colonic epithelial cells becomes unregulated, colon cancer may result.

Cancer is a form of unregulated cell growth in which growing cells invade surrounding tissue. Such a growth is said to be malignant. Colon cancer results when there are certain changes in the genes that control normal cell replication. In most cases, when cell growth becomes abnormal, a visible growth (**lesion**) protrudes into the colon's lumen and is termed an adenomatous polyp (or **adenoma**). The polyp is not yet cancerous but may become so, at which point it is called carcinoma. This process, in which normal tissue becomes cancerous, is known as the adenoma-carcinoma sequence and may take between ten and fifteen years.

lesion damage

adenoma a tumor (cell mass) of gland cells

Major Genes Involved

The genes altered in the adenoma-carcinoma sequence normally play a role in regulating the cell cycle, controlling the division and turnover of epithelial cells. DNA mutations result in a loss of function of the gene and subsequent unregulated cell growth. While many genes have been studied, the ones most commonly associated with the majority of colon cancers are *APC*, *p53* and *K-ras*.

Colorectal cancer may develop in an individual who has a strong inherited risk. This occurs, for example, in patients with familial adenomatous

polyposis (FAP) and hereditary nonpolyposis colon cancer (HNPCC). These patients will usually have a specific genetic alteration.

APC. The Adenomatous Polyposis Coli (*APC*) gene is found on chromosome 5 and is a tumor suppressor gene. Both **alleles** of the gene must be inactivated for tumor growth to occur. In the normal cell, the *APC* gene plays a role in regulating the cycle of cellular division and replication, as well as in cell-to-cell communication, thereby suppressing tumor development. Mutations of *APC* result in a loss of gene function, thus allowing unregulated cellular proliferation. *APC* mutations are found in the majority of common colon polyps and cancers and in patients with FAP, and they may be one of the earliest genetic alterations in the adenoma-carcinoma sequence.

K-ras. The *K-ras* gene plays an active role in cellular signaling and promoting cell growth. The normal gene exists in both an active and inactive form. However, in the abnormal state, the active form predominates and results in a continually growth-stimulated state.

p53. The normal *p53* gene is responsible for regulating cells with damaged DNA by directing abnormal cells either to halt the cycle of cell division or to die as the result of a process called **apoptosis**. Like *APC*, the *p53* gene is a tumor suppressor. With the *p53* mutation, the gene no longer functions, and this permits the uninhibited proliferation of cells that may have damaged DNA. *p53* mutations are seen in more than half of colorectal cancers.

Familial Adenomatous Polyposis (FAP)

FAP is characterized by the presence of hundreds or thousands of colonic polyps, and it accounts for less than 1 percent of colon cancers. Affected individuals are at increased risk of colorectal polyps and malignancy, and they usually develop polyps by age thirty-five and cancer by age forty. Because **endoscopic** removal of all the polyps is impossible, patients are usually advised to consider **colectomy** at a relatively young age. In addition to occurring in the colon, polyps occur in the upper digestive tract, and a variety of tumors may develop outside the gastrointestinal tract. Because mutations in the *APC* gene can be identified in most cases, genetic testing is now available for affected families.

FAP is inherited as an **autosomal** dominant disorder with 95 percent penetrance, meaning 95 percent of those who inherit one mutated *APC* gene will develop FAP. At the cellular level, the *APC* mutation is actually recessive: As long as there is one copy that is not mutated, the cell cycle will remain controlled.

The apparent paradox of a recessive gene causing a dominant disorder is resolved when we consider how a gene defect predisposes a person to cancer. Of the many millions of cells lining the colon, it is highly likely that some will undergo spontaneous mutation in one of the two copies of the *APC* gene.

For a person not affected by the *APC* gene, a spontaneous mutation of one allele will not lead to cancer, because the other gene copy remains intact. However, for a person affected who inherits one mutated copy of the *APC* gene, each of the cells lining the colon begins with one bad gene copy. Any mutation to the remaining good copy will cause the cell to lose control of

alleles particular forms of genes

apoptosis programmed cell death

endoscopic describes procedure in which a tool is used to see within the body

colectomy colon removal

autosomal describes a chromosome other than the X and Y sex-determining chromosomes

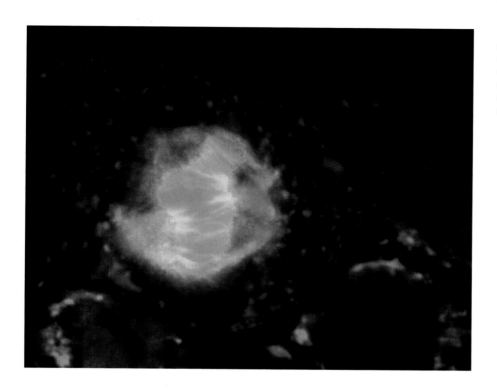

Digital image of multipolar spindles (green) separating the chromosomes (red) of a cancerous colonic epithelial cell in a late stage of its division.

its cell cycle and begin the process of polyp development. Given the number of cells involved, it is almost inevitable that this will occur in some of them. Hence FAP is a dominant condition.

Hereditary Nonpolyposis Colon Cancer

HNPCC is an autosomal dominant disorder that may be responsible for up to 5 percent of colon cancers. The genetic mutation leading to the abnormality is the mutation of DNA mismatch repair genes. Individuals with this mutation have up to an 80 percent chance of developing colon cancer. At least five genes are involved in this syndrome.

Malignancy in patients with HNPCC occurs at a younger age than in the general population, is more often located in the proximal colon (the portion nearest the small intestine), and may be associated with multiple tumors. HNPCC also carries an increased risk of tumors of the **endometrium**, ovaries, stomach, small intestine, bile ducts, bladder, renal pelvis, and ureters. Genetic testing is useful for HNPCC families.

endometrium uterine lining

Genetic Testing

Genetic testing may benefit patients and families affected by an inherited colon cancer syndrome. The genetic mutations in FAP and HNPCC can often be characterized. If a family member with colon cancer has an identified genetic abnormality, other family members can be tested to see if they have the same abnormality.

If the mutated gene is not found, the abnormal gene was not inherited and the family member is not at increased risk of developing cancer. Because screening (i.e., colon examination) is more frequently performed for those with FAP and HNPCC than others, genetic testing can be useful for determining who would benefit from intense surveillance.

Other Risk Factors

In addition to the well-described genetic syndromes of FAP and HNPCC, other factors that place an individual at increased risk include a personal or family history of colon cancer and the presence of inflammatory bowel disease (e.g., ulcerative colitis and Crohn's disease). Population studies support an association between the development of colon cancer and a high-fat, low-fiber diet, although a cause-and-effect relationship has not been proved.

Prevention

Inhibiting the development of polyps and cancers, finding and removing premalignant polyps, and testing individuals at high risk may reduce colon cancer–related morbidity and mortality. Colon cancer occurs less commonly in individuals whose diets are high in calcium and folate, who take multivitamins, and who maintain high-fiber and low-fat diets.

Non-steroidal anti-inflammatory medications like aspirin may reduce the numbers of polyps, particularly in families with FAP. Colonoscopy can identify polyps that may be premalignant and can facilitate polyp removal. It is recommended that all individuals have a colonoscopy at age fifty. High-risk patients, such as those with inflammatory bowel disease, FAP, or HNPCC, should have screening initiated at an earlier age and repeat exams at shorter time intervals. SEE ALSO APOPTOSIS; BREAST CANCER; CANCER; CARCINOGENS; CELL CYCLE; DNA REPAIR; GENETIC TESTING; MUTATION; ONCOGENES; TUMOR SUPPRESSOR GENES.

David E. Loren and Michael L. Kochman

Bibliography

Giardiello, Francis M., Jill D. Brensinger, and Gloria M. Petersen. "AGA Technical Review on Hereditary Colorectal Cancer and Genetic Testing." *Gastroenterology* 121 (2001): 198–213.

Yamada, Tadataka, et al., eds. *Textbook of Gastroenterology*, 3rd ed. Philadelphia, PA: Lippincott Williams & Wilkins, 1999.

Internet Resource

"The Burden of Gastrointestinal Diseases." *American Gastrointestinal Association.* May 2001. <http://www.gastro.org/pdf/burden-report.pdf>.

Color Vision

nanometer 10^{-9} meters; one billionth of a meter

Sight is a complex process that results when visible light, a narrow band of the electromagnetic spectrum between 400 and 700 **nanometers** (nm), is converted into signals that can be interpreted by the brain. This process involves special light-sensitive cells called photoreceptors that are located in the retina, a thin structure that lines the inside of the eye. These cells capture packets of light, called photons, and transform their energy into signals that are transported from the eye to the occipital cortex, the portion of the brain that allows us to interpret these signals as sight.

Normal human color vision is trichromatic (based on the perception of three primary colors) and requires three types of photoreceptor cells, called cones, each of which contains a different photopigment. Each photopigment

CLASSIFICATION AND INCIDENCE OF COLOR VISION DEFECTS

Table 1. Adapted from American Academy of Ophthalmology, 1995.

Color Vision	Inheritance	Incidence in Male Population (*Percent*)
I. Hereditary		
Trichromats		
1. Normal		92.0
2. Deuteranomalous	XR	5.0
3. Protanomalous	XR	1.0
4. Tritanomalous	AD	0.0001
Dichromats		
1. Deuteranopes	XR	1.0
2. Protanopes	XR	1.0
3. Tritanopes	AD	0.001
Monochromats (achromats)		
1. Typical (rod monochromats)	AR	0.0001
2. Atypical (cone monochromats)	XR	Unknown
II. Acquired		
1. Tritan (blue-yellow)		
2. Protan-deutan (red-green)		

absorbs particular wavelengths of light in the short (blue, 440-nm), middle (green, 545-nm), or long (red, 560-nm) wavelength region of the visible spectrum. About 7 percent of all cones are blue-sensitive, 37 percent are green-sensitive, and 56 percent are red-sensitive. These cones are the basic mediators of color vision. If one or more of their pigments is missing, color blindness results. Rod cells, unlike cones, detect light intensity but not color. The photopigment in rod cells is called rhodopsin.

The spectral sensitivity of the cone photopigments is intimately related to the structure of the pigment molecules. These are concentrated in the photoreceptor outer segment, the portion of the cell containing the photo-transduction machinery. Each pigment molecule consists of an opsin protein and a chromophore (11-cis-retinal), which is a derivative of vitamin A. Photon absorption by the pigment molecules causes a change in the shape of the chromophore, which initiates the processes that lead to vision.

The different opsins of the cone photopigments and of the rod photopigment are encoded by four separate genes, the *BCP* (blue cone pigment), *GCP* (green cone pigment), *RCP* (red cone pigment), and *RHO* (rhodopsin) genes. The genes encoding the blue cone and rod pigments reside on the long arms (called the q arms) of chromosome 7 and chromosome 3, respectively. The genes encoding the red and green cone pigments reside on the q arm of the X chromosome.

Color vision defects may be divided into two groups, hereditary and acquired. Hereditary color vision defects are almost always "red-green" and affect 8 percent of males and 0.5 percent of females. Acquired defects are more often "blue-yellow," and affect males and females equally. Hereditary defects are typically bilateral (affecting both eyes), while acquired defects may affect one eye only and are often asymmetric. Hereditary color vision defects tend to remain stable throughout life and are usually not associated with other retinal or optic nerve **pathology**. Acquired defects, however, may have a more variable course and are frequently associated with observable

pathology disease process

COLOR VISION TESTS

Test	Sensitivity/ Quantification	Ease of Administration
AO-HRR	Will miss very mild R-G/good classification	Excellent for all ages
Farnsworth-Munsell 100 hue	Extremely sensitive/ classify by error scoring	Tedious to administer
Ishihara	Extremely sensitive/ nil	Difficult for pre-school children and low-IQ patients
Farnsworth's Panel D-15	Will only detect severe anomalous trichromats and dichromats/good classification	Easy to administer
Nagel's anomaloscope	Very sensitive/classify by anomaly (R-G) quotient	Good
Sloan's achromatopsia test	Grossly sensitive/ very incomplete achromatopsia pass	Easy to administer

NOTE: All tests, with exception of Nagel's anomaloscope, are to be administered under an illuminant C source such as provided by Macbeth easel lamp.

ocular pathology. A common cause of acquired color-vision loss is optic nerve disease, such as optic neuritis.

Inherited color blindness usually results from the loss of one of the photopigments and reduces color vision to two dimensions, or dichromacy. Other less common conditions reduce color vision to one dimension (monochromacy), or may completely extinguish it (achromacy). Vision in this last circumstance is purely dependent on the rods, which function primarily in dim conditions and do not contribute to color vision.

The most common forms of hereditary color blindness are protanopia/anomaly and deuteranopia/anomaly, both of which are caused by defects in the red (L) and green (M) cones. Also known as red-green color vision deficiencies, they typically demonstrate an X-linked recessive pedigree pattern. The incidence of X-linked color-vision defects varies between human populations of different racial origin, with some of the highest rates appearing in Europeans and some populations in India.

The incidence of these common forms of color blindness is much lower in females than in males because the defects are inherited as X-linked recessive traits. Males, who have only one X chromosome, are hemizygous (meaning that they have only one gene present for the trait) and they will always manifest a color vision deficiency if they inherit an abnormal gene from their mother. Females, on the other hand, have two X chromosomes, one inherited from each parent, so they will not usually show a complete manifestation of the typical color defect unless they are homozygous, though a partial manifestation of color blindness may be present in heterozygotic carriers. A variety of special tests are used to screen for these red-green color-vision defects. SEE ALSO INHERITANCE PATTERNS; MOSAICISM; SIGNAL TRANSDUCTION; X CHROMOSOME.

Eric A. Postel

Bibliography

American Academy of Ophthalmology. *Retina and Vitreous: Basic and Clinical Science Course.* San Francisco: American Academy of Ophthalmology, 1995.

Benson, William E. "An Introduction to Color Vision." In *Clinical Opthalmology*, vol. 3, T. D. Duane and E. A. Jaeger, eds. Philadelphia: Harper & Row, 1987.

Connor, Michael, and Malcolm Ferguson-Smith. *Essential Medical Genetics*, 5th ed. Oxford U.K.: Blackwell Science, 1998.

Gegenfurtner, Karl R., and Lindsay T. Sharpe, eds. *Color Vision: From Genes to Perception.* Cambridge, U.K.: Cambridge University Press, 2001.

Combinatorial Chemistry

Combinatorial chemistry is a technology for creating a multitude of different compounds by reacting different combinations of interchangeable chemical "building blocks." The compounds are then screened for their ability to carry out a specified function, most commonly to act as drugs to treat a disease. Combinatorial chemistry allows the rapid synthesis and testing of many related compounds, greatly speeding the pace of drug discovery. Automated synthesis and screening systems are key to this approach.

The Combinatorial Approach

There are two general approaches for finding the correct answer to a question (besides asking someone who knows). One way is to learn everything relevant to the topic and then to use your knowledge to arrive logically at the answer. Scientists, and most other people, almost always use this method. A second approach is to keep guessing until you've guessed right! This seems like a foolhardy strategy, and usually is. What if it took a million guesses before you stumbled upon the right answer? But what if you could make a million or a billion guesses all at once? Through combinatorial chemistry, scientists can make and test millions, billions, or even quadrillions (10^{15}) of guesses about which chemical compound might have a desirable function, such as the ability to bind to a specific molecule, or to serve as a drug.

Many chemicals are pieced together through combinations of smaller building blocks. For example, benzene is a chemical consisting of six carbon atoms connected in an aromatic ring structure, with a hydrogen atom bound to each carbon. Substituting one of the hydrogens with a hydroxyl (-OH) group forms the chemical phenol. Substituting a methyl ($-CH^3$) group instead forms toluene, and substituting an amino ($-NH^2$) group forms aniline. Because of their different "functional groups," or side groups, all of these compounds have very different physical and chemical properties. More variations can be synthesized by substituting additional side groups with more than one of the hydrogens. By substituting one of just these three groups (or by not adding any groups) for any of the six hydrogens in a benzene ring, there are 4^6, or 4,096, possible combinations (the number of different compounds is much smaller, because benzene is symmetrical, and many of the combinations represent equivalent structures).

Side groups can also be placed onto other side groups. For example, a single chlorine atom can substitute for one of the hydrogens of the methyl group in toluene to form benzyl chloride. By using a moderately sized collection of side groups, placing them onto a "scaffold" molecule that is more complex than

The substitution of side groups on simple molecules can create many new molecules.

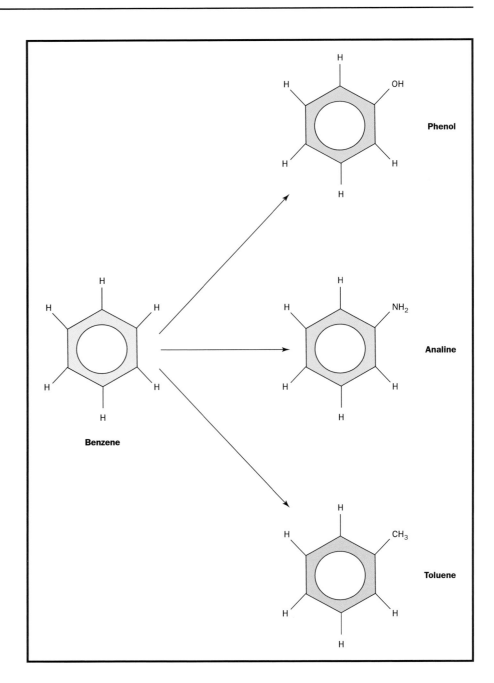

synthesize create

benzene (such as cholesterol, which has three six-carbon rings and a five-carbon ring), and by using additional levels of side groups, combinatorial chemists can **synthesize** vast numbers of distinct but related compounds.

Although the utility of combinatorial chemistry was not fully appreciated by scientists until the 1980s, nature uses this strategy over and over. Genes, after all, are composed of different combinations of only four different nucleotides, and just twenty different amino acids form the building blocks of all proteins. In the immune systems of mammals, B lymphocytes use an elaborate scheme for mutating and combining different segments of antibody genes to generate a diverse pool of antibody molecules that can recognize and bind a wide array of alien molecules that enter the body with a pathogen infection.

Drug Targets

Combinatorial chemistry is most often used to synthesize "small molecules," in contrast to **macromolecules** such as DNA, RNA, proteins, and polysaccharides, which are polymers containing long chains of monomer subunits. Because of their enormous size, macromolecules cannot easily enter cells, which is an important requirement for compounds intended for use as drugs.

In many cases the combinatorial chemist is looking for a compound that will bind tightly and specifically to a cellular molecule, such as the catalytic, or "active," site on the inside of an **enzyme**. Small molecules can fit into the holes and crevasses leading to the active site. By binding the enzyme, the synthetic compound may prevent it from binding to its natural substrate or from carrying out its catalytic reaction. Defective enzymes that resist normal cellular restraints on their activities are responsible for many diseases, including certain cancers. Chemical inhibitors of such rogue enzymes hold promise as powerful drugs. Alternatively, binding of a small molecule to an enzyme could enhance the enzyme's normal activity. Such molecules have potential as drugs for diseases caused by insufficient activity of a crucial enzyme.

For two molecules to bind to one another, they must have a proper fit, like a key in a lock. The fit depends on the shapes of the two molecules as well as on the chemical interactions between them. For example, two positively charged side groups will repel each other, but negative and positive groups can attract. Not surprisingly, a synthetic compound that binds a particular molecule often has chemical properties and a shape mimicking the natural ligand for the molecule. Such compounds are termed analogues.

When the drug target is known, its structure can be used as a **template** to create analogues with complementary shapes. Alternatively, if an analogue is already in hand that binds the target but has undesirable properties (such as weak binding, poor solubility, or serious side effects), this structure can be used as a starting point. Even without such clues, the speed and automation of the combinatorial approach makes it feasible to randomly synthesize and test millions of compounds.

High-Throughput Screening

A library of a billion or more different molecules is only useful if the molecules can be quickly and economically screened for the desired function. "High-throughput" techniques have been developed that automate most of the steps required to combine the molecules with their targets and evaluate the extent of any reaction.

Typically, the molecules are arrayed on a solid surface and the target is added. Unbound target is washed away. Fluorescent tags are often added to the target, to allow easy (and automated) visualization of the results. Robotic systems controlled by computers can react and evaluate billions of separate compounds in the time it would take a human to screen a dozen. One such approach is used in DNA microarrays, in which thousands of genes from a DNA library are attached to a solid base. These are reacted with messenger RNAs from a cell, and the results are visualized fluorescently.

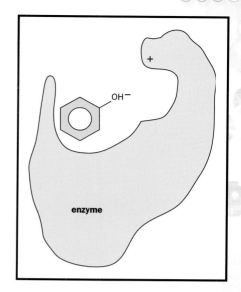

Interaction of small molecules with target enzymes depends on shape and charge.

macromolecules large molecules such as proteins, carbohydrates, and nucleic acids

enzyme a protein that controls a reaction in a cell

template a master copy

A high-throughput screening assay can quickly assess the target-binding ability of millions of compounds.

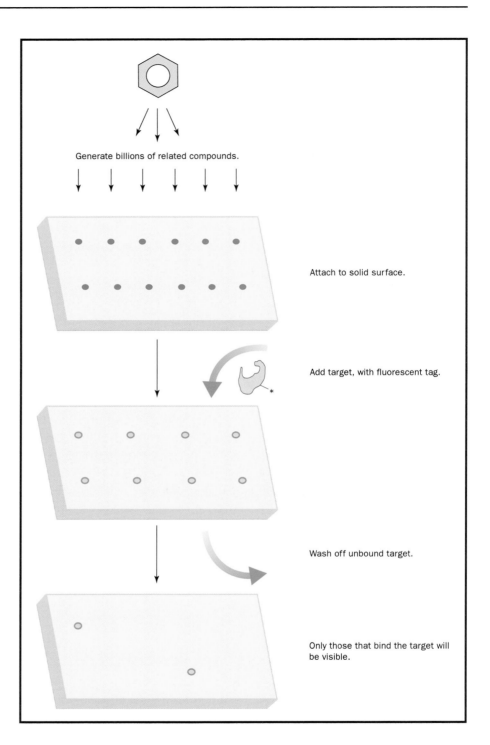

Generate billions of related compounds.

Attach to solid surface.

Add target, with fluorescent tag.

Wash off unbound target.

Only those that bind the target will be visible.

SELEX

In addition to its use in drug development, combinatorial chemistry can be applied to other areas of biomedical research, such as the design of molecules for diagnosing medical conditions. Compounds for these applications can be larger than pharmaceutical compounds, and do not have to be designed to enter the body. Using a novel combinatorial chemistry method called in vitro selection or SELEX (systematic evolution of ligands by exponential enrichment), a short DNA or RNA molecule (termed an oligonucleotide) with a desired property, such as the ability to specifically recognize and bind a molecule associated with a particular disease, can be selected in a single experiment from a library containing approximately 10^{15} different compounds. First, a library of oligonucleotides is created in a machine called an oligonucleotide synthesizer. This apparatus can make oligonucleotides with either a defined or random sequence.

Oligonucleotides for SELEX are designed to have a central region containing random sequence and outer, flanking regions with defined sequences. These defined sequences will be used as primer-binding sites for the polymerase chain reaction (**PCR**). The oligonucleotide library is prepared as a mixture, usually containing about 10^{14} to 10^{15} different sequences. These specialized oligonucleotides, termed aptamers, are then exposed to target ligand molecules, which are typically attached to a solid support, such as a filter membrane. The unbound aptamers are then washed away, leaving only the rare aptamers that can bind the ligand adhering to the filter. These aptamers can then be recovered from the filter by washing it with a solution that disrupts the binding.

These binding candidate aptamers represent a minuscule fraction of the original library. Some may bind the target ligand tightly, but others may bind weakly. Since all the aptamers have defined primer-binding sites on the ends, this much-reduced population can now be amplified exponentially by PCR. After amplification, the aptamers can be subjected to another round of ligand binding, now using more stringent washing conditions, in which only the tightest-binding molecules will stay bound. These high-affinity binders can be recovered again subjected to still more cycles of PCR amplification, binding, washing, and recovery, until the population of aptamers consists exclusively of very tightly binding molecules.

For some applications, these molecules are useful directly. They can also be studied to design non-DNA molecules that have similar shapes but that will have more potential as drugs. SEE ALSO DNA LIBRARIES; DNA MICROARRAYS; HIGH-THROUGHPUT SCREENING.

Paul J. Muhlrad

PCR polymerase chain reaction, used to amplify DNA

Bibliography

Alberts, Bruce, et al. *Molecular Biology of the Cell*, 4th ed. New York: Garland Science, 2002.

Borman, Stu. "Combinatorial Chemistry." *Chemical and Engineering News* 75, Feb. 24, 1997.

Complex Traits

Complex traits are those that are influenced by more than one factor. The factors can be genetic or environmental. This is in contrast to simple genetic traits, whose variations are controlled by variations in single genes. Examples of simple traits include Huntington's disease and cystic fibrosis. Each of these traits is caused by a **mutation** in a single gene that alters or destroys the function of that gene. There are several thousand disorders caused by single genes, but these are almost always quite rare in the population, often occurring in less than one in five thousand individuals.

mutation change in DNA sequence

Almost any trait that is not simple is considered complex. If there are just a few genes that affect a trait, it may be called oligogenic. If there are many different genes that affect a trait, it may be called polygenic. If other, nongenetic factors are involved, it may be called multifactorial.

Diseases inherited as complex traits are often much more common in the population and include heart disease, Alzheimer's disease, and diabetes. There are many factors that can affect a complex disease. Perhaps most commonly, these traits have multiple genes, where variations in those genes can influence the risk of developing the disease.

How Genes Are Involved in Complex Disease

Because there are multiple factors involved in complex traits, such traits are difficult for scientists to study and even more difficult to understand. This is because the different factors may not all act equally or independently on the trait.

For example, if there are three genes involved in a trait, the simplest hypothesis is that each gene will contribute about one-third of the genetic effect on the trait. If the effect of the variations in the genes can be added together, then this is called an additive effect. However, that is not the only way multiple genes can have an effect. Rather than being added together, the effects of variations in the genes may have to be multiplied together. If this is the case, this is called a multiplicative effect. Both parts of figure 1 show examples of additive and multiplicative effects.

Both additive and multiplicative effects imply that a variation in each gene has an effect and that the overall effect gets larger when additional genes are involved. However, in some cases two or more gene variants may need to occur together before any effect is seen. An example of this is shown in Figure 2. If multiple genetic variants must occur together, then this is called an epistatic interaction. In epistasis, a particular form of one gene must be present for the effect of a second to be felt.

The overall effect of genes on a trait can be even more complicated, because genes may act in combinations of additive, multiplicative, and epista-

Figure 1A. In this model two-gene system, the risk of developing a disease varies between 0 and 8. It is determined by adding risk (2) for each T a person has at both Gene 1 and 2.

	Variation in Gene 1		
Variation in Gene 2	GG	GT	TT
GG	0	2	4
GT	2	4	6
TT	4	6	8

Variation in Gene 2 \ Variation in Gene 1	GG	GT	TT
GG	1	3	9
GT	3	9	27
TT	9	27	81

Figure 1B. Here, the risk of developing a disease varies between 0 and 81. It is determined by multiplying the risk (3) for each T a person has at both Gene 1 and Gene 2.

tic ways. For example, if there are six different genes whose variations can influence the risk of developing the disease, Genes 1, 2, and 3 may act additively, Genes 4 and 5 may act multiplicatively, and Gene 6 may act in an epistatic manner with Gene 2. Another level of complexity can occur because different variations within the same gene may act differently. For example one variation (**allele**) in Gene 1 may act additively with Gene 2, whereas another may act multiplicatively with Gene 2.

allele a particular form of a gene

In simple traits, the variations within the gene usually create major changes in the way the gene's product (the protein it codes for) acts. In most cases, these changes (mutations) are considered causative, because having them is enough to cause the disease. In other words, having the mutation is sufficient to get the disease. These are called causative genes. For instance, in Huntington's disease, the presence of the expanded form of the *huntingtin* gene is sufficient to cause the disease.

In contrast, in complex traits, the variations in any one of the genes are not usually enough to cause the trait. These variations may simply increase (or decrease) the probability of developing the disease. Thus these genes, and the variations within them, are often called susceptibility genes and susceptibility alleles. It is a particular combination of susceptibility alleles across multiple genes, and possibly including environmental factors, that causes a complex disease.

How the Environment Is Involved in Complex Disease

Genes are not the only things that can affect a complex trait. Often environmental factors can also be involved. The type of environmental factors can be very different for different traits. One obvious example of this is lung cancer. Smoking cigarettes greatly increases the risk of developing lung cancer. Smoking also seems to have an effect on other diseases, including some eye diseases (such as age-related macular degeneration). However, not every chronic smoker will develop lung cancer or eye disease: The presence of particular alleles of susceptibility genes is also a risk factor, as discussed below.

Figure 2. Here, the risk of developing a disease is either 0 or 9. It is determined by having at least one T at both Gene 1 and Gene 2.

Variation in Gene 2 \ Variation in Gene 1	GG	GT	TT
GG	1	1	1
GT	1	9	9
TT	1	9	9

Other environmental factors, which may be more difficult to identify directly, can still be identified by measuring other risk factors, such as gender, age, occupation, level of education, use of alcohol, and other measures of lifestyle. For example, the effect of estrogen may not be measured directly, but could be identified by finding an effect of gender on the risk of developing the disease. The strength of an environmental effect may vary by how much exposure a person has to that effect. For example, the number of years someone has smoked or how many cigarettes per day someone smokes may change the effect of exposure to smoke.

Genes and Environmental Factors May Interact

Just as with the genes themselves, which may act independently or together to cause a complex trait, genes and environmental factors may act independently or together to cause a complex trait. The actions can be additive, multiplicative, or epistatic.

One example of a multiplicative effect occurs in lung disease. Some people have a variation in the *alpha-antitrypsin* gene that moderately increases the chance that they will develop lung disease. However, if individuals with this variation also smoke regularly, they are at a greatly increased risk of getting lung disease, more than just smoking alone or just having the variation alone. Other cancer-related genes include tumor suppressor genes, whose particular alleles can influence the risk of developing breast cancer, for instance.

An Example of a Complex Trait

Alzheimer's disease is characterized by a gradual loss of brain function, usually starting with increasing loss of memory. Four different genes have been identified that can play a role in causing Alzheimer's disease. Rare mutations in each of three different genes (on different chromosomes) can each cause Alzheimer's disease. A common variation in a fourth gene, called *ApoE*, which occurs in approximately 35 percent of the population, increases the risk of developing Alzheimer's disease about threefold if a person has one copy of the allele (called e4), and about tenfold if they have two copies of the allele.

Taken together, these four genes account for less than half of all the genetic effects in Alzheimer's disease, indicating that additional genes that have not yet been identified are also important. In addition, environmental risk factors can have an effect. For example, taking certain anti-inflammatory medications, such as ibuprofen, reduces the risk of developing Alzheimer's disease, whereas severe head injury increases it. SEE ALSO ALZHEIMER'S DISEASE; BREAST CANCER; CARDIOVASCULAR DISEASE; DIABETES; EPISTASIS; INHERITANCE PATTERNS.

Jonathan L. Haines

Bibliography

Haines, Jonathan L., and Margaret A. Pericak-Vance, eds. *Approaches to Gene Mapping in Complex Human Diseases.* New York: John Wiley & Sons, 1998.

Strachan, Tom, and Andrew P. Read. *Human Molecular Genetics,* 2nd ed. New York: John Wiley & Sons, 1999.

Computational Biologist

A computational biologist is a scientist who develops and utilizes computational tools to analyze biological data. The Human Genome Project and other large sequencing projects have generated an extraordinary amount of data. Biologists are now faced with the challenge of extracting meaning from linear sequences composed of billions of **base pairs**. The work of computational biologists is indispensable for this task and for many other biological problems that lend themselves to computational solutions.

base pairs two nucleotides (either DNA or RNA) linked by weak bonds

A basic knowledge of molecular biology and genetics is important, enabling the computational biologist to understand the issues, to interpret the meaning of results, and to communicate with research biologists who work at the laboratory bench. The actual job duties, apart from proficiency in oral and written communication and reading the literature, are very much centered around working with computers. It is, therefore, paramount that a computational biologist be highly skilled in computer technology, with expertise in hardware, software, and the principles of programming. Daily tasks range from accessing public databases and using publicly available and commercial tools for analysis, to developing novel methods for solving problems.

Until recently, there were no formal educational opportunities in computational biology at the graduate level. Therefore, most of the current practitioners and authorities in the field have a combination of degrees at the graduate (master's or doctorate) and undergraduate (baccalaureate) levels in computer science and biology. Still others hold degrees in one of the two fields and are self-taught in the discipline for which they lack formal training. Most often such individuals will have an advanced degree in computer science and will be self-taught in biology, although the converse can also be true. Nonetheless, the ability to program in a high-level language such as C++ or PERL is a major qualification.

With the release of the working draft of the human genome in 2001, computational biology has come of age. Highly qualified individuals are in demand at academic, private, and government research institutes alike. Academic institutions have taken notice and have begun implementing certificate and

graduate programs in computational biology and bioinformatics. Many more programs are in the planning stages. Thus, it will probably be desirable for future practitioners to take advantage of these specialized formal training initiatives if they wish to remain competitive.

Working environments can be fairly diverse, but be it in an office, a cubicle, a computer lab, or a corner in a **wet lab**, it invariably will consist of a desk and a secure networked UNIX workstation. Depending on the structure of the organization, one may work as part of a team or independently. It is common to have frequent contact with both biologists and programmers. The computational biologist often acts as a liaison between the two. Job opportunities range from entry level to team leader to project manager or principle investigator.

Helping biologists extract meaning from their findings is extremely rewarding. Interesting findings should be published to help research move forward. As with most scientific fields, a computational biologist enjoys the opportunity to travel to international meetings. The salary ranges from approximately $45,000 to approximately $150,000, depending on experience and educational attainment, and on whether one chooses to work in the public or private sector. SEE ALSO BIOINFORMATICS; INFORMATION SYSTEMS MANAGER; INTERNET; STATISTICAL GENETICIST.

Judith E. Stenger

wet lab laboratory devoted to experiments using solutions, cell cultures, and other "wet" substances

Bibliography

Gibas, Cynthia, Per Jambeck, and James M. Fenton. *Developing Bioinformatics Computer Skills.* Sebastopol, CA: Oreilly & Associates, 2001.

Setubal, Joao C., and Joao Meidanis. *Introduction to Computational Molecular Biology.* Monterey, CA: Brooks/Cole Publishing, 1997.

Waterman, Michael S. *Introduction to Computational Biology: Maps, Sequences, and Genomes.* Boca Raton, FL: Chapman & Hall/CRC Press, 1995.

Conjugation

Conjugation is one of several mechanisms that bacteria use to transfer DNA, and hence new genetic information, between two cells. The other primary mechanisms are transformation, in which free DNA is transported across the cell membrane, and transduction, in which DNA is carried into the recipient cell by a bacterial virus.

The Role of Plasmids

conjugation a type of DNA exchange between bacteria

Conjugation is about as close as single cells come to engaging in sex, and some of the terminology used to describe the process reflects that similarity. Conjugation, or mating, is a process of genetic transfer that requires cell-to-cell contact. The genetic instructions for conjugation are encoded on a double-stranded, circular piece of DNA. The circular DNA exists in the bacterial cell entirely separate from the much larger bacterial chromosome. Scientists refer to this specialized, extrachromosomal piece of DNA as a conjugative plasmid or a "fertility factor." Cells that possess it are donor or "male" cells, and those that lack a conjugative plasmid are recipient or "female" cells.

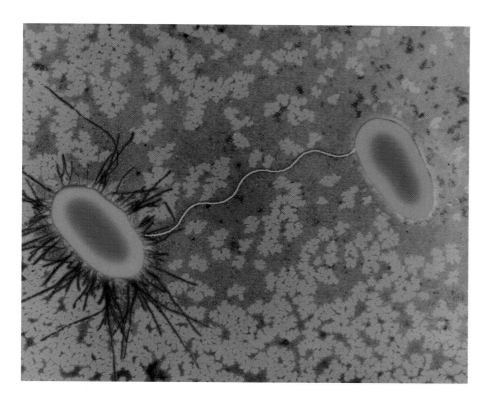

Conjugating *Escherichia coli* bacteria transfer DNA through cell-to-cell contact, which is made possible by the thread-like pilus that attaches to and reels in other cells.

There are multiple genes involved in the process of conjugation. Some of the genes code for a surface structure found on donor cells, the sex pilus. This is a threadlike tube made of protein. The sex pilus recognizes a specific attachment site on a recipient cell. When the donor cell comes near a recipient, the sex pilus attaches to the specific site and begins to retract, pulling the two cells together. This is a bit like throwing out a fishing line, hooking a fish, and pulling it into shore. The fishing analogy ends here, however. As the two cells draw close, their connection stabilizes and their outer membranes fuse together to allow the transfer of DNA from one cell to the other.

Only one of the two strands of DNA making up the plasmid passes through the fused membranes into the recipient cell. Thus DNA synthesis must occur in both donor and recipient to replace the missing strand in each. The genes encoding the enzymes responsible for this part of the conjugative process are also found on the plasmid. Once passage and synthesis are successfully completed, both donor and recipient cells contain a whole double-stranded, circular, conjugative plasmid. Thus there are now two donor cells when before there was only one. This process is so efficient that it can quickly change an entire population to donor cells. Some types of conjugative plasmids are transferred only between cells of the same species. Other types can be transferred across species; scientists call them promiscuous plasmids.

Large-Scale Gene Transfer

One of the two scientists who first described conjugation, Joshua Lederberg, ultimately won the Nobel Prize in medicine in 1958 for his discoveries concerning the organization of genetic material in bacteria. In 1946

"Rolling replication" transfers the F (fertility) plasmid between bacteria. The F plasmid contains the gene for formation of pili, the hairlike projection through which the plasmid passes. Adapted from Curtis, 1989.

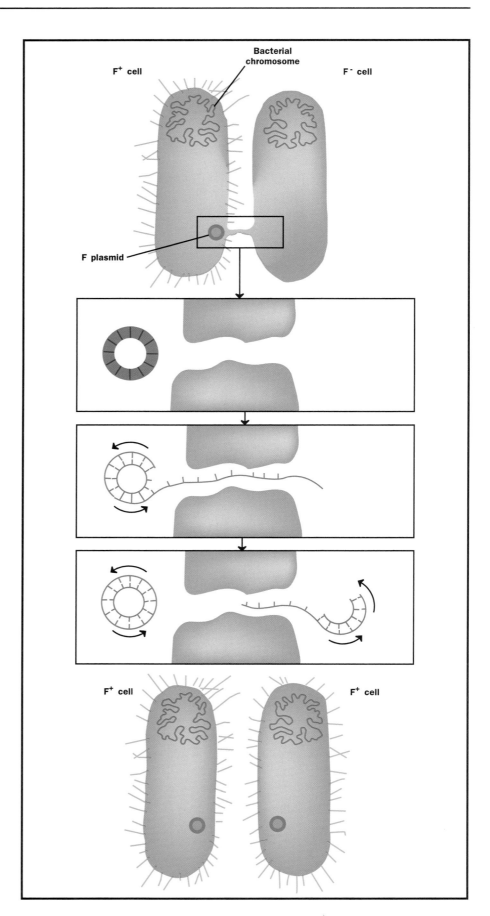

Lederberg and his colleague E. L. Tatum set out to determine whether a sexual process might occur in bacteria. The bacterial species he used in the experiments was *Escherichia coli*. This was fortuitous, as it turned out, because *E. coli* often contains a special kind of conjugative plasmid that has the ability to insert itself into the cell's chromosome. Once this happens, the donor cell can transfer to a recipient not only plasmid genes but also large numbers of chromosomal genes.

Lederberg worked with two different nutritional mutants of *E. coli*. One strain required biotin and methionine to grow; the other strain required threonine and leucine. Lederberg mixed the two strains together and then attempted to grow them without supplying any of the four nutrients. His hypothesis was that any cell able to grow without the four nutrients would have all four genes intact, and would thus have received the functioning genes from the other strain and incorporated them into its chromosome. The incorporation of the genes in this manner is called genetic recombination.

As he predicted, Lederberg's experiment yielded cells that did not require any of the nutrients to grow. In a second set of experiments, Lederberg showed that cell-to-cell contact was necessary for genetic recombination to occur. Over several years, he and other scientists discovered the mechanics of the entire process that we now call conjugation.

Antibiotic Resistance

From the human perspective, one of the significant consequences of a bacterium's ability to pass genetic information along to other cells via conjugation is its link to the widespread incidence of antibiotic resistance. The genes that encode for resistance to a variety of antibiotics like penicillin and tetracycline are commonly found on plasmids. When a population of susceptible bacteria is exposed to a given antibiotic, most of them will be killed. However, if the population contains cells with conjugative plasmids bearing the genes for resistance, they can rapidly spread the trait throughout the population. These plasmids are large and are often promiscuous, so that transfer of antibiotic resistance genes need not be restricted to cells of like species. In some cases, this has resulted in disease-causing bacteria that are resistant to almost every antibiotic available. For instance, antibiotic resistant tuberculosis bacteria are a significant public health threat in some metropolitan areas. SEE ALSO ANTIBIOTIC RESISTANCE; *ESCHERICHIA COLI* (*E. COLI* BACTERIUM); PLASMID; RECOMBINANT DNA; TRANSDUCTION; TRANSFORMATION.

Cynthia A. Needham

Bibliography

Curtis, Helena, and Sue Barnes. *Biology*, 5th ed. New York: Worth, 1989.

Madigan, Michael T., John M. Martinko, and Jack Parker. *Brock Biology of Microorganisms*, 9th ed. Upper Saddle River, NJ: Prentice Hall, 2000.

Robinson, Richard, ed. *Biology*. New York: Macmillan Reference USA, 2001.

Snyder, Larry, and Wendy Champness. *Molecular Genetics of Bacteria*. Washington, DC: ASM Press, 1997.

Conservation Biology: Genetic Approaches

Conservation biology is a multidisciplinary field dedicated to protecting global **biodiversity** and critical habitats. It incorporates biological approaches such as ecology, evolution, and behavior studies, as well as other disciplines, such as political science, law, economics, and cultural anthropology. One of the major goals of conservation biology is preserving critical habitats and the species that inhabit them. Genes can tell us something about how a particular habitat is used by species and populations. Genetic approaches are also used to identify and classify organisms and evaluate the extent of genetic diversity within a particular population.

biodiversity degree of variety of life

Categories of Threatened Populations

The International World Conservation Union (IUCN) provides definitions of terms used to describe the status of a species in the wild, based on a number of factors, including the size of a particular population, whether the population is declining in number, and, if so, the extent to which the trend will continue, as well as the threats the population faces. Genetic approaches can play an essential role in helping to evaluate populations, species, and species designations.

The following categories currently cover the range of definitions for species status according to IUCN: Extinct, Extinct in the Wild, Critically Endangered, Endangered, Vulnerable, Lower Risk, and Data Deficient.

Threatened. Populations are considered critically endangered, endangered, or vulnerable when there is considerable concern, based on available evidence or a high level of uncertainty, that the population will survive. With any of these classifications, the species or population of concern is considered to be facing a high to very high risk of extinction in the wild.

Conservation Genetics Applications

The practical applications of conservation genetics include analyzing fragmented populations in nature, determining units of conservation in nature, and monitoring captive populations. In general, conservation genetics integrates these types of information on particular species and populations to help prioritize areas for conservation.

Conservation genetics also plays a major role in guiding relocation and reintroduction efforts, in prioritizing species for conservation, and in designing captive-breeding programs. Identifying natural units based on systematics and population genetics allows researchers and wildlife officials to track organisms in the wild and in zoos, and it lets them identify parts or products of endangered and threatened organisms that are used in illegal trade. Conservation geneticists may use genetic techniques to determine, for example, if certain individuals in the pet-trade were illegally taken from the wild versus bred from permitted captive breeding programs.

Some of the most common issues addressed by genetic techniques in conservation are those confronting small or fragmented populations. Genetic approaches in these cases allow researchers to assess the variability in these populations, as well as to assess whether there is any history or future danger of loss of genetic variability. Genetics can help conservation biologists do viability analyses (tests of how likely that a population will survive over time) by testing **hypotheses** concerning how long genetic variation might persist into the future. This might be done by examining current levels of genetic variation in a species or population, and integrating these pieces of information with demographic and life history models to examine what happens to genetic variation over time.

The use of a conservation genetics approach may be an effective way for assessing the status of populations and species in the wild. Populations that decrease in number while becoming increasingly fragmented by loss of habitat in the wild can experience a loss of genetic variation that could have a severe impact on their fitness and survival. Conservation genetics permits scientists to assess the impacts of habitat fragmentation and loss in the wild using both theoretical and empirical methods. Results from these studies allow managers to evaluate the viability of populations and design protected areas for conservation.

Sometimes conservation initiatives are also concerned with the translocation or reintroduction of animals to areas where they have been extirpated or severely depleted. Such reintroduction or translocation measures require a detailed understanding of the genetics of the populations being reintroduced in order to ensure there is compatibility between populations as well as to maximize genetic variation and minimize the chance of inbreeding among related animals.

Determining the extent of genetic variation among captive populations in zoological parks and botanical gardens is also essential, because captive populations must have sufficient genetic variation so that they persist into the future without suffering from reduced fitness due to inbreeding and other effects associated with small populations. In some cases, captive populations may be viewed as a source for improving genetically or numerically depleted wild populations. However, they must be managed to minimize the effects of inbreeding. Accredited zoos, aquariums, and botanical gardens work to manage populations and establish conservation programs that strive to carefully manage the breeding of a species in captivity. The primary goal is to maintain a healthy and self-sustaining captive population that is both genetically diverse and demographically stable. Captive breeding specialists usually attempt to maximize the genetic health of a population by reconstructing pedigrees of the animals in the captive populations in order to understand and minimize how much inbreeding might occur.

The Tools of Conservation Genetics

The technique that revolutionized modern molecular genetics is the polymerase chain reaction (PCR). PCR has had major implications for conservation genetics. This technique allows the amplification of minute amounts of DNA, which can then be used for analysis. Amplification is critical for the study of endangered species, because biological samples may be obtained from nontraditional sources, such as hair, feathers, sloughed skin, or feces from which only small amounts of DNA are generally available. Once DNA has been obtained, conservation geneticists are able to use a wide arsenal of tools to characterize the genetics of endangered and threatened species and populations.

hypotheses testable statements

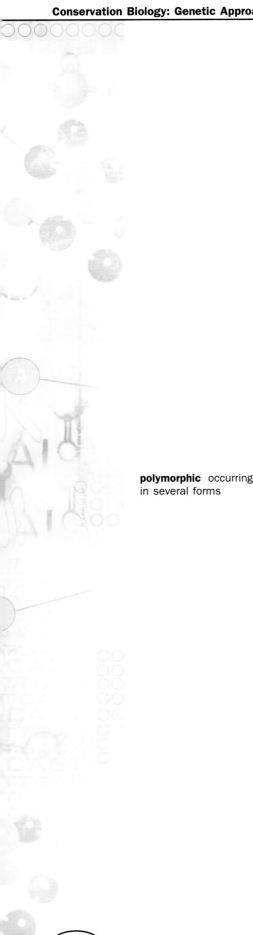

When conservation genetics is used to decipher the evolutionary relationships among species, DNA sequence comparisons are often made. Sequencing a region of a gene and properly analyzing the data may lead to novel findings. Based on analyzing underlying genetic variation, there could be evidence to suggest that a revision of numbers of species might be warranted. What was once considered a single species with two populations, for example, might actually merit consideration for separate species status. The level of genetic differentiation detected could have significant implications for how they are protected in the wild and the measures that must be taken by local, state, and federal authorities. Alternatively, the data may indicate that these populations are not sufficiently genetically different to merit separate species designations. DNA sequences are also used to aid in diagnosing natural units for conservation in the wild. Detecting fixed nucleotide characters among DNA sequences between well-sampled populations can provide sufficient evidence for defining units of conservation that potentially merit separate species status under the Phylogenetic Species Concept. Several studies have used DNA sequence polymorphism shared by certain units to the exclusion of other groups to unequivocally define or diagnose species.

Population-level analyses use DNA sequences as well as another set of molecular markers, called microsatellites, a type of repetitive DNA element. Microsatellites are used to address many conservation genetics questions. They are short, tandem-repeated motifs of DNA sequences, such as a dinucleotide repeat (e.g., $(AT)_n$), that usually vary in the number of repeats in a particular stretch of DNA. They are distributed throughout the genomes of plants and animals, are inherited in a Mendelian fashion, and have been found to be highly **polymorphic**. These genetic markers have proven to be useful in population studies for such purposes as estimating gene flow between populations, describing the genetic variation within and between populations, and examining the effects of hybridization between species. They are used in pedigree analysis to identify individuals based on a DNA sample, and they are used to decipher mating strategies and degrees of relatedness among members of a population.

polymorphic occurring in several forms

Implication of Genetics for Conservation in the Wild

In the wild, populations that once were large and widespread are increasingly being reduced to small and fragmented isolates due to human activities. Habitat loss and fragmentation trigger processes that further threaten populations. Small populations often face greater demographic and genetic risks relative to large populations. When populations become fragmented and small, the genetic diversity of a population may be greatly affected. Conservation geneticists focus on the impact of such severe reductions, called bottlenecks, on endangered species.

When a bottleneck occurs, there is an increased chance of breeding among close relatives. This is termed inbreeding, and it may result in a reduction in fitness due to the expression of deleterious genes, in a process known as inbreeding depression. Inbreeding and the loss of genetic variation in small populations can lead to a genetically reduced or homogeneous population that is more sensitive to diseases and to the effects of habitat alteration. The interaction between genetic and demographic declines has been termed "extinction vortex." We include below several real examples of the use of genetics in conservation biology.

Bottlenecks, Cheetahs, and Right Whales. In small populations, inbreeding depression may be more common because random mating is less likely and breeding among related animals may have a greater cumulative effect. Low genetic diversity and inbreeding is not always deleterious, and some small populations may be stable while permanently maintaining low levels of genetic diversity. In general, however, avoidance of inbreeding is a major goal in the management of small populations, since it has been shown to cause a reduction in fitness in captive populations of endangered species.

Perhaps the most famous case of a putative bottleneck being examined in conservation genetics is the cheetah, as examined by Stephen J. O'Brien, a molecular geneticist at the National Cancer Institute. In this 1980s study, cheetahs were shown to have extremely low levels of genetic diversity, which the researchers attributed to a bottleneck that happened less than ten thousand years ago and that may have left only a few females alive. The bottleneck was so extreme that even the usually highly diverse genes of our immune system, genes of the major histocompatibility complex, showed amazingly low levels of diversity. The extreme loss of genetic diversity was attributed to difficulties associated with the species' breeding in captivity and in the wild, abnormal sperm counts, and susceptibility to disease. While it remains controversial whether cheetah populations went through a bottleneck and the extent to which reproductive issues can be attributed to reduced genetic variation, the example remains one of the most prominent in the field of conservation.

Other species, such as the North Atlantic right whale, have faced demographic decline and extremely low levels of genetic diversity since the end of legal commercial hunting for whales at the beginning of the twentieth century. The North Atlantic right whale has maintained a low level of genetic diversity since the 1930s, but recent studies suggest that some additional genetic variation may eventually be lost.

Units of Conservation and DNA Sequences. Using genetic data to evaluate or define species and/or units of conservation can also lead to novel findings and enhance conservation management. Right whales are found in the North Atlantic, North Pacific, and southern oceans. For over a hundred years, they have been considered as two species—one in the north (in the Pacific and Atlantic Oceans) and one in the south. DNA diagnosis methods have corroborated that the southern right whales are distinct from all the others. However, the DNA data also clearly demonstrate that the North Pacific and North Atlantic whales are distinct from each other and warrant distinct species status. The ramifications for the conservation plan of these whales have been taken into consideration by the appropriate management authorities, who have developed a revised plan for naming and protecting three distinct species of right whales.

DNA Detectives and Endangered Species. As discussed above, conservation genetics can aid in the identification of endangered and threatened animals that are traded illegally as commercial products. Researchers at the Wildlife Conservation Society used species-identification methods to detect caiman crocodile tissue in leather products and thus thwart their illegal importation into the United States. Other scientists have used species identification methods to detect whale meat in Japanese fish markets and thus have had an impact on the policing of whale hunting.

Rob DeSalle, a curator at The American Museum of Natural History, and colleagues have recently used species identification methods to verify or reject the labeling of caviar origin. Such tests have been instrumental in getting sturgeons (the source of caviar) listed on the Convention on International Trade in Endangered Species of Wild Fauna and Flora. Prior to the test's development, there was no way to verify the contents of a container of caviar, leading authorities to be wary of prosecuting the illegal importation of caviar. With the development of species-identification procedures based on analyzing the DNA from single caviar eggs, enforcement of importation regulations became possible. SEE ALSO CONSERVATION BIOLOGIST; DNA PROFILING; GENE FLOW; POPULATION BOTTLENECK; REPETITIVE DNA ELEMENTS.

Howard C. Rosenbaum and Rob DeSalle

Bibliography

DeSalle, R., and V. Birstein. "PCR Analysis of Black Caviar." *Nature* 381 (1996): 97–198.

Conservation Geneticist

A career path in conservation biology can be taken in many different directions, many of which are not centered around genetics. In general, the science of conservation biology draws upon many traditional academic disciplines, combining them in a science dedicated to maintaining Earth's **biodiversity**. In recent years, however, genetics has taken on an increasingly important role in the field, for a genetic approach can be used to address questions concerning conservation of populations, species, and their habitats.

A Variety of Career Directions

Conservation biologists, and more specifically conservation geneticists or molecular ecologists, are often multi-disciplinary scientists who combine training in ecology, evolutionary biology, conservation, and molecular biology. This unique combination of skills enables these scientists to design field studies aimed at collecting biological specimens needed for research involving genetics, or to advise others who are planning such studies.

Genetics-based conservation involves both research and the implementation of that research's findings. Some conservation geneticists focus their research on identifying natural population units for conservation based on genetic criteria, in an effort to maximize genetic diversity. Others seek to establish **taxonomical** or population priorities for conservation efforts. Alternatively, a conservation geneticist may monitor trade in endangered species, guide captive-breeding programs, or work on the re-introduction of selected species to habitats from which they are in danger of disappearing. Results from their studies are often central for management decisions regarding the viability and protection of threatened or endangered populations and the designation of critical habitats for conservation.

The discovery of the polymerase chain reaction has enabled conservation biologists to study **endangered** or **threatened** species and

biodiversity degree of variety of life

taxonomical derived from the science that identifies and classifies plants and animals

endangered in danger of extinction throughout all or a significant portion of a species' range

threatened likely to become an endangered species

their habitats in ways that were previously unimaginable. The use of non-invasively collected samples, such as feces or hair, enables the conservation geneticist to acquire an extraordinary amount of information about life history, **demography**, distribution, and diversity of rare species without directly observing them. In addition to such studies, with research drawn from skilled geneticists, ecologists, and behavioral biologists, conservation geneticists are now faced with a new set of high-tech approaches derived from the emerging field of genomics. The challenge for these biologists will be to evaluate whether these advanced technologies can be used effectively to promote biodiversity, conservation, and habitat protection.

Conservation biologists are likely to lead projects such as this whooping crane breeding program located in Patuxent, Maryland.

demography aspects of population structure, including size, age distribution, growth, and other factors

Becoming a Conservation Biologist

To begin a career in conservation biology, a student should expect to carry out advanced study in one or more of the relevant sciences at the graduate level. Many researchers have completed doctorates in biology, genetics, or conservation, followed by several years of post doctoral training. Projects that are initiated as part of graduate work may develop into entire research programs in this field.

The key to success for a conservation geneticist goes far beyond designing and carrying out a research program. Equally important is how the program implements the results of that research, and it is often this

implementation that is used as a benchmark in evaluating the program's value. Research in this discipline may be carried out at academic centers, museums, nongovernmental organizations, and government institutions.

The rewards of a career in conservation biology are many. First, the multidisciplinary approach affords a researcher the chance to develop a unique background and breadth of diverse skills. Second, the work has a direct and clearly visible value for biodiversity conservation programs. Salaries for a conservation geneticist depend on the researcher's level of education and the type of institution in which he or she works, but most are comparable to the salaries offered to biology professors teaching at the university level. SEE ALSO CONSERVATION BIOLOGY: GENETIC APPROACHES; POLYMERASE CHAIN REACTION.

Howard C. Rosenbaum and Rob DeSalle

Bibliography

Gerber, Leah R., Douglas P. DeMaster, and Simona P. Roberts. "Measuring Success in Conservation." *American Scientist* 88 (2000) 4: 316–324.

Crick, Francis

British Biophysicist
1916–

Francis Crick is the co-discoverer, with James Watson, of the structure of DNA. He has remained a significant contributor to theoretical biology since that discovery.

Education and Training

Crick was born in Northampton, England, in 1916. He studied physics at University College in London until the outbreak of the Second World War. He then joined the British Admiralty Research Laboratory, where he contributed to the development of radar for tracking enemy planes, and magnetic mines used in naval warfare.

During this time, Crick read *What is Life?*, a book by the physicist Erwin Schrödinger. Schrödinger's book popularized the work of physicist Max Delbrück, who had begun to apply the analytical tools of physics to inquire what a gene was and how it might behave. Like many other physicists at that time, Crick was excited by Delbrück's approach, and turned his attention to biochemistry and biological physics. While he knew a great deal of physics, he knew very little chemistry or biology at that time. In 1949 he began research at the Cavendish Laboratory in Cambridge, England, using **X-ray crystallography** to study the three-dimensional structures of proteins. At that time, Crick wrote that he was interested in "the borderline between the living and the nonliving, as typified by, say, proteins, viruses, bacteria and the structure of chromosomes. The eventual goal, which is somewhat remote, is the description of these activities in terms of their structure, i.e., the spatial distribution of their constituent atoms" (Judson, 88).

X-ray crystallography use of X rays to determine the structure of a molecule

The Structure of DNA

Almost ten years earlier, it had been shown that genes encode proteins, but the chemical nature of the gene remained unknown. Genes were presumed to be composed of DNA (deoxyribonucleic acid), at least in part, but how DNA might encode hereditary information, and whether it acted alone or in partnership with proteins, was a complete mystery. Crick saw that the solution to the mystery lay in discovering the structure of DNA, whose linearity he guessed corresponded to the linear **amino acid** chains of which proteins are made.

amino acid a building block of protein

In 1951 a 23-year-old American named James Watson joined the Cavendish Laboratory. Watson and Crick got along well, and they decided to work together on the structure of DNA. DNA was known to be composed of nucleotide subunits, each of which had a sugar (deoxyribose), a phosphate, and a nitrogenous base. The sugars were known to alternate with phosphates to make long strands, off of which the bases projected. The bases came in four types: adenine, thymine, cytosine, and guanine (A, T, C, and G). Shortly before Crick and Watson began to collaborate, American biochemist Erwin Chargaff had discovered that across a wide range of species, the amount of adenine in an organism's DNA always equaled the amount of thymine, and the amount of cytosine always equaled the amount of guanine.

Crick and Watson proceeded to build models of the nucleotides, which they attempted to fit together in accordance with what was known from experimental data. The most important data came from X-ray images of DNA that had been generated by Rosalind Franklin, who also worked at the Cavendish. Using this information, they constructed a model in which the two sugar-phosphate strands wind around each other to form a double helix, their bases projecting inward, like the stair treads of a broad spiral staircase. The two strands are held together and stabilized by the hydrogen bonding between the bases across the interior. These weak chemical attractions, they discovered, are strongest when adenine projects across to meet a thymine, and guanine a cytosine, explaining the ratios discovered by Chargaff. They published their model in 1953. Watson and Crick received the Nobel Prize in physiology or medicine in 1962 for this work, along with Maurice Wilkins of the Cavendish Lab.

After the publication of DNA's structure, Crick turned his attention to understanding the coding function of DNA. He and Watson proposed that the order of bases in a gene encoded the order of amino acids in a protein. Over the next decade, the details of this insight were worked out by a large group of scientists, including Crick, Watson, Sydney Brenner, George Gamow, Seymour Benzer, Marshall Nirnberg, and Har Gobind Khorana. As part of this work, Crick hypothesized the existence of an "adaptor" that intervened between DNA and amino acids. This led to the discovery of messenger RNA and transfer RNA, which serve this function.

Later Work

Crick received his Ph.D. in 1954. He remained with the Medical Research Council at the Cavendish Laboratory, and became head of the Division of Molecular Genetics in 1962, continuing to work closely with Sydney Brenner. He turned his attention to embryology in the mid-1960s, and in 1975 he moved to the Salk Institute in La Jolla, California, to pursue

neurobiology, an interest that had vied with molecular biology from the very beginning of his career. At the Salk Institute, in collaboration with Christof Koch, he studies the neural correlates of conscious visual experience, seeking to understand how neuron firing patterns correspond to the conscious experience of seeing. SEE ALSO DELBRÜCK, MAX; DNA; DNA STRUCTURE AND FUNCTION, HISTORY; RNA; WATSON, JAMES.

Richard Robinson

Bibliography

Crick, Francis. *What Mad Pursuit: A Personal View of Scientific Discovery*. New York: Basic Books, 1988.

Judson, Horace F. *The Eighth Day of Creation*, expanded ed. Cold Spring Harbor, NY: Cold Spring Harbor Press, 1996.

Crossing Over

Crossing over, or recombination, is the exchange of chromosome segments between nonsister chromatids in meiosis. Crossing over creates new combinations of genes in the **gametes** that are not found in either parent, contributing to genetic diversity.

gametes reproductive cells, such as sperm or eggs

Homologues and Chromatids

All body cells are diploid, meaning they contain pairs of each chromosome. One member of each pair comes from the individual's mother, and one from the father. The two members of each pair are called homologues. Members of a homologous pair carry the same set of genes, which occur in identical positions along the chromosome. The specific forms of each gene, called alleles, may be different: One chromosome may carry an allele for blue eyes, and the other an allele for brown eyes, for example.

Meiosis is the process by which homologous chromosomes are separated to form gametes. Gametes contain only one member of each pair of chromosomes. Prior to meiosis, each chromosome is replicated. The replicas, called sister chromatids, remain joined together at the centromere. Thus, as a cell starts meiosis, each chromosome is composed of two chromatids and is paired with its homologue. The chromatids of two homologous chromosomes are called nonsister chromatids.

Meiosis occurs in two stages, called meiosis I and II. Meiosis I separates homologues from each other. Meiosis II separates sister chromatids from each other. Crossing over occurs in meiosis I. During crossing over, segments are exchanged between nonsister chromatids.

Mechanics of Crossing Over

The pairing of homologues at the beginning of meiosis I ensures that each gamete receives one member of each pair. Homologues contact each other along much of their length and are held together by a special protein structure called the synaptonemal complex. This association of the homologues may persist from hours to days. The association of the two chromosomes is called a bivalent, and because there are four chromatids involved it is also called a tetrad. The points of attachment are called chiasmata (singular, chiasma).

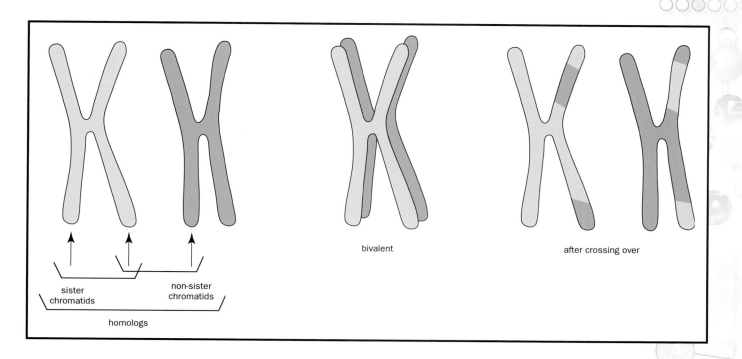

sister chromatids

non-sister chromatids

homologs

bivalent

after crossing over

The pairing of homologues brings together the near-identical sequences found on each chromosome, and this sets the stage for crossing over. The exact mechanism by which crossing over occurs is not known. Crossing over is controlled by a very large protein complex called a recombination nodule. Some of the proteins involved also play roles in DNA replication and repair, which is not surprising, considering that all three processes require breaking and reforming the DNA double helix.

One plausible model supported by available evidence suggests that crossing over begins when one **chromatid** is cut through, making a break in the double-stranded DNA (recall that each DNA strand is a double helix of nucleotides). A nuclease enzyme then removes nucleotides from each side of the DNA strand, but in opposite directions, leaving each side with a single-stranded tail, perhaps 600 to 800 nucleotides long.

One tail is then thought to insert itself along the length of one of the nonsister chromatids, aligning with its complementary sequence (i.e., if the tail sequence is ATCCGG, it aligns with TAGGCC on the nonsister strand). If a match is made, the tail pairs with this strand of the nonsister chromatid. This displaces the original paired strand on the nonsister chromatid, which is then freed to pair with the other single-stranded tail. The gaps are filled by a DNA **polymerase enzyme**. Finally, the two chromatids must be separated from each other, which requires cutting all the strands and rejoining the cut ends.

The Consequences of Crossing Over

A chiasma occurs at least once per chromosome pair. Thus, following crossing over, at least two of the four chromatids become unique, unlike those of the parent. (Crossing over can also occur between sister chromatids; however, such events do not lead to genetic variation because the DNA sequences are identical between the chromatids.) Crossing over helps

During crossing over, homologous chromosomes pair up to form a bivalent. Nonsister chromatids exchange segments.

chromatid a replicated chromosome before separation from its copy

polymerase enzyme enzyme complex that synthesizes DNA or RNA from individual nucleotides

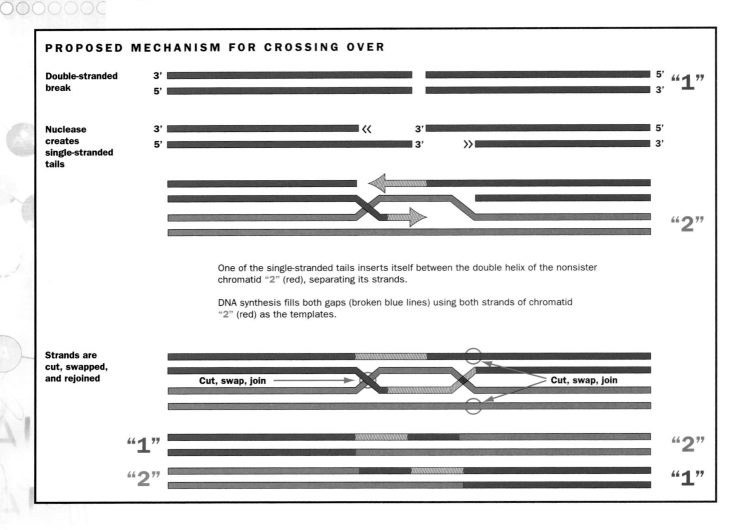

PROPOSED MECHANISM FOR CROSSING OVER

Double-stranded break · 3' · 5' · 5' · 3' · "1"

Nuclease creates single-stranded tails · 3' · 5' · 3' · 3' · 5' · 3'

"2"

One of the single-stranded tails inserts itself between the double helix of the nonsister chromatid "2" (red), separating its strands.

DNA synthesis fills both gaps (broken blue lines) using both strands of chromatid "2" (red) as the templates.

Strands are cut, swapped, and rejoined · Cut, swap, join · Cut, swap, join

"1" · "2"
"2" · "1"

The exact mechanism for crossing over is unknown. It is believed to involve cutting of one chromatid, removal of nucleotides from both strands, insertion of one strand into the matching region of the nonsister chromatid, and further cutting, swapping of segments, and rejoining of the strands. Adapted from <http://www.ultranet.com/~jkimball/BiologyPages/c/crossingover.html>.

to preserve genetic variability within a species by allowing for virtually limitless combinations of genes in the transmission from parent to offspring.

The frequency of recombination is not uniform throughout the genome. Some areas of some chromosomes have increased rates of recombination (hot spots), while others have reduced rates of recombination (cold spots). The frequency of recombination in humans is generally decreased near the centromeric region of chromosomes, and tends to be greater near the telomeric regions. Recombination frequencies may vary between sexes. Crossing over is estimated to occur approximately fifty-five times in meiosis in males, and about seventy-five times in meiosis in females.

X–Y Crossovers and Unequal Crossovers

The forty-six chromosomes of the human diploid genome are composed of twenty-two pairs of autosomes, plus the X and Y chromosomes that determine sex. The X and Y chromosomes are very different from each other in their genetic composition but nonetheless pair up and even cross over during meiosis. These two chromosomes do have similar sequences over a small portion of their length, termed the pseudoautosomal region, at the far end of the short arm on each one.

The pseudoautosomal region behaves similarly to the autosomes during meiosis, allowing for segregation of the sex chromosomes. Just proximal to the pseudoautosomal region on the Y chromosome is the *SRY* gene (sex-determining region of the Y chromosome), which is critical for the normal development of male reproductive organs. When crossing over extends past the boundary of the pseudoautosomal region and includes this gene, sexual development will most likely be adversely affected. The rare occurrences of chromosomally XX males and XY females are due to such aberrant crossing over, in which the Y chromosome has lost—and the X chromosome has gained—this sex-determining gene.

Most crossing over is equal. However, unequal crossing over can and does occur. This form of recombination involves crossing over between nonallelic sequences on nonsister chromatids in a pair of homologues. In many cases, the DNA sequences located near the crossover event show substantial sequence similarity. When unequal crossing over occurs, the event leads to a deletion on one of the participating chromatids and an insertion on the other, which can lead to genetic disease, or even failure of development if a crucial gene is missing.

Crossing Over as a Genetic Tool

Recombination events have important uses in experimental and medical genetics. They can be used to order and determine distances between **loci** (chromosome positions) by genetic mapping techniques. Loci that are on the same chromosome are all physically linked to one another, but they can be separated by crossing over. Examining the frequency with which two loci are separated allows a calculation of their distance: The closer they are, the more likely they are to remain together. Multiple comparisons of crossing over among multiple loci allows these loci to be mapped, or placed in relative position to one another.

loci sites on a chromosome (singular, locus)

Recombination frequency in one region of the genome will be influenced by other, nearby recombination events, and these differences can complicate genetic mapping. The term "interference" describes this phenomenon. In positive interference, the presence of one crossover in a region decreases the probability that another crossover will occur nearby. Negative interference, the opposite of positive interference, implies that the formation of a second crossover in a region is made more likely by the presence of a first crossover.

Most documented interference has been positive, but some reports of negative interference exist in experimental organisms. The investigation of interference is important because accurate modeling of interference will provide better estimates of true genetic map length and intermarker distances, and more accurate mapping of trait loci. Interference is very difficult to measure in humans, because exceedingly large sample sizes, usually on the order of three hundred to one thousand fully informative meiotic events, are required to detect it. SEE ALSO DNA POLYMERASES; DNA REPAIR; LINKAGE AND RECOMBINATION; MEIOSIS; MENDEL, GREGOR.

Marcy C. Speer

Bibliography

Strachan, Tom, and Andrew P. Read. *Human Molecular Genetics.* New York: Wiley-Liss, 1996.

Cycle Sequencing

DNA sequencing is the process of determining the order of **nucleotides** on a segment of DNA. Cycle sequencing is a method used to increase the sensitivity of the DNA sequencing process and permits the use of very small amounts of DNA starting material. This is accomplished by using a temperature cycling process similar to that employed in the polymerase chain reaction.

nucleotides the building blocks of RNA or DNA

The Chain Termination Method

The most popular approach to sequencing is the chain termination method, developed in 1977 by Fred Sanger. This technique makes use of a DNA synthesis reaction and a unique form of a base, called a dideoxynucleotide, that lacks the 3′ hydroxyl chemical group involved in forming the link between nucleotides in a DNA chain. A dideoxynucleotide can be added to a growing chain, but, once incorporated, no further nucleotides can be linked to it. Thus, chain growth is terminated.

In a reaction, the concentration of dideoxynucleotides is optimized so that all possible chain lengths are generated. In automated DNA sequencing, the newly formed chain fragments are marked with four fluorescent dyes, each of which corresponds to one of the four DNA nucleotides (A, C, T, and G). The fragments are then separated by a technique known as gel electrophoresis, during which the dyes are excited and detected by the automated DNA sequencing instrument. In this way, the identity of each successive terminating nucleotide is determined, revealing the sequence of the entire chain.

In a sequencing reaction, all the components needed for the synthesis of DNA are present. These include the DNA to be sequenced (the template) and a short (17 to 28 bases in length), single-stranded piece of DNA (the primer), which attaches itself to a specific site on the template and acts as a starting point for the synthesis of a new DNA strand.

complementary matching opposite, like hand and glove

In both a DNA synthesis reaction and in the chain termination method of DNA sequencing, one strand of the template DNA is copied to form a new, **complementary** strand. An A base on the template strand, for example, directs the addition of its complementary base T into the growing chain. Likewise, a C base on the template strand directs the addition of its complementary base G into the corresponding position of the new chain. During chain growth, an enzyme called DNA polymerase links one nucleotide to the next, extending the new strand until it either reaches the end of the template strand, or, in DNA sequencing, until a dideoxynucleotide is incorporated.

The Cycle Sequencing Technique

electrophoresis technique for separation of molecules based on size and charge

In cycle sequencing, a reaction is taken through several steps designed to prepare the template for copying, allow for initiation of DNA synthesis, and generate the terminated DNA chains needed for **electrophoresis** and sequence determination. The first step in the process is called denaturation, in which double-stranded template DNA is converted into its single-stranded form. This is accomplished by heating the template to between 94 °C and 98 °C, a temperature high enough to break the hydrogen bonds between the complementary bases holding the two strands together.

As a single-stranded molecule, the template's bases are now exposed, and are free to interact with the sequencing primer. The primer, in a step called annealing, locates and attaches itself to its complementary site on the template. Thus, Ts bind to As and Cs bind to Gs. However, primer annealing will only occur at a temperature where **hydrogen bonds** can form between the primer and **template** strands, usually between 40 °C and 65 °C. Because the high temperature used for the **denaturation** step in each cycle would destroy most DNA polymerase enzymes, a special heat-stable enzyme must be used in the annealing stage, one that remains active even after repeated exposure to very high temperatures. The enzyme most commonly used at this point in cycle sequencing is *Taq*, isolated from *Thermus aquaticus*, a bacterium that lives in the hot springs of Yellowstone National Park.

In the final step of the reaction, DNA polymerase extends the annealed primer by sequentially adding on to its end bases that are complementary to those on the template. It is during this extension step of a DNA sequencing reaction that random incorporation of a dideoxynucleotide can occur, terminating chain growth. All three of these steps, taken together, represent one round, or cycle, of a DNA synthesis reaction. By repeating a cycle over and over again, the amount of each fragment made in the reaction can be substantially increased. Since each fragment carries fluorescent dyes, increasing the number of copied fragments also increases the strength of the fluorescent signal. Cycle sequencing, therefore, greatly improves the sensitivity of the sequencing reaction, and even very small amounts of starting DNA sample can be used as template. SEE ALSO AUTOMATED SEQUENCER; GEL ELECTROPHORESIS; POLYMERASE CHAIN REACTION; SANGER, FRED; SEQUENCING DNA.

Frank H. Stephenson and Maria Cristina Abilock

hydrogen bonds weak bonds between the H of one molecule or group and a nitrogen or oxygen of another

template master copy

denaturation destruction of the structure of

Bibliography

Craxton, Molly. "Linear Amplification Sequencing, a Powerful Method for Sequencing DNA." In *METHODS: A Companion to Methods in Enzymology* 3 (1991): 20-26.

Cystic Fibrosis

Cystic fibrosis (CF) is one of the most common genetic diseases and one of the best known to the general public. There are approximately one thousand new cases of CF in the United States every year, and approximately thirty thousand people in the country are currently affected. In many ways, it has come to be viewed, along with sickle-cell disease, as a prototypical recessive genetic disorder, one that can teach us a great deal about the molecular basis of disease, population genetics, and delivery of genetic screening services.

Clinical Features

CF is a multisystem disease, affecting a number of different organs and tissues throughout the body. Most of these manifestations have in common the production of abnormally viscous secretions from glands and surface **epithelial cells**. In the lung the mucus secretions from the bronchial epithelial cells are unusually thick. They are difficult to clear properly from the

epithelial cells one of four tissue types found in the body, characterized by thin sheets and usually serving a protective or secretory function

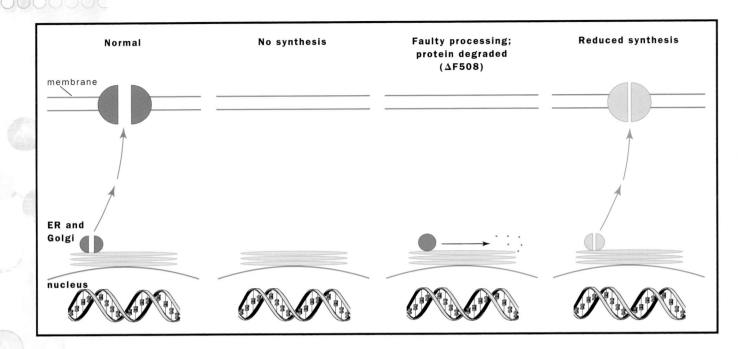

There are three types of defects in CFTR production that can cause cystic fibrosis. The most common mutation, ΔF508, causes faulty protein processing and degradation of the protein before it reaches the cell membrane. Other defects are also possible.

airway passages and, instead, tend to collect and obstruct the bronchial tree, while providing a perfect culture medium for dangerous bacteria. Over time, repeated bacterial infections damage and destroy the lung tissue, leading to chronic breathing problems and, eventually, to the loss of viable lung function. Indeed, the pulmonary manifestations of the disease are the main cause of death in most CF patients.

Obstruction in other organs is also seen. In the pancreas it leads to an insufficiency of pancreatic enzymes and malabsorption during digestion; in the nose and sinuses, it produces chronic sinusitis; and in the intestines, in a small minority of newborn infants with CF, it produces an often fatal condition called meconium ileus. Altered secretions also occur in the sweat glands, so that the sweat of CF patients has an abnormally high salt content. In fact, in the early days of medicine, the diagnosis of CF was often made by licking the skin and tasting the sweat! One additional mysterious clinical feature of CF occurs only in men with the disorder: They are infertile due to blockage or congenital absence of the vas deferens, the tube through which sperm pass prior to ejaculation.

Laboratory Diagnosis

Since the clinical symptoms of CF are so varied, diagnosis is aided by laboratory tests. For many years the only definitive test available was sweat chloride analysis, which detects and quantifies the abnormally high salt concentrations in the sweat of CF patients. Since the causative gene was discovered in 1989, as described below, patients can now also be diagnosed by DNA testing, which detects mutations in the gene.

Mode of Inheritance

CF is a classical autosomal recessive disease, in which affected patients are born to parents who are both carriers—that is, they have a mutation in one

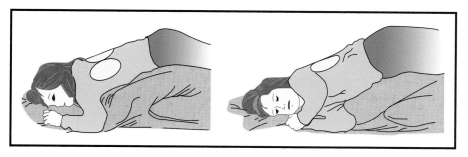

Chest physical therapy is the traditional airway clearance technique for cystic fibrosis. It includes postural drainage (positioning to allow movement of secretions) and percussion and vibration on the chest wall to loosen secretions. The circles indicate the sites of percussion. Adapted from <http://www2medsch.wisc.edu/childrenshosp/CF/cfpages/cpt2.html>.

of their two CF genes. Carrier couples have a one-in-four risk of producing an affected child with each pregnancy. The carriers themselves are completely asymptomatic and even have normal sweat chloride tests.

Treatment

There is still no cure for this disease, but supportive therapies have improved markedly in the last twenty years. The recurrent lung infections are now much better controlled with powerful, new-generation antibiotics, though, unfortunately, many patients eventually develop infections with drug-resistant bacteria. A variety of approaches are used to break up and remove the thick mucus secretions in the lung, including mucolytic (mucus-breaking) agents, bronchodilators, and chest physical therapy. One type of mucolytic treatment is DNase, a DNA-destroying **enzyme** that breaks up the long, sticky DNA strands left by dying cells. Some patients with end-stage lung disease can be rescued by lung transplantation. Patients with pancreatic obstruction can be managed with pancreatic enzyme supplements, and affected men with vas deferens obstruction have been able to father children by a technique called sperm aspiration, followed by **in vitro** fertilization.

enzyme a protein that controls a reaction in a cell

in vitro "in glass"; in lab apparatus, rather than within a living organism

The ultimate hope for the cure of CF lies in gene replacement therapy. A number of clinical trials are under way, most attempting to deliver the normal CF gene to the bronchial epithelium by aerosol spray, using a viral **vector** (usually adenovirus, a common respiratory virus that naturally targets the desired tissue). Thus far the attempts have not been completely successful, as most patients develop an immune response against the virus during the course of therapy. But with the median life expectancy of CF patients now at thirty years just through conventional therapies, the hope is that many CF patients alive today will survive long enough to avail themselves of gene therapy once it is perfected.

vector carrier

The Cystic Fibrosis Gene and CFTR Protein

The identification of the causative gene for CF in 1989 represented one of the great triumphs of molecular genetic research up to that time. With the gene's identification having preceded the official start of the Human Genome Project by one year, the search for the CF gene proceeded without the benefits of the fully mapped genome that we have today. Hence,

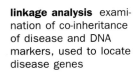

linkage analysis examination of co-inheritance of disease and DNA markers, used to locate disease genes

positional cloning the use of polymorphic genetic markers ever closer to the unknown gene to track its inheritance in CF families

some of the techniques used to identify the CF gene, such as "gene walking" and "gene jumping," are no longer used extensively.

The CF gene was identified through **linkage analysis** and **positional cloning**. Whereas nowadays the map of the human genome is saturated with these markers, which serve as convenient "signposts" in gene mapping studies, this was not the case when the CF mapping was done, and more laborious, brute-force techniques such as those mentioned above had to be employed. Thus, it was dramatic news indeed when the causative gene was found on chromosome 7.

As might have been expected based on the secretory defects in the disease, the gene, dubbed "cystic fibrosis transmembrane conductance regulator" (*CFTR*), encodes an ion-channel protein in epithelial cell membranes. The gene is quite large—250,000 nucleotides—and the spectrum of mutations in CF patients continues to grow. At the time of this writing, more than 950 different mutations have been reported. Most of these are quite rare and may only be found in individual families. A few are more common, most notably a three-nucleotide deletion of codon 508, called ΔF508, which is found in approximately 70 percent of Caucasian CF carriers. Several others are present in 1 percent to 3 percent of carriers, while the remainder are very rare, except for some that are found at higher frequency in particular ethnic and racial groups (such as W1282X in the Ashkenazi-Jewish population and 3120+1G→A in the African-American population).

The ΔF508 mutation in the *CFTR* gene deletes a phenylalanine amino acid from the final protein. Like other membrane proteins, *CFTR* is made at the endoplasmic reticulum in the interior of the cell, and must be transported to the plasma membrane to function. The absence of this amino acid results in improper folding of the *CFTR* protein within the endoplasmic reticulum, which causes it to be degraded by the cell's protein-recycling machinery before it reaches the membrane. Some of the less common mutations prevent any protein synthesis by introducing a stop codon into the gene, while others allow the protein to reach the membrane but without functioning properly.

The *CFTR* protein forms a pore to allow chloride ions to pass through the plasma membrane. The full range of functions served by this pore is not known, but the sticky secretions of CF are believed to result when chloride ions in the salty fluid secreted by the epithelial cells cannot be recovered by the membrane protein.

Cystic Fibrosis DNA Testing and Screening

The discovery of the *CFTR* gene raised hopes that the detection of mutations at the DNA level could supplement the traditional sweat test for CF diagnosis and, more importantly, might be used to identify carriers in the general population so that they could be offered genetic counseling. Unfortunately, these goals have been hampered by the large number of possible mutations in the gene, since present-day DNA tests can detect only a small subset of them. As in most recessive diseases, the vast majority of carriers have no family history of the disorder and do not discover that they are carriers until they happen to have a child with another carrier, giving birth to their first affected child.

CF is an appealing target for population carrier screening simply because of the relatively high carrier frequency in the general population. One in twenty-nine Caucasians, one in forty-six Hispanics, one in sixty-five African Americans, and one in ninety Asian Americans are carriers. It is not known why the mutation frequency is so high, especially in European populations. Some have proposed, using the analogy of the sickle-cell gene conferring relative resistance to malaria, that the mutations must have a protective effect against some disease appearing in European history, such as cholera or tuberculosis.

But all of this is just speculation. DNA screening of the entire adult population could potentially identify those couples at risk, who could then be offered prenatal diagnosis, affording couples the opportunity to consider their options. After several pilot studies and much debate at the national level, it has now been recommended that screening for the twenty-five most frequent *CFTR* mutations be offered to all couples expecting a child or planning a pregnancy. So most of the students reading this book will eventually be offered this DNA test! SEE ALSO CELL, EUKARYOTIC; GENE DISCOVERY; GENE THERAPY; GENETIC COUNSELING; HETEROZYGOTE ADVANTAGE; HUMAN DISEASE GENES, IDENTIFICATION OF; INHERITANCE PATTERNS; POPULATION SCREENING; PROTEINS.

Wayne W. Grody

Bibliography

Welsh, Michael J., and Alan E. Smith. "Cystic Fibrosis." *Scientific American* 273 (1995): 53–59.

Internet Resources

"Airway Clearance Techniques." University of Wisconsin Medical School. <http://www2medsch.wisc.edu/childrenshosp/CF/cfpages/cpt2.html>.

Cystic Fibrosis Foundation. <http://www.cff.org>.

Grody, Wayne W., et al. "Laboratory Standards and Guidelines for Population-Based Cystic Fibrosis Carrier Screening." *Genetics in Medicine* 3 (2001): 149–154. <http://www.acmg.net>.

Delbrück, Max

Physicist, Molecular Biologist
1906–1981

Max Delbrück made major contributions to the understanding of replication and viral function. Raised in Berlin in a distinguished family of German intellectuals, Delbrück trained as a physicist with Niels Bohr and other leaders in the field of quantum mechanics. In the early 1930s his interests turned toward biology and the nature of the gene. This was only thirty years after the rediscovery of Mendel's work and twenty years before Watson and Francis Crick discovered the structure of DNA. With two colleagues, he published a theoretical paper on quantum mechanical restrictions on gene structure. These ideas were popularized by the physicist Erwin Schrödinger in the book *What is Life?*, which inspired many young midcentury scientists to join the quest to understand the gene.

Max Delbrück.

bacteriophages viruses
that infect bacteria

Delbrück moved to the United States in 1937 to pursue genetics and escape the increasingly repressive atmosphere of Nazi Germany. He first went to Columbia University in New York, where he joined Thomas Hunt Morgan's group to study *Drosophila*. Soon, however, he became interested in **bacteriophages**. It was in the understanding of this model system that Delbrück made his greatest contribution.

Bacteriophages are among the simplest genetic systems, and thus provided Delbrück with an elegant tool for exploring fundamental processes of reproduction and mutation. Delbrück collaborated with Salvador Luria and Alfred Chase to work out the fundamental mechanisms of viral replication and to explore the genetics of mutation in this system. This loosely allied trio, and the ever-widening circle of scientists with whom they collaborated, became known as "the phage group." This group conducted training courses at Cold Spring Harbor Laboratory in New York, where they introduced many other biologists to this model system, while inculcating in them their own rigorous and quantitative approach. Watson was one of Delbrück's students in the phage course. The phage group began and shaped the field of molecular genetics, and Delbrück is usually considered the father of this discipline. Delbrück, Luria, and Hershey were awarded the Nobel Prize for physiology or medicine in 1969 for their discoveries in phage genetics.

Later in his life, Delbrück turned his attention to the cellular physiology underlying perception, but the model system he chose for this research, a light-sensitive fungus, had too little in common with animals to make the research strongly relevant to animal perception. He died in 1981. SEE ALSO CRICK, FRANCIS; MORGAN, THOMAS HUNT; WATSON, JAMES; VIRUS.

Richard Robinson

Bibliography

Fischer, Ernst Peter, and Carol Lipson. *Thinking About Science: Max Delbrück and the Origins of Molecular Biology.* New York: W. W. Norton, 1988.

Judson, Horace Freeland. *Eighth Day of Creation: Makers of the Revolution in Biology.* New York: Simon & Schuster, 1979.

Schrödinger, Erwin. *What is Life?* New York: Cambridge University Press, 1992.

Development, Genetic Control of

Development is the process through which a multicellular organism arises from a single cell. During development, cells become specialized, or differentiated, taking on different functions and forms. The organism develops a characteristic three-dimensional shape, the parts of which (such as limbs and organs) continue to maintain the same relationship to each other even as the organism grows. How the genes in a single fertilized egg dictate the creation of a complex multicellular creature is the central question in the genetic control of development.

While we are often most curious about human developmental processes, very little is known about the genetics of human development specifically, because experimentation on human embryos is forbidden by law and ethics. Instead, the details of genetic control are best understood in several model organisms, including the roundworm (*Caenorhabditis elegans*), the fruit fly

(*Drosophila melanogaster*), the zebrafish, and the mouse. Each organism differs in the details, and in some cases the overall logic, of genetic control. The understanding of developmental control is not complete for any of these organisms, but scientists have come to understand several mechanisms that contribute to, but do not entirely explain, development.

The Importance of Transcription Factors

With few exceptions, every cell in a multicellular organism contains the same set of genes as every other cell. Despite this genetic equivalence, cells differ greatly in form, function, longevity, and many other characteristics. These differences are due to the differential expression of genes within each cell type. Thus, a nerve cell will express a certain subset of the entire **genome**, while a gut cell will express a different subset. (To express a gene means to use it to create its encoded product, usually a protein.) Cells become different from one another, therefore, by expressing different sets of genes. Thus, the problem of development can be addressed by understanding how initially identical cells come to express different sets of genes.

genome the total genetic material in a cell or organism

The beginning of the answer to this question lies in understanding gene transcription and transcription factors. Transcription is the process in which an enzyme called RNA polymerase binds to a gene to make an RNA copy; this is the first step in expressing the gene. Transcription factors are proteins that bind to regulatory regions of the gene, thereby influencing how easily RNA polymerase attaches to it. Different genes require different sets of transcription factors, and when these factors are in low supply, expression of that gene is slowed or stopped.

Since transcription factors are proteins, they are encoded by their own genes, which are regulated by yet other transcription factors. As we will see, many of the "master" genes that control development encode transcription factors that are expressed early in development. The sequential activation of these genes, in a domino-like fashion, is one way that the overall developmental program is carried out.

The European Way and the American Way

A central question in development is whether a particular cell is predestined to become a specific cell type from the moment of its creation, or whether its fate is less determinate, depending on a variety of cues it receives from its local environment as development proceeds. The developmental geneticist Sydney Brenner dubbed these two alternatives the European way (what matters is who your ancestors are) and the American way (what matters is what your environment is and who your neighbors are). While no multicellular organism displays either alternative exclusively, the roundworm *C. elegans* operates primarily according to the European plan, with each of its exactly 959 cells largely following a set developmental path, and with few decisions made through interaction with neighboring cells. *Drosophila*, and mammals such as mice and humans, largely develop according to the American plan, with most cells only gradually taking on a final identity, through repeated communication and competition with neighboring cells.

The difference can be seen in transplant experiments, in which cells of the early embryo are moved from one region to another. In the roundworm, the transplanted cell generally follows its original developmental plan, regardless of the environment in which it finds itself. In fruit flies and mammals, the transplanted cell generally takes on the identity of the region into which it is transplanted, switching from one developmental pathway to another, such as from bone cell to gut cell. This change is not absolute and, most importantly, it is time-dependent. Cells transplanted later in development tend to remain committed to the pathway they were on, despite their new surroundings, leading to the aberrant development of bone cells in the gut, for instance.

Morphogens and Gradients

One problem has intrigued embryologists for many decades: In the absence of strictly defined developmental fates, how does a cell "know" where it is in an embryo, in order to know what to become? An early suggestion, which has been borne out by experiments, is based on the concept of a concentration gradient—a variation in concentration of a substance across a region of space.

A concentration gradient is formed whenever a substance is created in one place and moves outward by diffusion. When this occurs, there will be a high concentration of the substance near its point of origin, and increasingly lower concentrations further away. This provides positional information to a cell anywhere along the gradient. Cells pick up this chemical signal, and its strength (concentration) determines the cell's response. Typically, the signal is a transcription factor, and the response is a change in gene transcription.

Because such a signalling substance helps to give form to the embryo, it is termed a morphogen ("morph-" meaning "form," "-gen" meaning "to give rise to"). Morphogen gradients are a key means by which originally identical cells are exposed to different environments and thus sent along different developmental paths. In humans and other mammals, one morphogen that acts early in development is retinoic acid, a relative of vitamin A.

Gradients Determine the Axes of the Fruit Fly Embryo

We can see how such a morphogen acts by considering the development of the anterior-posterior axis in the fruit fly embryo. In the fly egg case, the oocyte, or fertilized egg, is accompanied by "nurse cells" at what will become the head end of the fly. This is called the anterior end; the tail end is posterior. Nurse cells create messenger RNA for a protein called bicoid, which they transport to the oocyte. Because these **mRNAs** originate in the anterior end, their concentration is highest there, and is lower towards the posterior end. Once the oocyte begins to divide, the mRNA is translated, and the bicoid protein is synthesized. Anterior cells have more of it than posterior cells, and the difference in concentration sets each cell group down its own developmental pathway, with anterior cells developing head structures, and posterior cells tail structures. Note that, in keeping with the "American plan," the fate of each cell is determined not by its ancestry, but by the environment it is in.

Retinoic acid is used as an acne treatment for humans, and must be avoided during pregnancy.

mRNA messenger RNA

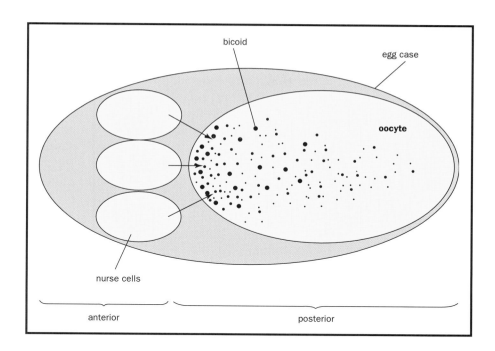

A concentration gradient of bicoid mRNA (and later bicoid protein) helps establish the anterior-posterior axis within the fruit-fly oocyte.

The effect of bicoid can be seen in transgenic flies, which have too many or too few copies of the gene. With extra bicoid, a higher-than-normal concentration exists further back in the oocyte, and anterior structures develop further back on the fly. With no bicoid, the anterior structures don't develop at all.

As we might expect, the bicoid protein is a **transcription factor**, which helps regulate expression of other genes. Other gradients of other transcription factors also exist at this stage, and together, these overlapping gradients establish the dorsal-ventral (back-belly) axis and map out the body segments that characterize all insects. While the details are complex, the fundamental idea is that of combinatorial control: At each position, it is the combination of transcription factors and their concentrations that determines which genes will be expressed, and therefore what the identity of the cells will be.

transcription factor protein that increases the rate of transcription of a gene

As segmentation becomes more firmly established and segments begin to take on their unique identities, gradients become less important. Instead, local gene control and cell to cell interactions create the increasingly fine level of spatial patterning.

Homeotic Genes and Segment Identity

Once segmentation is established, another important and remarkable set of genes turns on: the homeotic selector genes. These genes control development within each segment. For instance, the thorax region of the fly contains three segments, each with one pair of legs (the reason insects are six-legged). The homeotic selector gene *antennapedia* is normally expressed only in the thoracic segments, leading to the creation of a pair of legs.

Note that *antennapedia* does not itself "code for" legs. Instead, its protein product is a transcription factor. By regulating expression of many other

In place of its antennae this fly grew legs. Normally expressed in the thoracic region of a fly, the homeotic selector gene *antennapedia* "kick starts" a series of events leading to leg growth. A mutated form of this gene allowed for its expression in the head region of the fly.

apoptosis programmed cell death

genes, it sets off a cascade of events that results in the creation of legs. Remarkably, however, this single gene is sufficient by itself to turn on the leg-producing program, and its absence keeps the program silent. It can even turn it on in other segments. For instance, when *antennapedia* is mutated to allow it to be expressed in head segments, a pair of legs develops in place of the normal appendages, antennae (hence the gene name, which means "antenna foot").

Intriguingly, the sequence of homeotic selector genes along the fly chromosome matches the order of segments in which each is expressed. That is, the genes expressed in head segments come first, followed by those expressed in thoracic segments, then the abdomen, then the tail. The way in which this correspondence is exploited during development is still unknown, but the arrangement is clearly not accidental. Related genes have been found in vertebrates, including humans, and the same pattern holds: Genes expressed more anteriorly precede those expressed toward the posterior.

Homeotic Genes in Other Species

Homeotic genes also control development in other species, from yeast to humans, although the details are not as clear as they are for the fruit fly. Homeotic genes, called Hox genes, control the development of segments in the mammalian hindbrain, for example, and help establish the anterior-posterior and dorsal-ventral axes in the limbs. Vertebrates have duplicate copies of the Hox genes on several chromosomes, all of which function together to specify, for example, limb development. The multiple copies provide a redundancy not found in the fruit fly, thus making the effect of individual genes harder to detect. Nonetheless, by "knocking out" multiple versions of a particular Hox gene, researchers have shown their dramatic effects. For example, mice that are missing two copies of Hox 11 have no forelimbs, and the wrist is fused directly to the elbow.

The Homeobox

As transcription factors, the homeotic gene products must bind to DNA. Sequence analysis of both gene and protein has revealed that all share a 180-nucleotide DNA stretch, termed the homeobox, the sequence of which has remained almost unchanged over many millions of years of evolution. It codes for a 60-amino acid long DNA binding region, called the homeodomain. The homeodomain sequence of *antennapedia* and the mouse homeotic gene Hox B-7 differ by only two amino acids, despite having diverged several hundred million years ago.

Programmed Cell Death: Apoptosis

Development of a multicellular creature requires not only cell differentiation, but in some cases, cell death. **Apoptosis** helps create the spaces between the fingers, for instance. During brain development, nerve connections are sculpted through the apoptotic death of billions of cells. In *C. elegans*, exactly 131 cells die by apoptosis.

Cells can be directed to the apoptotic pathway if they fail to receive appropriate signals from their neighbors. In this way, it is thought that cells in the wrong location—a bone cell in the gut, for instance—might be terminated to prevent damage to the organism. The death program itself is carried out within the cell by activation of specific genes that ultimately trigger proteases, which are enzymes that break down cell contents, including the chromosomes. SEE ALSO APOPTOSIS; FRUIT FLY: *DROSOPHILA*; ROUNDWORM: *CAENORHABDITIS ELEGANS*; TRANSCRIPTION FACTORS; TRANSGENIC ORGANISMS; ZEBRAFISH.

Richard Robinson

Bibliography

Alberts, Bruce, et al. *Molecular Biology of the Cell*, 3rd ed. New York: Garland Publishing, 1994.

De Robertis, E. M., G. Oliver, and C. V. Wright. "Homeobox Genes and the Vertebrate Body Plan." *Scientific American* 263, no. 1 (1990): 46–52.

Gehring, Walter F. *Master Control Genes in Development and Evolution*. New Haven: Yale University Press, 1998.

Gilbert, Scott F. *Developmental Biology*, 5th ed. Sunderland, MA: Sinauer Associates, 1997.

Diabetes

Diabetes mellitus is a group of diseases characterized by elevated levels of **glucose** in the blood. Diabetes is caused by problems producing or responding to the hormone insulin. Insulin is produced in the pancreas by specialized cells called beta cells, in response to the presence of glucose absorbed through the gastrointestinal tract following a meal. Insulin promotes the uptake of glucose into muscle and fat cells, and it promotes the storage of excess glucose in the liver.

glucose sugar

Excess blood glucose over time damages organs, particularly the eyes, kidneys, nerves, heart, and blood vessels. It is the leading cause of adult blindness, end-stage kidney disease, and lower limb amputations, and it is a major risk factor for heart attacks and strokes. Diabetes is classified into four major groups: type 1 diabetes (T1DM), type 2 diabetes (T2DM), other specific types, and gestational diabetes (GDM), occurring during pregnancy. Approximately 5 percent to 8 percent of the people of the industrialized world have diabetes, mostly (approximately 90 percent) type 2, which at least 16 million Americans have.

Type 1 Diabetes

Type 1 diabetes is caused by beta cell destruction, leading to insulin deficiency. T1DM was previously called insulin-dependent diabetes mellitus (IDDM), because patients who have it require insulin for survival. It was also called juvenile-onset diabetes mellitus, because most type 1 diabetics are children or young adults. At the time of diagnosis, about 85 percent to 90 percent of people with type 1 diabetes have **antibodies** directed against components of their beta cells, indicating that the immune system is responsible for the progressive and irreversible beta cell destruction.

antibodies immune-system proteins that bind to foreign molecules

Current evidence indicates a genetic component to T1DM. HLA (histocompatibility leukocyte antigen) genes are a group of genes on chromosome 6 that encode proteins that are part of the immune system. Normally the immune system defends the body against disease by destroying foreign

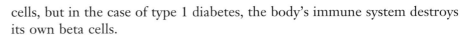

cells, but in the case of type 1 diabetes, the body's immune system destroys its own beta cells.

Certain types of HLA genes are strongly associated with type 1 diabetes, and other types protect against its development. However, these HLA genes are neither necessary nor sufficient to cause or protect from type 1 diabetes. T1DM is therefore a "complex" genetic disorder, in which several genes interact with the environment to result in the disease. Scientists are currently working to identify these other genes, as well as environmental factors (e.g., toxins and viruses) that provoke the development of T1DM.

Type 2 Diabetes

Type 2 diabetes is itself a group of disorders caused by some combination of insulin resistance—which occurs when cells' ability to *respond to* insulin is compromised—and insulin deficiency, which occurs when the beta cells' ability to *make* insulin is compromised. T2DM has, in the past, been called adult-onset diabetes, because most people with T2DM were adults. It was also called non-insulin-dependent diabetes mellitus (NIDDM), because people with type 2 diabetes usually do not require insulin injections. In the Unites States, T2DM is especially prevalent among certain ethnic minorities, including African Americans, Mexican Americans, Asian Americans, and Native Americans.

Obesity is a potent risk factor for T2DM. In the last thirty years, due to increased caloric intake and physical inactivity, both of which contribute to obesity, there was an explosion in the prevalence of T2DM, and it started occurring at younger ages—even in children. In addition to its association with an unhealthy lifestyle, T2DM is known to have a strong genetic component.

Scientists have been searching throughout the genome for T2DM-susceptibility genes. One such gene, *calpain 10 protease*, was identified on chromosome 2. A common variant of this gene may predispose certain individuals to T2DM; however, the true significance of this gene variant remains to be determined. In addition, several candidate genes have shown some evidence of being involved in T2DM. However, the effect of any single candidate gene variant on the risk of developing T2DM is modest. A candidate gene is a gene for which prior knowledge of its function leads researchers to assess whether chemical variation in it is associated with a disease.

As of 2002 there was no clinically available genetic test to predict the onset of type 2 diabetes, but it is anticipated that with a better understanding of the roles of various genes in T2DM, it will eventually be possible to use multiple genetic tests to identify individuals at risk for T2DM and to predict which treatments will be most helpful in specific patients. Although genetic susceptibility plays an important role in determining the risk of developing T2DM, studies have shown that the disease can often be prevented through diet, physical activity, and weight loss.

Other Specific Types of Diabetes

The third category of diabetes, containing other specific types, includes nongenetic forms as well as single-gene forms of diabetes. One group of single-gene diabetes disorders are genetic defects in beta cell function. The most common of the genetic beta cell defects are the disorders known as MODY, or maturity onset diabetes of the young. MODY constitutes no more than 2 percent to 5 percent of all cases of diabetes. It often occurs in children and young adults and is characterized by decreased but not absent insulin production. It is inherited in an autosomal dominant manner, which means that an affected person has a 50 percent chance of passing on the disease-version of the gene with each pregnancy. Most, but not all, people receiving a MODY gene do develop diabetes.

There are at least six different genetic forms of MODY. MODY2 is caused by a mutation in a gene on chromosome 7 that makes a protein called glucokinase, which is an enzyme in beta cells that helps to provide a chemical signal needed for insulin release. The other MODYs involve mutations in genes that encode proteins called transcription factors, which allow beta cells to develop and function properly. These are hepatocyte nuclear factor 4-alpha (*HNF4-alpha*, causing MODY1, on chromosome 20), *HNF1-alpha* (causing MODY3, on chromosome 12), insulin promoter factor 1 (*IPF1*, causing MODY4, on chromosome 13), *HNF1-beta* (causing MODY5, on chromosome 17) and *NeuroD1/beta2* (causing MODY6, on chromosome 2).

A very rare genetic insulin secretion disorder is maternally inherited diabetes and deafness (MIDD), caused by changes in the DNA of the mitochondria. The **mitochondria** are the energy powerhouses of the cell and the only part of the cell to contain DNA other than the nucleus, where most DNA is contained. MIDD and other mitochondrial disorders are maternally inherited because the fertilized egg has only mitochondria derived from the mother. The clinical features of MIDD can be similar to type 2 diabetes, and the hearing loss can be mild or even undetectable, except by special tests.

mitochondria energy-producing cell organelle

Another group of rare genetic diabetes types is characterized by extreme insulin resistance, which is defined as occurring when the ability of the body's cells to respond to insulin is severely compromised. Disorders of extreme insulin resistance include type A syndrome, leprechaunism, and Rabson-Mendenhall syndrome, and they are caused by inherited defects in the gene on chromosome 19 that makes the insulin receptor, a protein that allows cells to respond to insulin. Without properly functioning insulin receptors, insulin cannot work effectively. In addition to diabetes, individuals with insulin receptor defects may also have dental, genital, skin, and growth abnormalities. Most insulin receptor gene defects manifest in an autosomal recessive manner. That is, two defective copies of the gene are required for disease expression, and couples in which each partner has one defective copy (and in which neither is therefore affected) have a 25 percent chance of having an affected child, with each pregnancy.

Familial partial lipodystrophic diabetes (FPLD) is a rare condition in which children develop an unusual fat distribution at puberty, with little or no fat on their arms, legs, and trunk. They also develop insulin-resistant diabetes. FPLD is an autosomal dominant condition caused by mutations in the *lamin A/C* gene on chromosome 1. Another rare form of lipodystrophic diabetes is congenital (i.e., present at birth) generalized lipodystrophic (CGL) diabetes, which is autosomal recessive, and in about half of cases is due to mutations in the gamma-3-like gene (*GNG3*; also called the seipin gene), on chromosome 11.

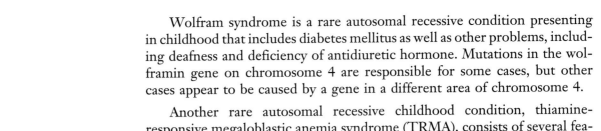

Wolfram syndrome is a rare autosomal recessive condition presenting in childhood that includes diabetes mellitus as well as other problems, including deafness and deficiency of antidiuretic hormone. Mutations in the wolframin gene on chromosome 4 are responsible for some cases, but other cases appear to be caused by a gene in a different area of chromosome 4.

Another rare autosomal recessive childhood condition, thiamine-responsive megaloblastic anemia syndrome (TRMA), consists of several features, including blood abnormalities, deafness, and diabetes. TRMA, which responds to treatment with thiamine (a form of vitamin B), is a disorder caused by mutations in the thiamine transporter gene *SLC19A2*, on chromosome 1.

Transient neonatal diabetes (TNDM) is a condition in which infants are born requiring injected insulin but are able to make sufficient insulin later in infancy. Later in childhood or in adulthood, they may again develop diabetes, which may or may not require insulin treatment. Most cases of transient neonatal diabetes appear to be caused by the inheritance of an extra copy of a region of chromosome 6 from the father.

Many known genetic disorders other than those mentioned previously are associated with an increased risk of diabetes. Among those most strongly associated are Friedreich's ataxia, cystic fibrosis, and hemochromatosis.

Gestational Diabetes Mellitus

Hormones associated with pregnancy may cause diabetes in susceptible individuals. Although the diabetes goes away after the pregnancy, individuals who have had GDM are at increased risk of developing T2DM. Currently very little is known about the genetic basis of GDM. It is possible that some of the same genes responsible for T2DM are also involved in GDM.

Genetic Susceptibility to Complications

As mentioned above, diabetes is associated with complications involving the eyes, kidneys, blood vessels, and heart. However, not all individuals with diabetes develop these complications. There is increasing evidence that there are genes other than those that increase susceptibility to developing the disease that may influence susceptibility to developing its complications. These genes are not yet identified, but they are likely to interact with other known risk factors for complications, including poor blood-sugar control and increased blood-pressure and blood-cholesterol levels. SEE ALSO COMPLEX TRAITS; DISEASE, GENETICS OF; GENE AND ENVIRONMENT; GENE DISCOVERY; IMMUNE SYSTEM GENETICS; MITOCHONDRIAL DISEASES.

Toni I. Pollin and Alan R. Shuldiner

Bibliography

Internet Resources

American Diabetes Association. <http://www.diabetes.org>.

Joslin Diabetes Center. <http://www.joslin.org>.

Juvenile Diabetes Research Foundation International. <http://www.jdrf.org>.

National Institute of Diabetes and Digestive and Kidney Diseases. <http://www.niddk.nih.gov>.

Online Mendelian Inheritance in Man. Johns Hopkins University, and National Center for Biotechnology Information. <http://www.ncbi.nlm.nih.gov/Omim>.

Disease, Genetics of

Genetics is believed to play a role in almost every human disease. Even for diseases traditionally described as environmental, such as tuberculosis and HIV, scientists are discovering that genetics is implicated either in the susceptibility to infection or in the severity of the disease. In some disorders a variation within a single gene is sufficient to cause disease, while in other disorders variations within a gene must interact with the environment and other genes to cause disease. The goal of human medical genetics is to identify all the genes that are involved in human disease and determine how the genes function to cause susceptibility to disease. This knowledge leads to the development of successful therapies that improve the quality of life of affected individuals and their families.

Mendelian and Complex Disorders

Geneticists typically classify genetic disorders into two main categories: Mendelian and complex disorders. Mendelian disorders, such as sickle-cell disease, cystic fibrosis, and Duchenne muscular dystrophy, are usually rare in the general population. These disorders have predictable, recognizable inheritance patterns (such as autosomal dominant and X-linked recessive), and variations in a single gene are sufficient to cause expression of the disorder. Furthermore, only individuals who carry a **mutation** in the causative gene are at risk for expressing the disorder.

mutation change in DNA sequence

In contrast, complex disorders, such as cardiovascular disease, diabetes, cancers, and psychiatric disorders, are common in the general population. In these disorders, genetics plays a significant role, but the biology of the disease is due to a tangled web of genetic and environmental interactions. Consequently, complex disorders generally do not display the distinct inheritance patterns seen in Mendelian disorders.

While the genetic variation at a single gene may contribute to the overall risk of developing a disease, it is not expected to be sufficient for expression of the disease. A well-known example of this is the association between Alzheimer's disease and the *APOE* gene. Individuals who carry the "4" **allele** of the *APOE* gene have a higher risk and earlier age-of-onset for Alzheimer's disease than those with other alleles. Furthermore, this association is dose-dependent. Individuals who have two copies of *APOE-4* are at greater risk for developing Alzheimer's disease than individuals who carry one copy of *APOE-4* and one copy of a different allele.

allele a particular form of a gene

How Important Are Genes?

Prior to searching for the genes involved in a disease and determining how those genes work in the various tissues of the human body, there must be clear evidence that genetics is involved in the disease. Genetic analyses are ethically and financially challenging, as well as quite laborious. Geneticists use several methods to evaluate whether or not genetics plays an important role in the **etiology** of a disorder before they begin a search for a gene. Some of these methods include familial aggregation, recurrence risks, and twin studies.

etiology causation of disease, or the study of causation

Familial aggregation can be established by obtaining a thorough family history on study participants. Individuals are simply asked if they have any

other family members who have the disease. If individuals with the disease have a higher frequency of affected relatives compared with individuals without the disease, there is evidence of familial aggregation. Although familial aggregation for a disorder is consistent with the involvement of genetics, it may also reflect the presence of a common environmental factor to which all family members have been exposed (such as pesticides or contaminated drinking water).

Relative risk ratios are another method commonly used to determine if there exists a genetic basis to a disease. For example, in cystic fibrosis, an autosomal recessive Mendelian disorder, the risk of disease in the siblings of an affected individual is 1 in 4. The **prevalence** of the disease is about 1 in 1,600 in the general population.

prevalence frequency of a disease or condition in a population

In 1990 Neil Risch demonstrated that by comparing the risk of a disease occurring in the relatives of an affected individual with the risk of the disease occurring in the general population, one could measure the significance of the genetic component of the disease. The risk ratio, labeled λ_R, where R is the type of relative (such as sibling), is the ratio of the risk of disease in the relative of type R to the prevalence of the disease in the general population. Thus in cystic fibrosis, $\lambda_s = 1/4 \div 1/1600 = 400$. This means that the risk of developing cystic fibrosis is four hundred times greater for siblings of an affected individual than for an individual in the general population. This clearly demonstrates the effect that genetics plays in the development of cystic fibrosis.

As a general rule, the larger the λ_R, the stronger the genetic component. However, this ratio is extremely sensitive to the frequency of the disease in the general population. The more common the disorder, the lower the λ_R, although this does not necessarily preclude a genetic component to the disorder.

Twin and adoption studies provide a unique opportunity to tease apart the role of genetics and the influence of a common familial environment. Because twins were born at the same time, the environments to which they were exposed are very similar. This holds true for the prenatal environment, and often for the childhood environment, but rarely for adult environments.

Monozygotic (MZ), or identical, twins share 100 percent of their genetic makeup, and dizygotic (DZ), or fraternal, twins share on average 50 percent of their genetic makeup. Twin studies compare disease "concordance" in MZ twins with DZ twins. When both twins in a pair have the disease, the twins are said to be concordant. When one twin has the disease and the other does not, the twins are said to be discordant. If the disorder has a genetic component, then MZ twins will be concordant more often than DZ twins. The difference between the MZ and DZ concordance rates can be used to assess the strength of the genetic component.

In summary, geneticists are finding that some disorders have a large genetic effect and others have a small genetic effect. There are several methods geneticists can use to determine the size of the genetic effect of a disease. However, ultimately, researchers trying to fully understand genetic disorders must identify the genes that are involved and determine their function. The revolution in human genetics, primarily due to the successes of the Human Genome Project, has made and will continue to make an impact on scientists' ability to define the role of genetics in human disease. SEE ALSO ALZHEIMER'S DISEASE; CANCER; CARDIOVASCULAR DISEASE; COMPLEX TRAITS; CYSTIC FIBROSIS; DIABETES; EPIDEMIOLOGIST; GENE DISCOVERY; HEMOGLOBINOPATHIES; INHERITANCE PATTERNS; MUSCULAR DYSTROPHY; TRIPLET REPEAT DISEASE; TWINS.

Allison Ashley-Koch

DNA

DNA (deoxyribonucleic acid) was discovered in the late 1800s, but its role as the material of heredity was not elucidated for fifty years after that. It occupies a central and critical role in the cell as the genetic information in which all the information required to duplicate and maintain the organism. All information necessary to maintain and propagate life is contained within a linear array of four simple bases: adenine, guanine, thymine, and cytosine.

DNA was first described as a monotonously uniform helix, generally called B-DNA. However, we now know that DNA can adopt many different shapes and conformations. Moreover, many of these alternative shapes have biological importance. Thus, the DNA is not simply an informational repository, from which information flows through RNA into proteins. Rather, structural information exists within the specific sequence patterns of the bases. This structural information dictates the interaction of DNA with proteins to carry out processes of DNA replication, transcription into RNA, and repair of errors or damage to the DNA.

The Components of DNA

DNA is composed of purine (adenine and guanine) and pyrimidine (cytosine and thymine) bases, each connected through a ribose sugar to a phosphate backbone. Many variations are possible in the chemical structure of the bases and the sugar, and in the structural relationship of the base to the sugar that result in differences in helical shape and form. The most common DNA helix, B-DNA, is a double helix of two DNA strands with about 10.5 **base pairs** per helical turn.

Bases and Base Pairs. The four bases found in DNA are shown in Figures 1 and 2. The purines and pyrimidines are the informational molecules of the genetic blueprint for the cell. The two sides of the helix are held together by hydrogen bonds between base pairs. **Hydrogen bonds** are weak attractions between a hydrogen atom on one side and an oxygen or nitrogen atom on the other. Hydrogen atoms of amino groups serve as the hydrogen bond donor while the carbonyl oxygens and ring nitrogens serve as hydrogen bond acceptors. The specific location of hydrogen bond donor and acceptor groups gives the bases their specificity for hydrogen bonding in unique pairs. Thymine (T) pairs with adenine (A) through two hydrogen bonds, and cytosine (C) pairs with guanine (G) through three hydrogen bonds (Figure 2). T does not normally pair with G, nor does C normally pair with A.

base pairs two nucleotides (either DNA or RNA) linked by weak bonds

hydrogen bonds weak bonds between the H of one molecule or group and a nitrogen or oxygen of another

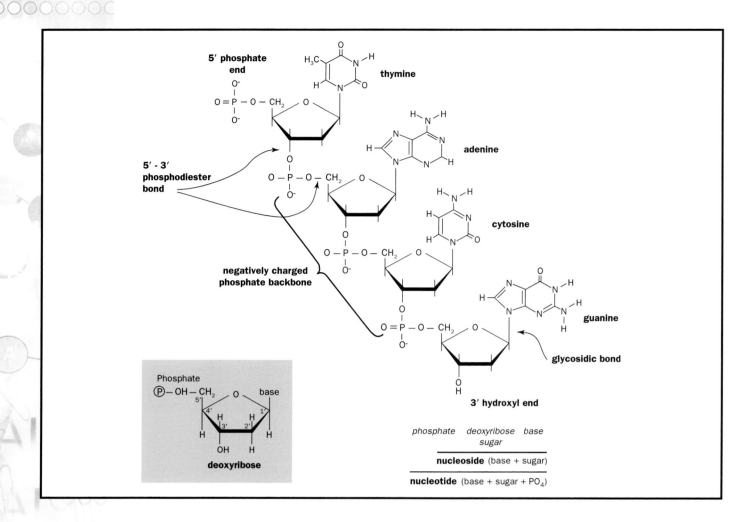

Phosphate

deoxyribose

phosphate	deoxyribose sugar	base
nucleoside (base + sugar)		
nucleotide (base + sugar + PO$_4$)		

Figure 1. DNA nucleotides include a base, a deoxyribose sugar, and a phosphate. The carbons on the sugar are numbered 1′ ("one prime") to 5′. The phosphate group links the 3′ end of one nucleotide to the 5′ end of the next. The phosphates are negatively charged, promoting their attraction to positively charged histone proteins (not shown).

β the Greek letter beta

nucleotide the building block of RNA or DNA

Deoxyribose Sugar. In DNA the bases are connected to a **β**-D-2-deoxyribose sugar with a hydrogen atom at the 2′ ("two prime") position. The sugar is a very dynamic part of the DNA molecule. Unlike the **nucleotide** bases, which are planar and rigid, the sugar ring is easily bent and twisted into various conformations (which exist in different structural forms of DNA). In canonical B-DNA, the accepted and most common form of DNA, the sugar configuration is known as C2′ endo.

Nucleosides and Nucleotides. The term "nucleoside" refers to a base and sugar. "Nucleotide," on the other hand, refers to the base, sugar, and phosphate group (Figure 1). A bond, called the glycosidic bond, holds the base to the sugar and the 3′–5′ ("three prime-five prime") phosphodiester bond holds the individual nucleotides together. Nucleotides are joined from the 3′ carbon of the sugar in one nucleotide to the 5′ carbon of the sugar of the adjacent nucleotide. The 3′ and the 5′ ends are chemically very distinct and have different reactive properties. During DNA replication, new nucleotides are added only to the 3′ OH end of a DNA strand. This fact has important implications for replication.

The Structure of Double-Stranded DNA

As mentioned above, the two individual strands are held together by hydrogen bonds between individual T·A and C·G base pairs. In DNA, the

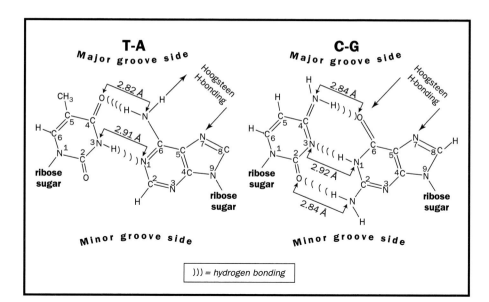

Figure 2. Hydrogen bonding between the bases. Note that the sets of distances between the base pairs are almost identical for the two base pairs, so that the distance across the double helix is unchanged. Hoogsteen hydrogen bonding can link a double helix to a third strand, to make a triple helix.

distance between the atoms involved is 2.8 to 2.95 angstroms (10^{-10} meters). While individually weak, the large number of hydrogen bonds along a DNA chain provides sufficient stability to hold the two strands together.

The stabilization of duplex (double-stranded) DNA is also dependent on base stacking. The planar, rigid bases stack on top of one another, much like a stack of coins. Since the two purine.pyrimidine pairs (A·T and C·G) have the same width, the bases stack in a rather uniform fashion. Stacking near the center of the helix affords protection from chemical and environmental attack. Both **hydrophobic interactions** and **van der Waal's forces** hold bases together in stacking interactions. About half the stability of the DNA helix comes from hydrogen bonding, while base stacking provides much of the rest.

Double-stranded DNA in its canonical B-form is a right-handed helix formed by two individual DNA strands aligned in an antiparallel fashion (a right-handed helix, when viewed on end, twists clockwise going away from the viewer). Antiparallel DNA has the two strands organized in the opposite polarity, with one strand oriented in the 5′–3′ direction and the other oriented in the 3′–5′ direction.

In the right-handed B-DNA double helix, the stacked base pairs are separated by about 3.24 angstroms with 10.5 base pairs forming one helical turn (360°), which is 35.7 angstroms in length. Two successive base pairs, therefore, are rotated about 34.3° with respect to each other. The width of the helix is 20 angstroms. An idealized model of the double helix is shown in Figure 3. As can be seen, the organization of the bases creates a major groove and a minor groove.

Adenine and thymine are said to be complementary, as are cytosine and guanine. Complementary means "matching opposite." The shapes and charges of adeninne and thymine complement each other, so that they attract one another and link up (as do cytosine and guanine). Indeed, one entire strand of duplex DNA is complementary to the opposing strand. During replication, the two strands unwind, and each serves as a template

hydrophobic interactions attraction between portions of a molecule based on mutual repulsion of water

van der Waal's forces weak attractive forces between two different molecules

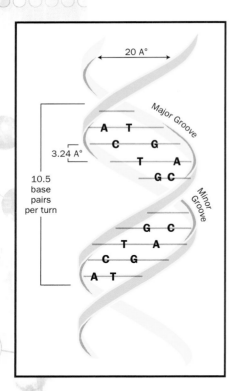

20 A°

Major Groove

3.24 A°

10.5
base
pairs
per turn

Minor
Groove

Figure 3. Canonical B-DNA double helix.

protonated possessing excess H$^+$ ions; acidic

for formation of new complementary strand, so that replication ends with two exact double-stranded copies.

Alternative DNA Conformations

While the vast majority of the DNA exists in the canonical B-DNA form, DNA can adopt an amazing array of alternative structures. This is the result of certain particular sequence arrangements of DNA and, in many cases, energy in the DNA double helix from DNA supercoiling, the property of DNA in which the double helix, in a high-energy state, becomes twisted around itself. Some alternative DNA conformations identified are shown in Figure 4.

Unwound DNA. Since A·T base pairs contain two hydrogen bonds and C·G base pairs contain three, A+T-rich tracts are less thermally stable that C+G-rich tracts in DNA. Under denaturing conditions (heat or alkali), the DNA begins to "melt" (separate), and unwound regions of DNA will form, and it is the A+T-rich sequences that melt first. In addition, in the presence of superhelical energy (a high-energy state of DNA resulting from its supercoiling, which is the natural form of DNA in the chromosomes of most organisms), A+T-rich regions can unwind and remain unwound under conditions normally found in the cell. Such sites often provide places for DNA replication proteins to enter DNA to begin the process of chromosome duplication.

Cruciform Structures. DNA sequences are said to be palindromic when they contain inverted repeat symmetry, as in the sequence GGAATT-AATTCC, reading from the 5′ to the 3′ end. Palindromic sequences can form intramolecular bonds (within a single strand), rather than the normal intermolecular (between the two complementary strands), hydrogen bonds. To form cruciforms ("cross-shaped"), the DNA must form a small unwound structure, and then base pairs must begin to form within each individual strand, thus forming the four-stranded cruciform structure.

Slipped-Strand DNA. Slipped-strand DNA structures can form within direct repeat DNA sequences, such as $(CTG)_n \cdot (CAG)_n$ and $(CGG)_n \cdot (CCG)_n$ (where "n" denotes a variable number of repetitions). They form following denaturation, after the strands become unwound, and during renaturation, when the strands come back together. To form slipped-strand DNA, the opposite strands come together in an out-of-alignment fashion, during renaturation. Expansion of such triplet repeats are features of certain neurological diseases.

Intermolecular Triplex DNA. Three-stranded, or triplex DNA, can form within tracts of polypurine·polypyrimidine sequence, such as $(GAA)_n \cdot (TTC)_n$. Purines, with their two-ring structures, have the potential to form hydrogen bonds with a second base, even while base paired in the canonical A·T and G·C configurations. This second type of base pair is called a Hoogsteen base pair, and it can form in the major groove (the top of the base pair representations in Figure 2). Pyrimidines can only pair with a single other base, and thus a long Pu·Py tract must be present for triplex DNA formation. The important factor for triplex DNA formation is the presence of an extended purine tract in a single DNA strand. The third-strand base-pairing code is as follows: A can pair with A or T; G can pair with a **protonated** C (C$^+$) or G.

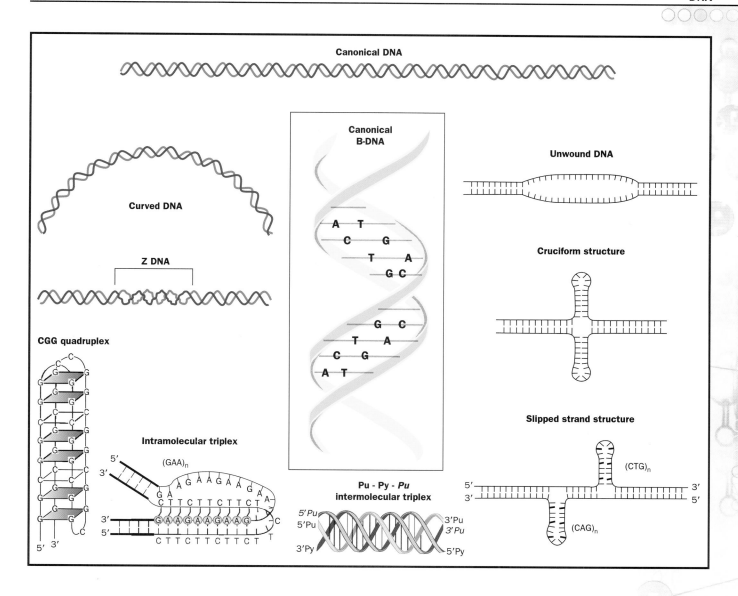

Figure 4. DNA can exist in a variety of conformations. "Canonical" DNA is the most common, but each of the other forms have particular functions or are found in certain conditions.

Intramolecular Triplex DNA. When a Pu·Py tract exists that has mirror repeat symmetry (5′ GAAGAG–GAGAAG 3′), an intramolecular triplex can form, in which half of the Pu·Py tract unwinds and one strand wraps into the major groove, forming a triplex. The structure in Figure 4 shows the pyrimidine strand (CTT) pairing with the purine strand (GAA) of a canonical DNA duplex. In an intramolecular triplex, one strand of the unwound region remains unpaired, as shown.

Quadruplex DNA. DNA sequences containing runs of G·C base pairs can form quadruplex, or four-stranded DNA, in which the four DNA strands are held together by Hoogsteen hydrogen bonds between all four guanines. The four guanines are aligned in a plane, and the successive rings of guanines are stacked one upon another.

Left-handed Z-DNA. Alternating runs of $(CG)_n \cdot (CG)_n$ or $(TG)_n \cdot (CA)_n$ dinucleotides in DNA, under superhelical tension or high salt (more than 3 M NaCl) (M, moles per liter) can adopt a left-handed helix called Z-DNA. In this form, the two DNA strands become wrapped in a left-handed helix, which is the opposite sense to that of canonical B-DNA. This can occur

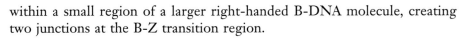

within a small region of a larger right-handed B-DNA molecule, creating two junctions at the B-Z transition region.

Curved DNA. DNA containing tracts of $(A)_{3-4} \cdot (T)_{3-4}$ (that is, runs of three or four bases of A in one strand and a similar run of T in the other) spaced at 10-base pair intervals can adopt a curved helix structure.

In summary, DNA can exist in a very stable, right-handed double helix, in which the genetic information is very stable. Certain DNA sequences can also adopt alternative conformations, some of which are important regulatory signals involved in the genetic expression or replication of the DNA. SEE ALSO CHROMOSOME, EUKARYOTIC; CHROMOSOME, PROKARYOTIC; DNA MICROARRAYS; GENE; GENOME; NUCLEOTIDE; SEQUENCING DNA; TRIPLET REPEAT DISEASE.

Richard R. Sinden

Bibliography

Sinden, Richard R. *DNA Structure and Function.* San Diego: Academic Press, 1994.

DNA Footprinting

DNA footprinting is a technique for identifying exactly where a protein binds to DNA. Knowing where a protein binds to DNA often aids in understanding how gene expression is regulated. Consequently, DNA footprinting is often part of a larger study to determine how a particular gene is controlled.

How It Works

DNA footprinting is based on the observation that when a protein binds to DNA, the DNA is protected from chemicals that would otherwise **cleave** it. In a typical DNA footprinting experiment, a DNA fragment with a suspected protein-binding site is first isolated, then labeled with a radioactive **nucleotide** or other chemical that will allow it to be detected later on.

Once labeled, the DNA is then mixed in a test tube with a DNA-binding protein and a chemical that cleaves DNA, such as the **enzyme** DNase I. In a separate test tube, more of the same labeled DNA is mixed with the same cleaving chemical, but without the binding protein. The DNA fragments in each tube are allowed to incubate long enough for the molecule to cleave once, and then are separated by size (fractionated) in a DNA sequencing gel.

The reactions in the two test tubes (one with the binding protein and one without) are then compared. If the DNA actually contains protein-binding sites, these will have been protected from cleaving in the test tube that contains DNA-binding protein, and a "footprint" of those sites where no DNA cleavage occurred will be observed. By comparison with a sequencing reaction run on the same gel, one can determine the exact location where a protein has been bound to the DNA. A related technique, called gel retardation, can also be used to test for protein binding to DNA, but this method is less precise than DNA footprinting.

cleave split

nucleotide the building block of RNA or DNA

enzyme a protein that controls a reaction in a cell

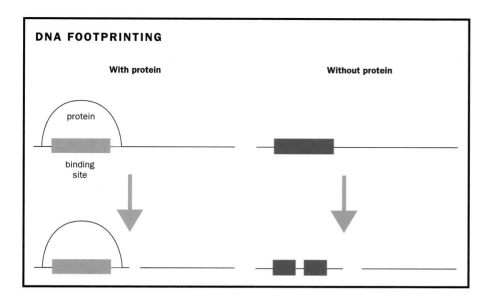

DNA FOOTPRINTING

With protein   Without protein

protein

binding
site

The protein protects the
binding site region from
cleavage by DNase or
another treatment.

Uses in Research

DNA footprinting is often used to locate the binding site for proteins that
regulate **transcription**. For example, a researcher may suspect that a particular protein binds to a particular DNA fragment and inhibits transcription. After conducting a DNA footprinting experiment, the researcher will
know the location of the exact sequence of DNA bound by that protein. If
that sequence matches the sequence of a **promoter** the DNA footprinting
experiment can help explain how that DNA-binding protein carries out its
function.

Modified DNA footprinting experiments can also be performed to
detect where proteins bind to DNA in a living cell. In these experiments,
cells are grown under conditions where the protein of interest would be
expected to bind to DNA. The cells are then treated with a chemical that
causes proteins bound to DNA to become permanently attached to the
DNA. The resulting DNA-protein complexes are then purified from the
cell, and the DNA sequences are identified.

Since DNA footprinting is used to identify the specific sequences in
DNA where a protein binds, the technique is likely to be of continuing usefulness in genetic research. For example, DNA footprinting is likely to be
heavily used in characterizing the function of proteins identified in the
Human Genome Project and other genome projects, making it an important component of a molecular geneticist's toolbox. SEE ALSO GENE EXPRES-
SION: OVERVIEW OF CONTROL; NUCLEASES; SEQUENCING DNA.

Patrick G. Guilfoile

transcription messen-
ger RNA formation from
a DNA sequence

promoter DNA
sequence to which RNA
polymerase binds to
begin transcription

Bibliography

Guilfoile, Patrick. *A Photographic Atlas for the Molecular Biology Laboratory.* Englewood,
CO: Morton Publishing, 2000.

Internet Resource

DNA Footprinting Reveals the Sites Where Proteins Bind on a DNA Molecule. National
Center for Biotechnology Information. <http://www.ncbi.nlm.nih.gov/entrez/
query.fcgi?db=Books>.

DNA Libraries

DNA libraries, like conventional libraries, are used to collect and store information. In DNA libraries, the information is stored as a set of DNA molecules, each of which contains biological sequences that can be used for a variety of applications. All DNA libraries are collections of DNA fragments that represent a particular biological system of interest. By analyzing the DNA from a particular organism or tissue, researchers can answer a variety of important questions. The two most common uses for these DNA collections are DNA sequencing and gene cloning.

The Importance of Vectors

Several types of DNA libraries have been developed for specific purposes, but all share some common features. The DNA fragments that make up the library are attached to other DNA sequences that are used as "handles" to maintain the fragments. These "handles," called vectors, allow the DNA to be replicated and stored, typically within model organisms such as yeast or bacteria.

vectors carriers

plasmid a small ring of DNA found in many bacteria

Different types of **vectors** can be used to store DNA fragments of different lengths. For example, **plasmid** vectors can store small fragments (from a few hundred bases up to ten or twenty thousand bases of sequence), while viral vectors, or viral-plasmid hybrids such as cosmids, can store up to fifty thousand bases, and yeast artificial chromosome (YAC) vectors can store hundreds of thousands of bases. In general, plasmid-based vectors are the easiest to manipulate but store the smallest fragments. They are commonly used for applications that involve complex manipulations, such as cloning or gene expression, but that require only small DNA fragments (e.g., cDNA libraries, as described below).

Artificial Chromosome Vectors and Genomic Libraries

Yeast artificial chromosome vectors act like real chromosomes in yeast and can store much longer DNA fragments, some over 150 kilobases in size, big enough for several genes along with their regulatory sequences. However, YAC vectors are difficult to manipulate, are prone to spontaneous rearrangement, and have been supplanted by bacterial artificial chromosome (BAC) vectors.

Escherichia coli common bacterium of the human gut, used in research as a model organism

BAC vectors are derived from the F plasmid of *Escherichia coli*. This plasmid behaves like a chromosome and not like a typical plasmid. BACs can store very large DNA fragments—in excess of three hundred kilobases in some cases, although typical fragments are about half that size. The unique features of BAC vectors are very well suited to creating and maintaining DNA libraries. For example, once a BAC vector enters a cell, it will exclude all other BAC vectors, which means that a given *E. coli* clone will contain only one unique library fragment. Furthermore, *E. coli* cells are relatively easy to grow and store, and DNA purification from the bacterium is straightforward. BAC libraries played a key role in the massive sequencing efforts that made up the Human Genome Project.

Many of the large-format DNA libraries (YACs, BACs) are used exclusively to store genomic DNA for sequencing projects. Larger fragments per-

CREATING A cDNA LIBRARY

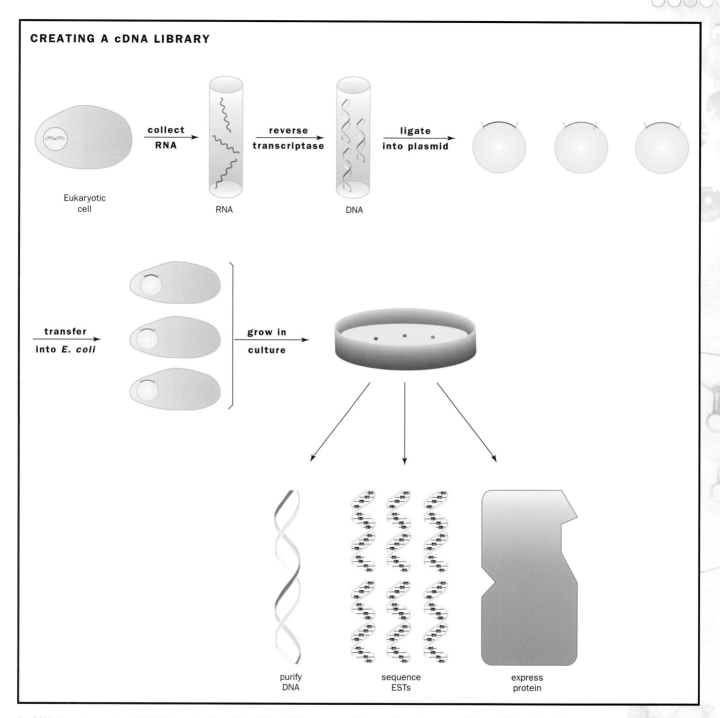

Eukaryotic cell — collect RNA → RNA — reverse transcriptase → DNA — ligate into plasmid →

transfer into *E. coli* → grow in culture →

purify DNA   sequence ESTs   express protein

A cDNA (complementary DNA) library begins with all the RNA expressed by a cell under a specific condition. The RNA is reverse transcribed into DNA, and then each DNA copy is inserted into a separate plasmid. These are taken up by *E.coli* bacteria, one to a cell, which are then grown in culture. At that point, the DNA can be purified for analysis or further cloning, "expressed sequence tags" for each can be sequenced, or the proteins can be re-expressed by the bacteria to yield enough protein for further study.

mit easier assembly of finished DNA sequence and require the maintenance of fewer clones, which is particularly important when sequencing large genomes. DNA sequencing, however, is only one application of DNA libraries.

DNA libraries are created by generating a set of DNA fragments of the desired size and then attaching those fragments to the appropriate vector sequence. For genomic DNA, the fragments are normally generated by either enzymatic digestion or simple mechanical shearing of all the DNA of the genome, including noncoding sequences. Fragments are then enzymatically attached to the vector sequences, in a reaction known as ligation. The collected fragments, now attached to vector sequences, are then moved into the appropriate host organism for growth and evaluation. Conditions are chosen so that only one fragment enters each organism, which can then be grown up into a colony whose individuals all carry the same fragment.

Complementary DNA Libraries

cDNA complementary DNA

RNA ribonucleic acid

Genomic DNA is not always the source of the fragments in a DNA library. A second major class of libraries uses **cDNA** which is generated by copying the messenger **RNA** from an organism or tissue of interest. Because it reflects the mRNA content of a biological system (or cell type) at a particular time and under particular conditions, a cDNA library can be considered a "snapshot" of gene expression in that system. This information can be of great value in understanding when and how certain genes are expressed in an organism or cell type. Additionally, cDNA, unlike genomic DNA, lacks **introns** and other noncoding segments of sequence and is relatively straightforward to clone and express. This greatly facilitates the analysis of gene products (proteins) in **eukaryotes**.

introns untranslated portions of genes which interrupt coding regions

eukaryotes organisms with cells possessing a nucleus

Creating a cDNA library is similar to creating a genomic DNA library, except that the starting material for cDNA libraries is mRNA, not DNA. The enzyme reverse transcriptase is used to copy the mRNA to DNA. The DNA fragments are then cloned into vectors (typically plasmids) by ligation and moved into a host organism, as with genomic libraries.

Often, cDNA libraries are constructed using plasmid vectors with sequences that allow the cloned cDNA fragments to be expressed as proteins. Such "expression libraries" can be searched with protein-finding tools such as antibodies, and then the gene coding for the protein can be isolated. cDNA libraries are also used for expressed sequence tag (EST) analysis, in which small portions of many cDNAs are sequenced to provide an overview of gene expression in a particular sample.

DNA libraries play important roles in modern molecular biology research. The many genome-sequencing projects that are revolutionizing our understanding of genetics are entirely dependent on genomic DNA library techniques. cDNA libraries are invaluable in the study of gene expression and protein function, and for EST analysis. Continued progress in the development of library techniques and a continued interest in their applications suggest that these tools will remain an important part of the field for years to come. SEE ALSO CHROMOSOMES, ARTIFICIAL; CLONING GENES; MODEL ORGANISMS; PLASMID; POLYMERASE CHAIN REACTION; RESTRICTION ENZYMES; REVERSE TRANSCRIPTASE.

Daniel J. Tomso

Bibliography

Bloom, Mark V., Greg A. Freyer, and David A. Micklos. *Laboratory DNA Science: An Introduction to Recombinant DNA Techniques and Methods of Genome Analysis.* Menlo Park, CA: Addison-Wesley, 1996.

DNA Microarrays

DNA microarrays are tools used to analyze and measure the activity of genes. Researchers can use microarrays and other methods to measure changes in gene expression and thereby learn how cells respond to a disease or to some other challenge.

Gene Expression

Humans have 30,000 to 70,000 genes, each consisting of a sequence of bases, the building blocks of the hereditary material DNA. Before they can carry out their function, genes are copied to make messenger RNA (mRNA), in a process called **transcription**. This molecule is in turn used as a template for the synthesis of a protein molecule (**translation**). This entire process, including transcription of RNA and translation of protein, is referred to as gene expression. Only a subset of the full set of genes is expressed in a given tissue at a given time. In fact, this differential pattern of gene expression is ultimately what distinguishes lung tissue from skin, liver, and muscle tissue.

transcription messenger RNA formation from a DNA sequence

translation synthesis of protein using mRNA code

Even within a given tissue type, different genes are expressed at different times. For example, there is a very tightly controlled sequence of gene expression during the course of embryonic development. Tissues also respond to metabolic and other challenges. The pattern of gene expression changes in the liver in response to the consumption of a large meal. Similarly, muscle gene expression changes in response to vigorous exercise or injury. Drugs can also affect gene expression. Researchers can use microarrays and other methods to measure these changes in gene expression, and from them learn about how cells respond to disease or to other challenges.

Hybridization

Microarrays measure gene expression by taking advantage of the process of **hybridization (molecular)**. DNA is made up of four bases: guanine, adenine, cytosine, and thymine, which are abbreviated G, A, C, and T, respectively. G and C can bind to one another, forming a base pair, as can A and T, but no other combinations of bases can form base pairs. G and C are said to be "**complementary**" bases, as are A and T.

hybridization (molecular) base-pairing among DNAs or RNAs of different origins

complementary matching opposite, like hand and glove

The bases on each of the two strands of DNA that make up a chromosome are complementary to the bases on the opposite strand. Long pieces of DNA will not bind to each other (or "hybridize") unless they are complementary. Hybridization allows researchers to test whether two pieces of DNA are complementary. If they bind to one another (hybridize) then they are opposite strands of a single gene. If they do not bind to one another, then they are unrelated.

Hybridization can be used to measure the levels of hundreds of different mRNAs within a given tissue, thereby providing a picture of gene

Tools like this DNA microarray aid researchers in measuring gene expression. Each color represents a different level of gene expression.

expression within that tissue. RNA is isolated from the tissue of interest and allowed to hybridize to a solid support to which many different DNA pieces, from many different genes, have been attached. Because the RNA is labeled with a fluorescent tag, the amount bound to a given spot can be measured. The fluorescent intensity of each spot is a measure of the level of that mRNA that was expressed in the original tissue. In this way, the levels of expression of up to 12,000 different genes can be measured with a single microarray.

There are two basic types of microarrays. One type is created by a company called Affymetrix. Affymetrix manufactures silicon and glass chips that resemble semiconductor chips and that are manufactured using the same photolithographic techniques. These chips have sets of very short (20 base-pair) stretches of DNA representing each gene. A second type of microarray is commonly called a printed array and is made by spotting small amounts of DNA on glass slides. These arrays frequently have smaller numbers of genes on each slide, but researchers can easily modify them for specific experiments.

Microarray Analysis

Microarrays produce enormous amounts of data, and the analysis of that data can be quite complex. The sheer volume of data requires special software and a database in which to store both the measurements and the results of the analyses. The exact form that the analysis takes depends on the nature of the experiment being performed. If just two samples are being directly compared (for example, gene expression in mouse heart tissue is compared with and without the administration of a drug), relatively straightforward statistical tests can be performed. If larger numbers of samples are being measured, the same tests can be performed between two samples at a time, but more sophisticated, "clustering" analyses can be performed as well.

Clustering analysis identifies groups of genes that react the same way across several different samples. For example, researchers might analyze gene expression in heart tissue from a set of mouse embryos that range in age from five to fifteen days. A clustering analysis would be able to detect a group of genes whose expression levels all increase slowly from days five to nine, peak at day ten and then fall to zero by day twelve. Only genes that have this precise pattern of expression would cluster together, in this type of analysis.

The Role of Bioinformatics

One of the tremendous difficulties in performing any kind of expression analysis is the manipulation of very large amounts of biological data, a field of study called bioinformatics. The usefulness of gene expression data depends on how much information is available for each identified gene. In other words, the identities of the genes associated with each spot on a microarray must be accessible as the analysis is done.

Descriptions and classifications of each gene on the array must be readily available, as no researcher can remember such details about the tens of thousands of genes that may be involved in the analysis. An analysis might be done many times, with slight changes in the parameters of the clustering algorithm each time. The genes that cluster together are examined at the end of each analysis, to look for reproducible patterns. This analysis must be done with the full understanding of the biology of the system being studied. Clusters of genes are most informative if they group in a biologically reasonable way. For this reason, microarray expression analysis is frequently exploratory. The results of the analysis are used to suggest additional, corroborative experiments.

Another bioinformatics challenge in gene expression studies is collecting information about the samples under analysis and storing the information in databases. If gene expression patterns of one hundred different **tumor** samples are being examined, it may be necessary to restrict the analysis to subgroups of the tumors in order to observe patterns in the data. This subgrouping or stratification of the samples is best performed on the basis of independently determined properties of those samples. For example, samples from only **metastatic** cancer cells could be grouped together for analysis and compared with those from nonmetastatic cancer cells, or the age of the patient at the onset of disease could be used to segregate the samples into different groups. Such subgroup analysis can only be done if complete information is collected and stored for all samples.

tumor mass of undifferentiated cells; may become cancerous

metastatic broken away cancerous cells from the initial tumor

Applications of Microarray Analysis

Microarrays are new enough that their applications are still being developed. Microarray expression analysis can be used to help study complex, multigenic diseases such as Parkinson's disease (PD). The great challenge in understanding the genetics of such disorders is identifying susceptibility genes, which are genes that increase a person's risk of developing the disease. Frequently, the first step in discovering a susceptibility gene is **linkage analysis**. This technique can identify regions of a chromosome that harbor such a gene, but the regions that are identified are frequently very large, containing hundreds of genes. Screening through all of these genes individually is tremendously slow and labor-intensive. **Expression analysis** using microarrays can help prioritize these genes for further analysis by providing independent lines of evidence that specific genes are involved in the disease process.

Brain tissue can be collected through anatomical donations from patients with Parkinson's disease and from unaffected individuals, for example. Regions of the brain that are especially affected in Parkinson's patients can be compared to the same regions from unaffected individuals. Genes whose levels of expression vary can be identified. Hundreds or thousands of genes may be identified in this way, but they can then be compared to those that are found, through linkage analysis, to be linked to Parkinson's disease. There may be only tens of genes common to both groups. These genes can be prioritized for detailed examination through other methods. The key here is that expression analysis and linkage analysis provide independent evidence of a given gene's involvement in a disease process. It is the synthesis of information from these two independent lines of evidence that makes this approach powerful.

Another very powerful application of microarray expression data is called classification analysis. This technique uses gene expression data to separate tissue samples into two or more groups. For example, one type of tumor may respond very well to an aggressive program of chemotherapy treatment, while another type may respond better to surgical removal followed by radiation therapy. Further, these two types of tumors may be difficult or impossible to tell apart under a microscope. Choosing the correct method of treatment and applying that treatment early in the course of disease could significantly improve a patient's chances of survival.

In such a case, expression analysis can be used to give a detailed picture of the genes that are expressed in the two types of tumor. A training set (a small set of samples in each category) can be used to find specific patterns of gene expression that are characteristic for each type of tumor. New tumors can then be analyzed, and their expression profiles can be used to predict the group to which they belong. These approaches are used with great success to refine the clinical management of cancer patients. A 2001 study by S. Dhanasekaran, "Delineation of Prognostic Biomarkers in Prostate Cancer," offers an example of this kind of work. Additional applications of microarrays are still being developed.

SAGE Analysis

Gene expression analysis can also be done using a powerful technique called serial analysis of gene expression (SAGE). Like microarrays, SAGE starts

linkage analysis examination of co-inheritance of disease and DNA markers, used to locate disease genes

expression analysis whole-cell analysis of gene expression (use of a gene to create its RNA or protein product) by isolating RNA from the tissue of interest. This RNA is then processed through a long series of steps resulting in the isolation of a set of very short sequences, called tags, from each **transcript** in the cell. These tags are converted into corresponding segments of DNA. These pieces of DNA, which are 14 base pairs long, are then linked together into long chains, and their sequence of bases is determined. Tens of thousands of these SAGE tags are sequenced from each tissue that is being studied. The tags corresponding to a given gene from one tissue are counted and compared to those from the same gene in another tissue.

transcript RNA copy of a gene

For example, a colon cancer tumor sample might generate 50,000 SAGE tags, thirty-three of which correspond to a specific gene. A second library made from normal colon cells might have fifty thousand tags, eleven of which correspond to the same gene. This would indicate that the gene is expressed at a level that is three times as great in tumor cells than it is in normal cells.

SAGE data is significantly more difficult and expensive to produce than microarray data, but it offers the advantage of providing very precise and quantitative measurements of expression levels. SAGE has the further advantage that it can detect genes that have not been previously characterized. Such unknown genes cannot be detected by microarrays, because researchers must first know their sequence before they can place them on the array. SAGE therefore can be used as a gene discovery tool.

SAGE has been used most extensively in cancer research. Investigators in the Cancer Genome and Anatomy Project have created more than one hundred SAGE libraries from normal and cancerous tissue. Analysis of these libraries has revealed a great deal about the way that gene expression changes in cancerous tissue, which in turn has provided insight into new diagnostic and treatment options.

SAGE has also been used as a tool to help calculate the total number of genes in the human body, as well as to describe the ways in which genes are regulated and processed at different times. Microarrays and SAGE analysis are only two of the many ways that scientists have examined gene expression. As these techniques become more refined, and as new techniques are developed, they will provide a powerful tool to investigate how the incredible diversity and complexity of our tissues can arise, even though every cell in our bodies contains exactly the same set of genes. SEE ALSO BIOINFORMATICS; CANCER; COMPLEX TRAITS; GENE DISCOVERY; *IN SITU* HYBRIDIZATION; LINKAGE AND RECOMBINATION; MAPPING.

Michael A. Hauser

Bibliography

Bloom, Mark V., Greg A. Freyer, and David A. Micklos. *Laboratory DNA Science: An Introduction to Recombinant DNA Techniques and Methods of Genome Analysis.* Menlo Park, CA: Addison-Wesley, 1996.

Dhanasekaran, S. M., et al. "Delineation of Prognostic Biomarkers in Prostate Cancer." *Nature* 412 (2001): 822–826.

Internet Resource

Cancer Genome Anatomy Project. National Cancer Institute. <http://cgap.nci.nih .gov>.

DNA Polymerases

DNA polymerases are proteins that synthesize new DNA strands using pre-existing DNA strands as templates. Before one cell divides to produce two cells, the DNA containing the genetic information in it must be duplicated for the new cell, in a process known as **polymerization**. In human cells, duplicating the DNA genome requires the polymerization of 2.91 billion nucleotides, the building blocks of DNA. In the bacterium *Escherichia coli*, the polymerization of 4.64 million nucleotides is necessary to duplicate the genome for the new cell. In all cells, the DNA polymerases are the protein catalysts that link together the nucleotide building blocks of the new DNA polymer in an accurate and timely process that occurs during **replication**.

The DNA polymerases are also required to repair the DNA of the genome. The genome's DNA can be damaged by highly reactive molecules that are either produced in the cell during normal metabolic processes or brought into the cell from external sources. The damage, if not repaired, could result in the production of **mutations** in the genome or possibly cell death. Several DNA repair processes occurring in the cell have been identified that preserve the integrity of the genome by removing the damaged nucleotides and resynthesizing DNA by the DNA polymerases.

The DNA Polymerase Mechanism

All DNA polymerases share a common mechanism for DNA chain synthesis. The polymerization of DNA occurs by the linkage of one nucleotide at a time to the end of a preexisting DNA chain. The sequence fluctuations of the nucleotides on the DNA template upon which the DNA polymerase is moving determines which nucleotide is added onto the end of the growing DNA chain. If a thymine (T) nucleotide is positioned in the DNA template, for example, then an adenine (A) is polymerized onto the DNA chain opposite the thymine in the DNA template. If a guanine (G) nucleotide is positioned in the template, a cytosine (C) is linked to the growing DNA chain opposite the guanine. This polymerization process results in the synthesis of a DNA chain that is **complementary**, rather than identical, to the template strand of DNA, and is sequenced according to the proper Watson-Crick nucleotide base pairing rules. During replication, both strands of the duplex DNA molecule serve as templates. The DNA strands are separated, and each of the DNA strands is copied by the DNA polymerases. This process results in two identical copies of the original duplex DNA molecule being produced for the two cells.

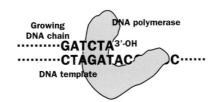

The DNA polymerase uses the **nucleoside triphosphate** form of the **deoxynucleotides** to build the DNA polymer. The monophosphate form of the deoxynucleotide is incorporated into the growing DNA chain, and a pyrophosphate molecule, a kind of salt, is released. The DNA polymerase can add nucleotides only to the 3′-OH end of the growing DNA chain (see

polymerization linking together of similar parts to form a polymer

replication duplication of DNA

mutations changes in DNA sequences

complementary matching opposite, like hand and glove

nucleoside triphosphate building block of DNA or RNA, composed of a base and a sugar linked to three phosphates

deoxynucleotides building blocks of DNA

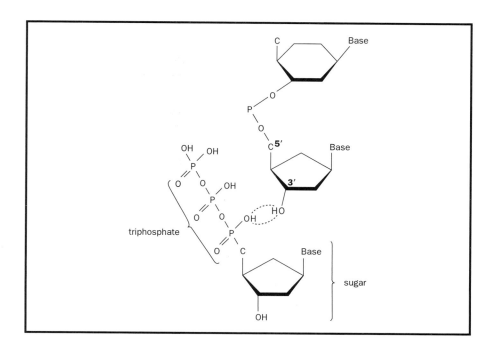

DNA polymerase attaches an incoming triphosphate nucleotide to the 3′ position on the growing chain. Removal of a water molecule between the two attaches them, and the two terminal phosphates break off.

above diagram). Therefore, DNA polymerization occurs in only one direction. Some DNA polymerases are highly processive, polymerizing many nucleotides to the 3′ end of the DNA chain before falling off the DNA template. Other DNA polymerases are distributive in nature, incorporating just one nucleotide and then falling off the DNA template.

Occasionally, the DNA polymerase will incorrectly polymerize a nucleotide onto the growing DNA chain. Removal of this misinserted nucleotide must be performed by a "proofreading" **exonuclease**, which is a substance that removes nucleotides from the 3′ end of the DNA molecule. The combined actions of DNA polymerases and proofreading exonucleases improve the accuracy of DNA synthesis and thus minimize introduction of errors into the genome.

exonuclease enzyme that cuts DNA or RNA at the end of a strand

The Variety of DNA Polymerases

Like all proteins, the DNA polymerases are encoded in genes. The genes that encode the human DNA polymerases are contained in the genomic DNA at various positions on several different chromosomes. In February 2001 the first "working draft" sequence of the human genome was published. Analysis of this sequence shows us that there are perhaps as many as fifteen different DNA polymerase genes in the human genome. Each of these genes encodes a different DNA polymerase protein. However, biochemists have not yet isolated all of these enzymes. A similar analysis of the *Escherichia coli* genome shows us that there are five different DNA polymerase genes present in this bacterium. The multitude of DNA polymerases in human and bacterial cells indicates a specialized role for the different enzymes in various aspects of DNA replication and repair, many of which have yet to be identified.

The human DNA polymerase α is encoded in the *POLA* (polymerase alpha) gene located on the human X chromosome. The DNA polymerases δ and ϵ are encoded in the *POLD1* (polymerase delta 1) and *POLE1*

mitochondria energy-producing cell organelle

catalytic describes a substance that speeds a reaction without being consumed

(polymerase epsilon 1) genes, which are located on chromosomes 19 and 12, respectively. These three DNA polymerases are most frequently associated with replication of the human genome. The DNA polymerase β is encoded by the *POLB* (polymerase beta) gene on chromosome 8 and is involved in DNA repair.

The DNA polymerase γ is encoded by the *POLG* (polymerase gamma) gene on chromosome 15 and replicates the DNA of the **mitochondria**. In *Escherichia coli* the DNA polymerase I is the most active. This enzyme functions in the bacterial cell to repair DNA, while the DNA polymerase III is responsible for replicating the genome. There are several additional DNA polymerases in human and in bacterial cells of which the precise function is not known. Some of these enzymes might be necessary to replicate genomic DNA that has been damaged. The ability to replicate damaged DNA could lead to mutations introduced into the genome but would preserve the life of the cell.

The amino acid sequences of the DNA polymerase proteins can be deduced from the genetic code contained in the DNA polymerase genes. Based on the amino acid sequences of the DNA polymerases, these proteins have been classified into several families. Analysis of these sequences reveals a relatively diverse collection of proteins with some very important similarities in specific amino acid regions along the length of the protein.

The similarities in amino acid sequences in certain parts of the DNA polymerase proteins tell us that these regions of the protein have been conserved throughout evolution. These specific amino acids are those that are important in the **catalytic** function of DNA polymerization by these proteins. Some of the similar amino acids are necessary for binding metal atoms that are needed by the DNA polymerase to carry out the polymerization reaction. Other amino acid sequences allow the DNA polymerase to hold on to the DNA and the four different deoxynucleoside triphosphates as the enzyme polymerizes the new DNA chain. Some DNA polymerases have the necessary amino acid sequences to generate a 3′ proofreading exonuclease domain (region), allowing the DNA polymerase to remove mistakes, or proofread, as it builds the DNA polymer.

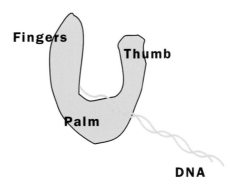

The DNA polymerases range in size from just over three hundred amino acids in length to more than two thousand amino acids in length. Three-dimensional studies of these enzymes have shown that the DNA polymerases have a common protein fold that resembles the shape of a "right hand" (see diagram). The "thumb," "fingers," and "palm" form a pocket along which the DNA can move. The DNA molecule interacts with specific amino acids located in the "palm" region of the DNA polymerase, and the "thumb" clamps down on the DNA to hold it as the DNA chain-elongation reaction proceeds. The amino acids that are common in many of the DNA polymerases are found in the regions where the enzyme contacts the DNA molecule. SEE ALSO DNA REPAIR; *ESCHERICHIA COLI* (*E. COLI* BACTERIUM); HUMAN GENOME PROJECT; NUCLEASES; NUCLEOTIDE; MUTATION; REPLICATION; X CHROMOSOME.

Fred W. Perrino

Bibliography

Alberts, Bruce, et al. *Molecular Biology of the Cell*, 4th ed. New York: Garland Science, 2002.

Baltimore, D. "Our Genome Unveiled." *Nature* 409 (Feb. 15, 2001): 814–816.

Blattner, F. R., et al. "The Complete Genome Sequence of *Escherichia coli* K–12." *Science* 277 (1997): 1453–1462.

Venter, J. C., et al. "The Sequence of the Human Genome." *Science* 291 (2001): 1304–1351.

Wood, R. D., M. Mitchell, J. Sgouros, and T. Lindahl. "Human DNA Repair Genes." *Science* 291 (2001): 1284–1289.

DNA Profiling

DNA profiling is a molecular testing method used to uniquely identify people and other organisms. In many ways, it is similar to blood typing and fingerprinting, and it is sometimes called "DNA fingerprinting." Because every organism's DNA is unique, DNA can be examined to identify people who might be related to each other, to compare suspected criminals to DNA left at the scene of a crime, or even to identify certain strains of disease-causing bacteria.

Blood Typing and the ABO Groupings

Before the development of the molecular biology tools that make DNA testing possible, investigators identified people through blood typing. This method hails from 1900, when Karl Landsteiner first discovered that people inherited different blood types. Several decades later, researchers determined that the basis for those blood types was a set of proteins on the surface of red blood cells.

The main proteins on the surface of red blood cells used in blood typing come in two varieties: A and B. Every person inherits from their parents either the genes for the A protein, the B protein, both, or neither. Someone who inherits the A gene from one parent and neither gene from the other parent has blood type A. If a person inherits both genes, they are AB. A person who inherits neither is type O. Another protein group found on red blood cells is referred to collectively as the Rh factor. People either have the Rh factor or they do not, regardless of which of the A and B genes they inherited. To type a person's blood, **antibodies** against these various proteins (A, B, and Rh) are mixed with a blood sample. If the proteins are present, the blood cells will stick together and the sample will get cloudy.

antibodies immune-system proteins that bind to foreign molecules

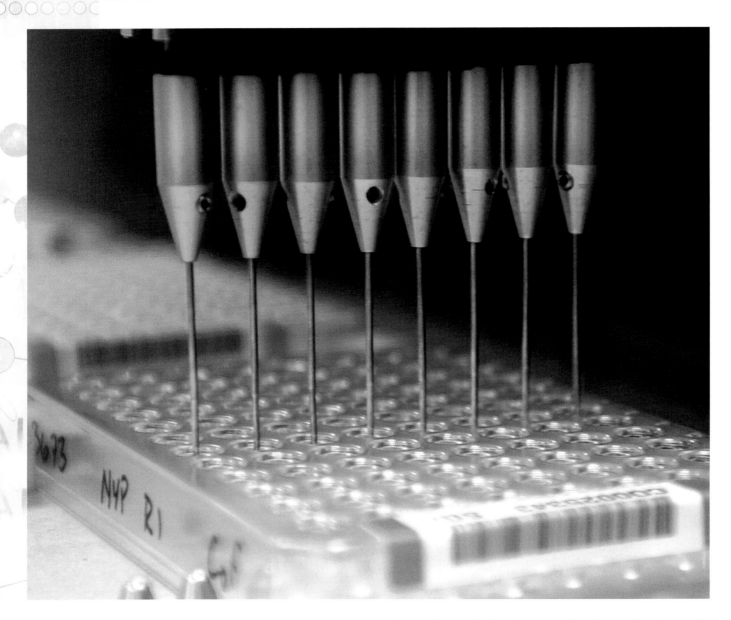

A robot used in DNA profiling adds solution and stirs DNA samples from tissues taken from September 11, 2001, New York terrorism victims. The tissue is being identified by matching DNA samples, which is the essence of DNA profiling.

Blood typing can be used to exclude the possibility that a blood sample came from a particular person, if the person's type does not match that of the sample. However, it cannot be used to claim that any particular person is the source of the sample, because there are so few blood types, and they are shared by so many people. About 45 percent of people in the United States are type O, and another 40 percent are type A. If four people were physically present at the scene of a murder, and the candlestick found nearby had type O blood spilled on it, chances are good that two of those individuals could be found guilty of the crime, based solely on the blood typing evidence. Most court cases, however, rely on more evidence than just blood or DNA typing, such as whose fingerprints are also found on the candlestick (see Statistics and the Prosecutor's Fallacy, below).

DNA Polymorphism Offers High Resolution

DNA is the molecule that contains all the genetic information of an individual. One person's DNA is made up of about three billion building blocks known as nucleotides or bases. Every organism in the world has a unique DNA sequence except for identical twins. Although identical twins accrue changes as they develop, they generally do not accumulate enough genetic differences for DNA typing to be useful. Portions of the DNA, called genes, encode proteins within the sequence of bases. Genes are separated by long stretches of noncoding DNA. Because these sequences do not have to code for functional proteins, they are free to accumulate more differences over time, and thus provide more variation than genes. Thus, they are more useful than gene sequences in distinguishing individuals.

Polymorphisms are differences between individuals that occur in DNA sequences which occupy the same **locus** in the chromosome. An individual will have only one sequence at a particular polymorphic locus in each chromosome, but if the population bears several to dozens of different possible sequences at the site in question, then the locus is considered "highly variable" within the population. DNA profiling determines which polymorphisms a person has at a small number of these highly variable loci. Because of this, DNA profiling can provide high resolution in distinguishing different individuals. The chances of one person having the same DNA profile as another are typically much less than the chances of winning a lottery.

polymorphisms occurring in several forms

locus site on a chromosome (plural, loci)

STR Analysis

The technology of DNA profiling has advanced from its beginnings in the 1980s. Today, DNA profiling primarily examines "short tandem repeats," or STRs. STRs are repetitive DNA elements between two and six bases long that are repeated in tandem, like GATAGATAGATAGATA. These repeat sequences often exist in a chromosomal region called heterochromatin, a largely unused portion of DNA found in each chromosome.

Different STR sequences (also called genetic markers) occur at different loci. While their positions are fixed, the number of repeated units varies within the population, from four to forty depending on the STR. Therefore, one genetic marker may have between four and forty different variations, and each variation is referred to as an allele of that marker. Each person has at most two **alleles** of each marker, one inherited from each parent. The two alleles for a particular marker may be identical, if both parents had the same form.

alleles particular forms of genes

The United States Federal Bureau of Investigation has designated thirteen of these sequences to use with STR analysis. These thirteen markers are all four-base repeats, and were chosen because multiple alleles of each exist throughout the population. The FBI system, called CODIS (Combined DNA Indexing System), has become the standard DNA profiling system in use today.

STR analysis begins with sample collection. Because of the often small samples involved and the legal weight that will be given to them, it is vital that the sample not be contaminated by other DNA. This may occur for instance if skin cells from the person collecting the sample are mixed with skin cells under the fingernails of a victim. Once the sample is collected, it must be kept secure at all times, to prevent any possibility of tampering.

In the laboratory, the DNA is isolated and purified, and then multiple copies of it are made using the polymerase chain reaction (**PCR**).

PCR polymerase chain reaction, used to amplify DNA

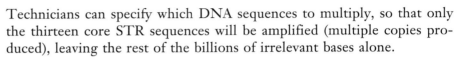

Technicians can specify which DNA sequences to multiply, so that only the thirteen core STR sequences will be amplified (multiple copies produced), leaving the rest of the billions of irrelevant bases alone.

In order to specify which DNA to amplify, "primers" are used. The primers are DNA sequences that recognize a nonrepeated sequence in the genetic markers, and which are used by the DNA polymerase that does the actual copying. After the DNA has been copied, the new DNA molecules are separated by size, by gel electrophoresis. A fluorescent molecule previously attached to each primer will send a light signal to the machine that measures the length of the molecule, or allele.

VNTR Analysis

An early form of DNA profiling, rarely used today, is based on VNTRs, or "variable number of tandem repeats." VNTRs requires extensive sample processing: The DNA is chopped up with restriction enzymes, separated by size, and probes are applied to the fragmented DNA to view only the relevant DNA pieces. In the DNA of two different individuals, different spacing between two cut sites for the restriction enzymes gives a unique pattern of DNA size fragments, called "restriction fragment length polymorphisms," or RFLPs.

Making a Match

To understand how DNA profiling is used to identify a person, imagine a sample of blood collected at a crime scene that doesn't match the victim's blood, and is presumably from the unknown perpetrator. DNA from the blood is isolated and its set of STRs are analyzed. The results will be a list of the alleles found at each of the markers (for example, VWA–12, 13; TH01–6, 7, and so on), where the initial symbol is the abbreviation for the markers and the last two are the numbers of the alleles found in the sample for that marker. The full set of thirteen markers may or may not be analyzed in each case. When a suspect is identified, his or her DNA can be analyzed for these same markers. If the set of alleles are different, the investigators can be sure that the two DNAs came from different sources, and the suspect is not the source of the blood. Since the introduction of DNA profiling, an absence of matching DNA has been used to free dozens of wrongly convicted prisoners.

If the samples do match, the question becomes whether the blood is actually from the suspect, or from someone else with the same set of alleles. As with blood typing, this is a matter of statistics, and depends on how frequently each allele occurs in the population. This information has been tabulated and is kept on file in the FBI CODIS database. If two samples share a very rare allele, that increases the likelihood they came from the same source.

Matching multiple alleles increases the certainty they came from the same source. Since the thirteen STRs are inherited independently of each other, the likelihood that one person's DNA will include specific alleles of all thirteen STR sites is the product of the individual allele frequencies. For example, if each allele a person carries occurs in 25 percent of the population, then the probability that all thirteen alleles will occur in one

13 CORE CODIS STR LOCI WITH CHROMOSOMAL POSITIONS

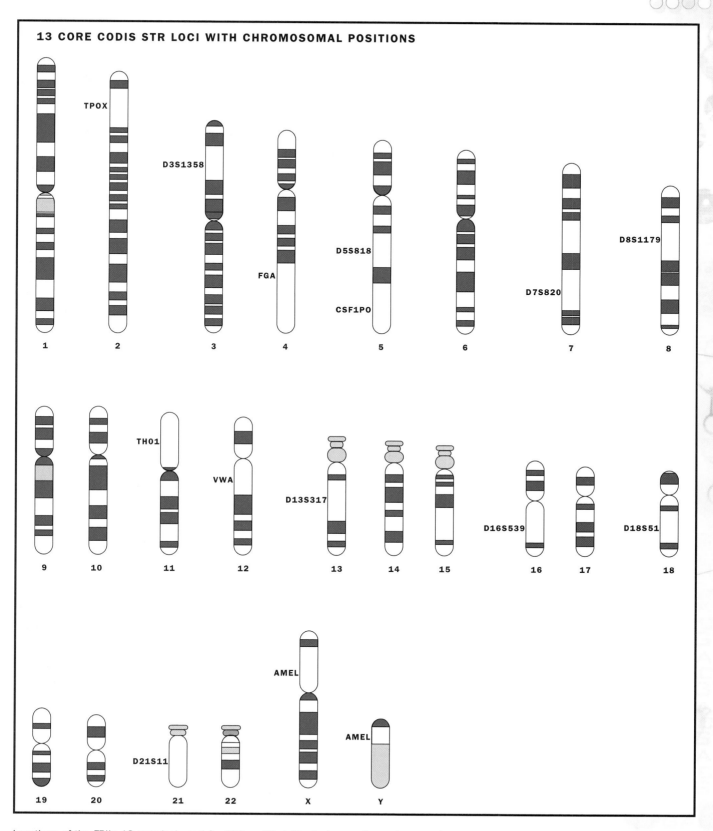

Locations of the FBI's 13 core loci used for DNA profiling. The loci were chosen to span the genome, and to have maximum variability among humans. Adapted from <http://www.cstl.nist.gov/biotech/strbase/images/codis.jpg>.

individual is (0.25 × 0.25 × 0.25 × 0.25 × 0.25 × 0.25 × 0.25 × 0.25 × 0.25 × 0.25 × 0.25 × 0.25) or 1 in more than 67 million. This analysis can discriminate between millions of people, far better than is possible using the four blood groups. Since many alleles are even rarer than 25 percent, their presence in both samples further increases the probability that they came from the same source.

Statistics and the Prosecutor's Fallacy

Despite the persuasiveness of such figures, it is quite possible to misuse DNA evidence to incorrectly argue that an innocent suspect must be the perpetrator of the crime, or that a guilty suspect should go free. Both defense and prosecution attorneys can—accidentally or otherwise—misinterpret data to make a highly likely event seem improbable, or a highly unlikely event seem probable. Jurors can be confused because DNA testing reveals the probability that an innocent person's DNA profile matches the sample at the scene of the crime. Jurors must decide, however, what the probability is that a person is innocent, if his DNA matches that sample. The prosecutor's fallacy occurs when investigators focus on the existence of the match, rather than the possibility that the match could be a coincidence.

Let's assume the DNA profile found at the crime scene—and the matching DNA of the suspect—is expected to occur once in every million people. The correct statement of probability arising from these facts is, "If the suspect is innocent, there is a one-in-one-million chance of obtaining this DNA match." The fallacy is to reverse these clauses, and state, "If the DNA matches, there is a one in one million chance that the suspect is innocent." To understand the logical fallacy, imagine the statement, "If it's Tuesday, it must be a school day." The reverse is not true—there are other school days besides Tuesday.

Similarly, there are other ways of misusing statistics in DNA profiling. Let's assume the suspect in the above case is actually guilty. If the suspect hails from a city with a population of ten million, there are ten people in the city whose DNA matches the DNA at the crime scene. Therefore, his defense lawyers could argue there is a 90 percent chance that the suspect is innocent, because he is 1 out of 10 individuals with that same DNA profile. If the defense can convince the jury to ignore other incriminating evidence, such as the suspect's bloody glove left behind at the scene, then the attorney may introduce reasonable doubt. Only by considering DNA typing within the context of other evidence can the probability of a DNA match improve the integrity of the justice system.

DNA Profiling Comes of Age

Although DNA profiling was viewed with some skepticism when it first made its way into the courts, DNA typing is now used routinely, in and out of the courthouse. It is commonly used in rape and murder cases, where the assailant generally leaves behind some personal evidence such as hair, blood, or semen. In paternity tests, the child's DNA profile will be a combination of the profiles of both parents. DNA profiling has also been used to identify victims in disasters where large numbers of people died at once, such as in airplane crashes, large fires, or military conflicts.

DNA testing can also used in organisms other than humans. For instance, it has been used to type cattle in a cattle-stealing case. It can also be used to identify pathogenic strains of bacteria to track the outbreak of disease epidemics. SEE ALSO BIOTECHNOLOGY AND GENETIC ENGINEERING, HISTORY; GEL ELECTROPHORESIS; POLYMERASE CHAIN REACTION; POLYMORPHISMS; REPETITIVE DNA ELEMENTS.

Mary Beckman

Bibliography

Bloom, Mark V., Greg A. Freyer, and David A. Micklos. *Laboratory DNA Science: An Introduction to Recombinant DNA Techniques and Methods of Genome Analysis.* Menlo Park, CA: Addison-Wesley, 1996.

Evert, Ian W., and Bruce S. Weir. *Interpreting DNA Evidence: Statistical Genetics for Forensic Scientists.* Sunderland, MA: Sinauer Associates, 1998.

Steward, Ian. "The Interrogator's Fallacy." *Scientific American* (September 1996): 172–175.

Internet Resources

"13 CODIS Core STR Loci with Chromosomal Positions." National Institute of Standards and Technology. <http://www.cstl.nist.gov/biotech/strbase/images/codis.jpg>.

The Biology Project. The University of Arizona. <http://www.biology.arizona.edu/human_bio/activities/blackett2/gifs/sample2.gif>.

FBI Core STR Markers. <http://www.cstl.nist.gov/biotech/strbase/fbicore.htm>.

The Innocence Project. <http://www.innocenceproject.org/>.

DNA Repair

When it was discovered that DNA is the **macromolecular** carrier of essentially all genetic information, it was assumed that DNA is extremely stable. Consequently, it came as something of a surprise to learn that DNA is actually unstable and subject to continual damage. When DNA damage is severe, the cell is unable to replicate and may die. Repair of DNA must be regarded as essential for the preservation and transmission of genetic information in all life forms. In this article, we will discuss various types of DNA damage and the DNA repair systems that have evolved to correct that damage.

macromolecular large molecule, composed of many similar parts

Sources of Damage

DNA is subject to spontaneous instability and decay. In addition to spontaneous damage, cellular DNA is under constant attack from reactive chemicals that the cell itself generates as by-products of **metabolism**. Moreover, the integrity of cellular DNA is assaulted by such environmental threats as X rays, ultraviolet radiation from the sun, and many chemical agents, some of which are products of our industrialized society.

metabolism chemical reactions within a cell

Since mutations can be introduced into DNA as a consequence of DNA damage, there is currently great interest and concern about the expanding list of chemicals released into the environment. In humans, damage to DNA has been implicated in many cancers as well as in certain aspects of aging. Genetic diseases such as cystic fibrosis and sickle cell disease can be caused by a single DNA mutation in one gene.

Figure 1. Four types of mutation: (A) complete loss of a base. (B) Loss of an amino group, converting a cytosine to a thymine. (C) Addition of a small alkyl group, such as –CH3. (D) Reaction with oxygen.

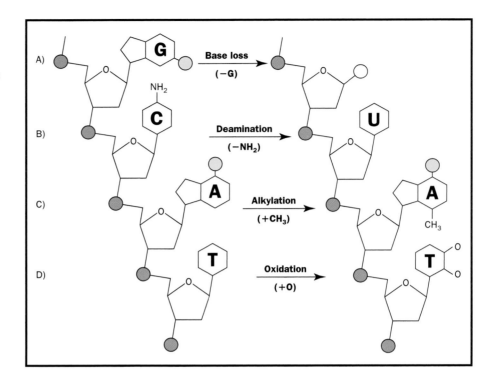

Types of DNA Damage

Damage to DNA can result from several different types of processes. Hydrolysis, deamination, alkylation, and oxidation are all capable of causing a modification in one or more bases in a DNA sequence.

Hydrolysis. DNA consists of long strands of sugar molecules called deoxyribose that are linked together by phosphate groups. Each sugar molecule carries one of the four natural DNA bases: adenine, guanine, cytosine, or thymine (A, G, C, or T). The chemical bond between a DNA base and its respective deoxyribose, although relatively stable, is nonetheless subject to chance **cleavage** by a water molecule in a process known as spontaneous **hydrolysis**. Loss of the "purine" bases (guanine and adenine) is referred to as depurination, whereas loss of the "pyrimidine" bases (cytosine and thymine) is called depyrimidination. In mammalian cells, it is estimated that depurination occurs at the rate of about 10,000 purine bases lost per cell generation. The rate of depyrimidination is considerably slower, resulting in the loss of about 500 pyrimidine bases per cell generation.

The baseless sugars that result from these processes are commonly referred to as AP-sites (apurinic/apyrimidinic). They are potentially lethal to the cell, as they act to block the progress of DNA replication, but are efficiently repaired in a series of enzyme-catalyzed reactions collectively referred to as the base excision repair (BER) pathway. In fact, AP-sites are intentionally created during the course of BER.

Deamination. The bases that make up DNA are also vulnerable to modification of their chemical structure. One form of modification, called spontaneous deamination, is the loss of an amino group (–NH$_2$). For example, cytosine (C), which is paired with guanine (G) in normal, double-stranded DNA, has an amino group attached to the fourth carbon (C4) of the base.

cleavage hydrolysis

hydrolysis splitting with water

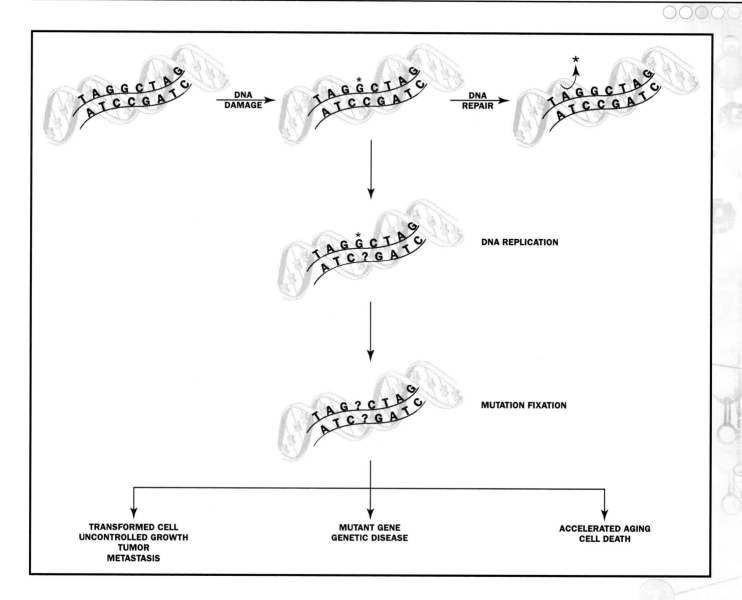

Figure 2. If damaged DNA (*) is not repaired, the mistake (mutation) can be replicated and become permanent (fixed) in the genome. This may cause severe problems.

polymerase enzyme complex that synthesizes DNA or RNA from individual nucleotides

When that amino group is lost, either through spontaneous, chemical, or enzymatic hydrolysis, a uracil (U) base is formed, and a normal C–G DNA base pair is changed to a premutagenic U–G base pair (uracil is not a normal part of DNA).

The U–G base pair is called premutagenic because if it is not repaired before DNA replication, a mutation will result. During DNA replication, the DNA strands separate, and each strand is copied by a DNA **polymerase** protein complex. On one strand, the uracil (U) will pair with a new adenine (A), while on the other strand the guanine (G) will pair with a new cytosine (C). Thus, one DNA double-strand contains a normal C–G base pair, but the other double-strand has a mutant U–A base pair. This process is called mutation fixation, and the mutation of the G to an A is said to be fixed (meaning "fixed in place," not "repaired"). In other words, the cell now accepts the new mutant base pair as normal. It is estimated that approximately 400 cytosine deamination events per genome occur every day. Clearly, it is very important for the cell to repair DNA damage before DNA replication commences, in order to avoid mutation fixation. One cause of

normal human aging is the gradual accumulation over time of mutations in our cellular DNA.

Alkylation. Another type of base modification is alkylation (Figure 2C). Alkylation occurs when a reactive **mutagen** transfers an alkyl group (typically a small hydrocarbon side chain such as a methyl or ethyl group, denoted as $-CH_3$ and $-C_2H_5$, respectively) to a DNA base. The nitrogen atoms of the purine bases (N_3 of adenine and N_7 of guanine) and the oxygen atom of guanine (O_6) are particularly susceptible to alkylation in the form of methylation. Methylation of DNA bases can occur through the action of exogenous (environmental) and endogenous (intracellular) agents. For example, exogenous chemicals such as dimethylsulfate, used in many industrial processes and formed during the combustion of sulfur-containing fossil and N-methyl-N-nitrosoamine, a component of tobacco smoke, are powerful alkylating agents. These chemicals are known to greatly elevate mutation rates in cultured cells and cause cancer in rodents.

Inside every cell is a small molecule known as S-adenosylmethionine or "SAM." SAM, which is required for normal cellular metabolism, is an endogenous methyl donor. The function of SAM is to provide an activated methyl group for virtually every normal biological methylation reaction. SAM helps to make important molecules such as adrenaline, a **hormone** secreted in times of stress; creatine, which provides energy for muscle contraction; and phosphatidylcholine, an important component of cell membranes. However, SAM can also methylate inappropriate targets, such as adenine and guanine. Such endogenous DNA-alkylation damage must be continually repaired; otherwise, mutation fixation can occur.

Oxidation. Oxidative damage to DNA bases occurs when an oxygen atom binds to a carbon atom in the DNA base (Figure 2D). High-energy radiation, like X rays and gamma radiation, causes exogenous oxidative DNA base damage by interacting with water molecules to create highly reactive oxygen species, which then attack DNA bases at susceptible carbon atoms. Oxidative base damage is also endogenously produced by reactive oxygen species released during normal respiration in mitochondria, the cell's "energy factories."

Humans enjoy a long life span; thus, it would seem that healthy, DNA repair-proficient cells could correct most of the naturally occurring **endogenous** DNA damage. Unfortunately, when levels of endogenous DNA damage are high, which might occur as the result of an inactivating mutation in a DNA repair gene, or when we are exposed to harmful exogenous agents like radiation or dangerous chemicals, the cell's DNA repair systems become overwhelmed. Lack of DNA repair results in a high mutation rate, which in turn may lead to cell death, cancer, and other diseases. Also, if the level of DNA repair activity declines with age, then the mutational burden of the cell will increase as we grow older.

Base Excision Repair

DNA bases that have been modified by the addition or loss of a small chemical group as described above are repaired by the BER pathway (Figure 3). The BER pathway begins with the **excision** of a damaged base by an enzyme called DNA glycosylase (Figure 3, step 1). DNA glycosylases bind to chemically altered (damaged) bases and **catalyze** the cleavage (hydrolysis) of the

mutagen any substance or agent capable of causing a change in the structure of DNA

hormone molecule released by one cell to influence another

endogenous caused by factors inside the organism

excision removal

catalyze aid in the reaction of

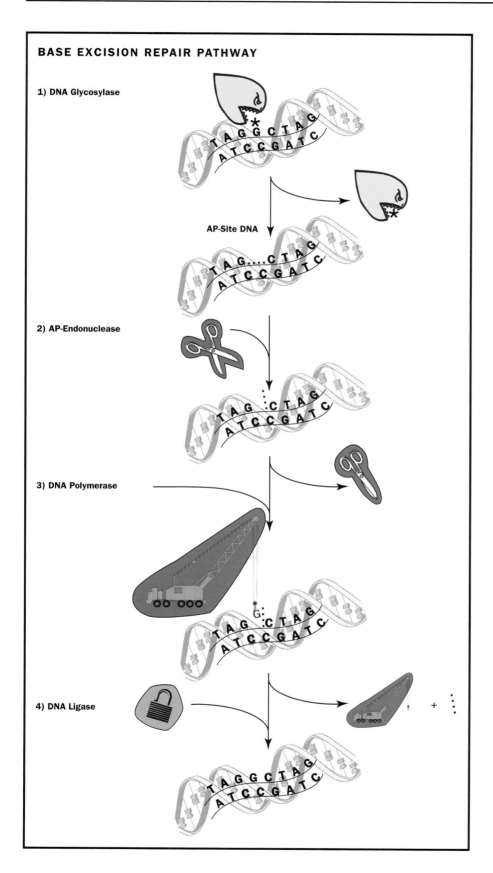

BASE EXCISION REPAIR PATHWAY

1) DNA Glycosylase

AP-Site DNA

2) AP-Endonuclease

3) DNA Polymerase

4) DNA Ligase

Figure 3. Schematic diagram of the base excision repair pathway. The damaged base (in this case, guanine) is removed by a glycosylase enzyme. This creates an AP-site, which is recognized and cleaved by AP-endonuclease. DNA polymerase fills in the gap. DNA ligase links the broken strand together again.

bond linking the modified base to its sugar, which results in the release of the modified base from the DNA chain and in the insertion of an AP-site. Several types of DNA glycosylases exist, each one specifically excising a different type of damaged base. It is important that a DNA glycosylase act only on damaged and not natural DNA bases, otherwise too many baseless sugars would be produced, weakening the integrity of the DNA chain.

Excision of the damaged base by a DNA glycosylase creates an AP-site, which in turn is acted upon by the second enzyme in the BER pathway, apurinic/apyrimidinic (AP) endonuclease (Figure 3, step 2). The most abundant AP-endonuclease in human cells cleaves (incises) the sugar-phosphate backbone on the left side of the baseless sugar to yield a one-nucleotide gap. On the left margin of the incision is a normal nucleotide (DNA base + sugar + phosphate); however, the right margin of the gap contains the baseless sugar-phosphate residue.

In order to fill the gap (replace the missing nucleotide), an enzyme specialized in synthesizing DNA, a DNA polymerase, will insert the correct nucleotide into the gap and link it to the normal nucleotide on the left margin by recognizing which base is opposite the gap on the complementary DNA strand. Figure 3, step 3 shows that the DNA polymerase recognizes that a G nucleotide is needed since the complementary base is a C. Note that an entire nucleotide is added here, not just a base. Before DNA polymerase is finished with the repair of the one-nucleotide gap, it removes the baseless sugar phosphate left behind by AP-endonuclease.

At this point, repair of the gap is almost, but not quite, finished, since there is a "nick" in the top DNA strand at the right margin of the former gap. Thus, the final step in the BER pathway is to **ligate** the DNA strands on both sides of the nick (Figure 3, step 4). If we examine the sugar phosphate DNA chain shown in Figure 2, we can see that the sugars that carry the DNA bases are linked together by phosphate groups. This type of linkage is referred to as a **phosphodiester bond**. The enzyme DNA ligase joins the strands by creating a phosphodiester bond between them, sealing the nick. In summary, the basic steps of the BER pathway are damage recognition and base excision, AP-site incision, DNA repair synthesis, and DNA ligation.

Nucleotide Excision Repair

DNA damage that involves particularly "bulky" molecules or chemical bonds between bases, or that significantly distorts the double-stranded structure of DNA, is subject to repair by the nucleotide excision repair (NER) pathway. For example, it has long been known that the ultraviolet (UV) light in sunshine can damage DNA by forming what are called photoproducts. UV radiation excites many types of molecules, causing them to react with each other and with DNA. In particular, UV light can catalyze the formation of chemical bonds between adjacent thymine and/or cytosine bases; these bonds are called intra-strand UV crosslinks (Figure 4A). These crosslinked bases distort the double-stranded structure of DNA and block DNA replication.

A second example of bulky DNA damage is that caused by large, organic molecules like aflatoxin, found in mold-contaminated peanuts, and benzo[*a*]pyrene (Figure 4B), a main component of smoke and soot. Both

ligate join together

phosphodiester bond the link between two nucleotides in DNA or RNA

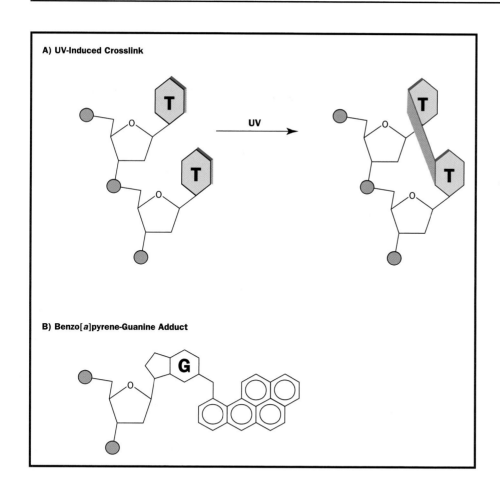

A) UV-Induced Crosslink

UV →

B) Benzo[*a*]pyrene-Guanine Adduct

Figure 4. Two types of mutations repaired by the nucleotide excision repair pathway. Ultraviolet light can trigger a chemical reaction that links adjacent thymines. Complex organic molecules such as benzo[*a*]pyrene can link on to a base such as guanine.

aflatoxin and benzo[*a*]pyrene are potent **carcinogens**. Ingestion or inhalation of these and similar compounds activates the body's detoxification systems, which convert the **hydrophobic** organic molecules into water-soluble forms for removal. However, the intermediate forms of aflatoxin and benzo[*a*]pyrene produced during the detoxification reaction happen to be very reactive with DNA purines, and form DNA base adducts (they "add on" to DNA). Specifically, such compounds tend to adduct guanine and, to a lesser extent, adenine. These large DNA adducts can cause mutations, and, since they block DNA replication, deletions of large segments of DNA can occur. Also, they activate the cell's damage surveillance systems, and, if not repaired, can cause cell death (apoptosis).

The mechanism of NER, involving some thirty proteins, is more complex than that of BER, but the basic principles are similar: damage recognition, damage excision, DNA repair synthesis, and DNA ligation (Figure 5). Damage recognition is obviously very important (Figure 5, step 1), but how can a single multiprotein complex detect so many different types of DNA damage? The answer is that the DNA damage must (1) distort the normal double-stranded structure of DNA, and/or (2) block transcription by RNA polymerase. Unusual kinks or twists in double-stranded DNA are recognized by the NER damage-recognition multiprotein complex. Also, when RNA polymerase stalls at a damaged DNA base, components of the NER damage-recognition complex are recruited to the site of damage.

carcinogens substances that cause cancer

hydrophobic "water hating," such as oils

DNA crosslinks also interfere with another vital cellular process: transcription of genes by RNA polymerase.

Next, the double-stranded DNA adjacent to the damage is unwound by a DNA unwinding enzyme called a helicase (Figure 5, step 2). Unwinding of the DNA allows repair proteins access to the site of damage. The DNA strand containing the damaged base is then cleaved a few nucleotides after the damage, and about twenty-five nucleotides before it, by specific endonucleases associated with the NER protein complex (Figure 5, step 3). Endonucleases are enzymes that cleave inside a segment of DNA.

Next, the DNA segment that contains damage is displaced by DNA polymerase and associated proteins, and a corresponding repair patch is synthesized (Figure 5, step 4). Lastly, DNA **ligase** seals the nick, joining the newly synthesized piece of DNA to the preexisting strand (Figure 5, step 5).

ligase enzyme that repairs breaks in DNA

DNA Mismatch Repair

The DNA mismatch repair (MMR) pathway has evolved to correct errors made by DNA polymerase during DNA replication. Such errors fall into two broad categories: base substitutions and insertions/deletions. A base substitution error occurs when DNA polymerase inserts an incorrect (noncomplementary) nucleotide opposite the template base, like a T opposite G instead of C, or A opposite C instead of G. These incorrect base pairs are referred to as mispairs or mismatches. Often, DNA polymerase will make a base substitution error when copying a base that has been damaged by alkylation. For example, DNA polymerase will frequently insert a T opposite O^6-methylguanine on the other strand.

An insertion error occurs when DNA polymerase adds one or more extra nucleotides (+1, +2, +3, and so on) to a sequence; a deletion error is made when one or more nucleotides (−1, −2, −3, and so on) are omitted from a sequence. Sequences that contain repeats of the same nucleotide (mononucleotide repeat), such as AAAAAAAA, are particularly vulnerable to +1 or −1 insertion/deletion errors when copied by DNA polymerase. Such sequences might be called "slippery," in that DNA polymerase can "slide" on the DNA and lose its place. Other repetitive sequences, like the dinucleotide repeat CACACACA and the trinucleotide repeat CTGCTGCTG, are prone to +2 and +3, or −2 and −3 insertion/deletion errors, respectively. These repetitive DNA sequences are called **microsatellites**.

microsatellites small repetitive DNA elements dispersed throughout the genome

Defects in DNA mismatch repair have been found in several types of cancer, notably colon cancer, and microsatellite sequences that are either shorter or longer than normal are a hallmark of defective MMR. Expansion of trinucleotide repeat sequences is associated with a number of hereditary neurological disorders, such as fragile X syndrome, myotonic dystrophy, and Huntington's disease.

The process of MMR, like the BER and NER pathways, comprises damage recognition, damage excision, DNA repair synthesis, and DNA ligation. First, a mismatch or insertion/deletion error must be recognized by a complex of proteins specialized for the particular type of damage (mismatch, or small or large insertion/deletion). Just how the mismatch recognition protein complex "knows" which DNA strand contains the "right" nucleotide and/or which DNA strand contains the "wrong" one has not yet been determined.

Next, a phosphodiester bond in the DNA strand containing the mismatched nucleotide is cleaved by an endonuclease, the strand is displaced

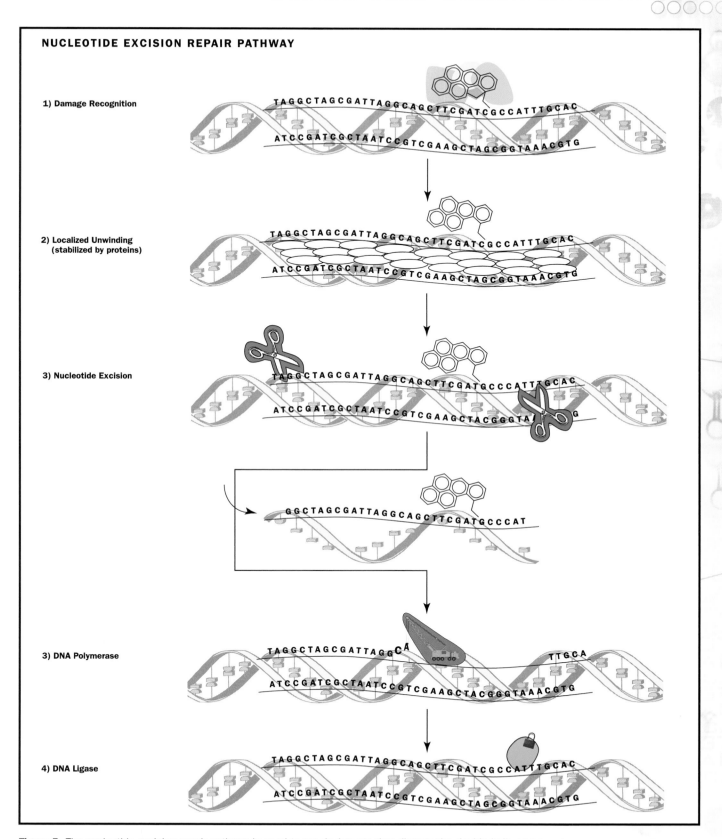

NUCLEOTIDE EXCISION REPAIR PATHWAY

1) Damage Recognition

2) Localized Unwinding
(stabilized by proteins)

3) Nucleotide Excision

3) DNA Polymerase

4) DNA Ligase

Figure 5. The nucleotide excision repair pathway is used to repair damage that distorts the double helix or prevents replication. After recognition, helicase unwinds the two strands, which are stabilized by single-strand binding proteins. Multiple nucleotides are removed from either side of the damage. DNA polymerase fills the gap, and ligase relinks the strand.

by DNA helicase, and a portion of the strand is removed by a combination of DNA exonuclease and DNA polymerase. Lastly, DNA polymerase carries out DNA repair synthesis, and DNA ligase restores the continuity of the sugar-phosphate-DNA backbone. The patch of DNA newly synthesized by the MMR DNA polymerase is relatively large, approximately 1,000 nucleotides long, compared to the DNA repair synthesis that takes place during BER, which typically replaces 1 nucleotide, or NER, which replaces approximately 30 nucleotides. MMR is especially important in tissues that are constantly regenerating, like the intestinal lining and the endometrium (the lining of the uterus), since growth requires DNA replication, which sometimes makes mistakes.

Future Directions

In addition to the three critical DNA repair pathways already discussed (BER, NER, and MMR), there are two additional types of DNA repair: double-strand break repair and recombinational repair. These are both complex phenomena, and scientists' understanding of them is still at an early stage. Also, many questions about BER, NER, and MMR still await answers. For example, since DNA damage that escapes repair leads to deleterious alterations of our DNA, could we prevent mutation by increasing the levels of DNA repair proteins? Could we live longer and healthier lives with more or better DNA repair? How are DNA repair pathways regulated by the cell? Is there such a thing as too much DNA repair? If repairs always took place whenever DNA damage occurred, would there be no evolution? Exactly how do the proteins and enzymes involved in DNA repair accomplish their jobs? These and many other exciting lines of inquiry are in store for future investigators. SEE ALSO APOPTOSIS; CANCER; CARCINOGENS; DNA POLYMERASES; FRAGILE X SYNDROME; MUTAGEN; MUTATION; NUCLEASES; NUCLEOTIDE; RNA POLYMERASES; TRIPLET REPEAT DISEASE.

Samuel E. Bennett and Dale Mosbaugh

Bibliography

Friedberg, E. C., G. C. Walker, and W. Siede. *DNA Repair and Mutagenesis.* Washington, DC: ASM Press, 1995.

Hanawalt, P. C. "DNA Repair: The Basis for Cockayne Syndrome." *Nature* 405, no. 6785 (2000): 415–416.

Kolodner, R. D. "Guarding Against Mutation." *Nature* 407, no. 6805 (2000): 687–689.

Marx, J. "DNA Repair Comes into Its Own." *Science* 266, no. 5186 (1994): 728–730.

Rennie, J. "Kissing Cousins: A DNA Repair System Stops Species from Interbreeding." *Scientific American* 262, no. 2 (1990): 22–23.

Wood, R. D., et al. "Human DNA Repair Genes." *Science* 291 (2001): 1284–1289.

DNA Structure and Function, History

DNA was discovered in the nineteenth century, but its significance as the physical basis of inheritance was not understood until midway through the twentieth. The realization that it was the molecule of heredity led to intensive efforts to determine its three-dimensional structure, and to understand how it stores and transmits genetic information. The discovery of the structure of DNA, and the elucidation of its function, ranks as one of the greatest achievements of science.

Discovery of DNA

Deoxyribonucleic acid (DNA) was first discovered in 1869 by Johann Friedrich Miescher (1844–1895), a young Swiss chemist studying in Tübingen, Germany. Miescher's interest in the biochemistry of the cell nucleus led him to collect used surgical bandages, from which he collected pus (white blood cells), which have very large nuclei. From these, he purified a new compound, which he termed "nuclein." Miescher showed that nuclein was a large molecule, acidic, and rich in phosphorus. Miescher continued to work with nuclein over the next two decades, turning for his source to salmon sperm, which have exceptionally large nuclei and were plentiful in the rivers near his laboratory. One of his students renamed the compound "nucleic acid."

In 1885 the German biologist Oskar Hertwig (1849–1922) suggested that nucleic acid might be the hereditary material, based on its presence in the nucleus and the growing certainty that the nucleus was the center of heredity. Despite this promising beginning, no further progress was made in understanding its true role until the 1940s.

The biochemistry of nucleic acid continued to be studied, however, and by 1900, scientists had learned that it was composed of three parts: a sugar, a phosphate, and a base (together termed a **nucleotide**). The five-carbon sugar is a ringed structure, and it forms an alternating chain with phosphate (a phosphorus atom surrounded by four oxygens). Also attached to the sugar is one of five different bases: adenine, cytosine, guanine, thymine, and uracil, usually abbreviated A, C, G, T, and U. The names of the bases are related to their historical origin: Guanine was isolated from bird guano, thymine from the thymus gland of calves, and adenine from calf pancreas ("adeno-" is a Greek root for gland); uracil is chemically related to urea; and "cyto-" means cell.

nucleotide a building block of RNA or DNA

In the 1920s it was discovered that there were two types of sugars, ribose and deoxyribose, differing by the presence or absence of one oxygen atom. DNA was shown to incorporate the bases A, C, G, and T, while RNA (ribonucleic acid) incorporates A, C, G, and U. Much of this work was carried out by the German biochemist Albrecht Kossel (1853–1927) and the American Phoebus Aaron Levene (1869–1940).

Also by the 1920s, Thomas Hunt Morgan and his colleagues had shown that genes, whatever they were made of, were carried on chromosomes. Chromosomes were shown to contain both protein and DNA, so the question of which of these two substances composed the genes took center stage. The more complex chemical nature of proteins gave them the theoretical edge, an opinion given great and, in hindsight, unfortunate weight by Levene, a widely respected biochemist.

Levene proposed the tetranucleotide hypothesis, which held that the structure of DNA was a monotonous repetition of the four nucleotides in succession. Levene's evidence was that DNA, which he had isolated from a variety of sources, had roughly equal amounts of A, C, G, and T. While the actual proportions he found were not exact, Levene attributed the differences to experimental error, rather than biochemical reality. Such a simple

Oswald Avery showed that DNA from dead S strain of *Pneumococcus* bacteria could transform a harmless strain into a deadly strain.

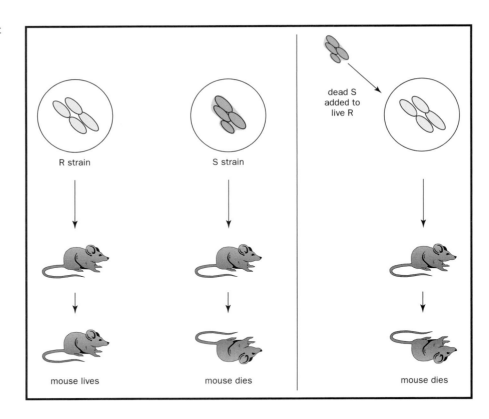

and highly regular molecule as the one Levene proposed could not account for the diversity of life, however, and so DNA was assumed to play only a structural role in the chromosome.

DNA Is the Transforming Factor

DNA was not again taken seriously as the hereditary material until 1944, when Oswald Avery (1877–1955) published a landmark paper outlining his experiments with two strains of *Pneumococcus* bacteria. The S ("smooth") strain was able to cause disease in mice, while the R ("rough") strain was not. Under the microscope, the S strain had a smooth, glistening surface, due to a sugar capsule it secreted. The R strain lacked the capsule.

Fifteen years earlier, Frederick Griffith had shown that injecting R bacteria plus heat-killed S bacteria into mice would cause disease just as surely as injecting live S bacteria, a result that Griffith attributed to a "transforming principle." When Avery grew R bacteria in a dish with the heat-killed S bacteria, he saw that the R bacteria were transformed into S bacteria, capable of making capsules and causing disease, just as Griffith had observed. Avery's group purified the components of the S bacteria, and showed that DNA alone could cause the transformation, while protein could not.

Though not immediately accepted by all scientists, Avery's discovery triggered intense interest among biochemists and geneticists, who turned their attention to discovering how DNA could be the genetic molecule. As in every other branch of biochemistry, the structure was presumed to hold the key to the function, and so the hunt was on for the structure of DNA.

One key was the discovery by Erwin Chargaff (1905–2002) that, contrary to Levene's conclusions, the nucleotide proportions were not all the same. Instead, in 1950 Chargaff showed that they varied from species to species, although within a species they were constant between tissues. Further, and most tantalizingly, he discovered that the amounts of adenine and thymine were equal to one another, and the amounts of cytosine and guanine were equal to one another; in other words, A = T, C = G. Chargaff initially did not understand the significance of this discovery, however.

Model Building

At the same time, other groups were trying to solve the DNA structure puzzle by building three-dimensional models. The group that succeeded was that of Francis Crick (born 1916) and James Watson (born 1928), in Cambridge, England. Watson and Crick built their models using data from X-ray crystallography, a technique for measuring interatomic distances through analysis of the scatter patterns made by X rays bouncing off a pure crystal. Rosalind Franklin (1920-1958), working in London, had made the best X-ray pictures of DNA, and Watson and Crick had been given access to these (without Franklin's permission) at a critical time in their model-building endeavors and saw features in it that Franklin had not yet discovered. Shortly after, Watson and Crick deduced the correct structure, and published their work in April 1953.

DNA is a double helix, in which two sugar-phosphate strands wind around each other, forming a structure that looks like a broad spiral staircase. The sugar molecule has a head and a tail, and each sugar-phosphate strand therefore has a direction, like a chain of arrows. The two helical strands point in opposite directions.

The bases project toward the inside like stair treads. In contrast to the monotony predicted by the tetranucleotide hypothesis, the bases along one strand can be in any sequence whatsoever. Critical to the stability of DNA is the hydrogen bonding between the bases across the interior. These weak chemical attractions form only when the base atoms are positioned just so—in particular, Watson and Crick discovered, only when adenine projects across to meet a thymine, and guanine a cytosine. Chargaff's ratios reflect this essential **base pairing**.

Replication

Given the DNA structure, three questions immediately arose: How is DNA copied to allow faithful inheritance of genes, how does DNA store information, and how does DNA use that information to determine the properties of the cell? Watson and Crick addressed the theoretical underpinnings of each of these issues in their next paper, published in May 1953. Regarding copying, they wrote:

> Now our model for deoxyribonucleic acid is, in effect, a *pair* of templates, each of which is complementary to the other. We imagine that prior to duplication the hydrogen bonds are broken, and the two chains unwind and separate. Each chain then acts as a template for the formation onto itself of a new companion chain, so that eventually we shall have *two* pairs of chains where we only had one before. Moreover, the sequence of the pairs of bases will have been duplicated exactly.

In outline, this is precisely correct, as confirmed in 1957 by Matthew Meselson (b. 1929) and Franklin Stahl (b. 1930). They fed bacteria

base pairing linking of two nucleotides (either DNA or RNA) linked by weak bonds

polymerase enzyme complex that synthesizes DNA or RNA from individual nucleotides

radioactive nucleotides, so that both DNA chains would be labeled with radioactivity. They then removed the radioactive nucleotide source and measured the dilution of the radioactivity in each round of DNA copying. After one round, each DNA molecule had half the amount of radioactivity. According to the Watson-Crick prediction, this meant that one strand of each was completely new. After the second round, half the DNA molecules maintained this level of radioactivity, and half had none at all, just as expected from the Watson-Crick model.

This process, in which one parental DNA strand is conserved, unchanged, and acts as a template to synthesize a new partner, is called semiconservative replication. The details of the copying process, called replication, are much more complex than this simple outline, though, and the entire process is still not fully understood in all its particulars. Central to it is DNA **polymerase**, a large multiprotein complex first discovered in 1957 by Arthur Kornberg (born 1918) and Severo Ochoa (1905–1993).

Coding

Regarding how DNA can act as a gene, storing information and directing activities of the cell, Watson and Crick wrote,

> The phosphate-sugar backbone of our model is completely regular, but any sequence of the pairs of bases can fit into the structure. It follows that in a long molecule many different permutations are possible, and it therefore seems likely that the precise sequence of the bases is the code which carries the genetical information.

This too is correct, as a series of experiments showed.

Proteins are the workhorses of the cell, controlling the rates of all the reactions within and providing much of the cell's structure as well. Therefore, it was quickly realized, genes must control the production of proteins, and the genetic information carried in the sequence of bases in DNA is a code for the sequence of amino acids in proteins. Proteins are made of twenty amino acids, linked together in varying sequences. The sequence determines the shape and chemical properties of the protein, and so specifying protein sequence is the essential role of DNA.

Since there are four bases and twenty amino acids, a single nucleotide is not enough to specify one amino acid. Even two are not enough, because two nucleotides will only give rise to sixteen unique combinations (AA, AC, AG, and so on). Therefore, it was immediately obvious that each amino acid must be coded for by at least three nucleotides.

Stating this must be so was quite a bit easier than working out the details of how DNA and amino acids interacted to form a protein. Some researchers suggested a solution in which amino acids lined up directly on the surface of DNA; other alternatives were also proposed. A suggestion by Crick that there was some type of adapter between the two was confirmed with the discovery of transfer RNA. In fact, DNA and amino acids never do interact during protein synthesis—instead, an RNA copy of DNA is made (messenger RNA), which links with transfer RNAs that carry amino acids. The code itself was worked out between 1961 and 1967, by several different groups, including Marshall Nirenberg (born 1927), Har Gobind Khorana (born 1922), and Johann Matthaei, who developed pioneering cell-free systems that allowed researchers to work without the complexity and constraints involved with living organisms.

These heroic discoveries marked the beginning of the molecular biology revolution. From them even deeper questions have arisen, about how **gene expression** is regulated, how genes control development, and how (and whether) genes can be modified to treat disease and improve human life. SEE ALSO CRICK, FRANCIS; DNA; GENETIC CODE; NUCLEOTIDE; REPLICATION; TRANSCRIPTION; WATSON, JAMES.

Richard Robinson

gene expression use of a gene to create the corresponding protein

Bibliography

Judson, Horace F. *The Eighth Day of Creation*, expanded edition. Plainview, NY: Cold Spring Harbor Press, 1996.

Olby, Robert C. *The Path to the Double Helix*. Mineola, NY: Dover, 1994.

Watson, James D. *The Double Helix*. New York: Penguin, 1968.

DNA Vaccines

A DNA vaccine uses foreign DNA to express an encoded protein and stimulate the body's immune system. It represents a new way to immunize against infectious disease that is potentially less expensive than classic vaccination forms.

Classic Vaccines

One of the greatest achievements in the history of medicine has been the development of vaccination. The use of vaccines has saved more lives than all other medical procedures combined, and represents one of the highest points in civilization's technical accomplishments. Vaccines are used to mobilize the immune system to prevent or combat infectious disease caused by exposure to viruses, bacteria, or parasites.

A vaccine works by mimicking an infectious agent and inducing a protective immune response in the host, without actually causing the disease. Successful vaccination provides protection for individuals by making them immune to the disease, and it protects whole populations by hindering the spread of the infectious agent.

Historically, vaccines have consisted of formulations using live, noninfectious (attenuated) microbes that resemble the original **pathogen**; whole organisms that have been killed; or purified by-products of the infectious agent. More recently, some vaccines have used recombinant DNA technology to genetically engineer purified proteins from infectious agents.

pathogen disease-causing organism

All of these classic vaccines are based on the principle of using a protein to stimulate the immune system. In other words, the individual is immunized with a protein vaccine consisting of either a modified pathogen or some protein or proteins derived from that pathogen. When the immune system encounters a foreign protein (called an **antigen**), it mounts a two-pronged defense. It produces proteins called antibodies, which can bind and neutralize the antigen, and it produces specific immune cells, which work to eliminate those host cells that have been infected by the microbe. Thus our immune system is capable of producing two distinct

antigen a foreign substance that provokes an immune response

253

Certain DNA vaccines have been tested on cattle.

types of responses to combat infectious microbes: An antibody response and a cell-mediated response. Typically, vaccines activate only the antibody response to an infectious microbe.

Advantages of DNA Vaccines

DNA vaccines represent a new approach to immunization, in that an individual is directly inoculated (injected) with DNA that genetically encodes one or more of the antigens associated with the infectious agent. In effect, the recipient of a DNA vaccine produces the immunizing protein (antigen) within his own cells as a result of the immunization process.

This revolutionary approach to vaccination offers many advantages over conventional vaccines. A major advantage is that DNA vaccination stimulates both the antibody and cell-mediated components of the immune system, whereas conventional protein vaccines usually stimulate only the antibody response. Furthermore, DNA vaccines are simpler to produce and store than conventional vaccines, and are therefore less expensive. Preliminary studies to date indicate that DNA vaccines appear to be very safe and to produce no side effects.

DNA Vaccination Techniques

DNA vaccination involves immunization with a circular piece of DNA, known as a plasmid, that contains the gene (or genes) that code for an

antigen. When injected into an individual, the plasmid is taken up by cells and its genetic information is translated into the immunizing protein. This enables the host immune system to respond to the antigen as it is presented to other cells.

In many respects, this process is reminiscent of what occurs during a viral infection, when viral proteins are expressed within host cells. Thus, a DNA vaccine is somewhat like a very simple, nonreplicating virus. However, plasmid DNA vaccines do not replicate within the host, and therefore do not infect neighboring cells, as occurs during a viral infection.

This innovation in vaccination strategy was discovered some years ago, but the active development of this technology only began after Stephen Johnston's group at the University of Texas Southwestern Medical Center demonstrated that **plasmid** DNA can induce the formation of antibodies against an encoded protein in 1992. Johnston's group was able to show that when mice are innoculated with plasmid DNA encoding human growth hormone, the mice produce antibodies against the hormone.

Shortly thereafter, another research group reported that a protective cell-mediated immune response against influenza virus followed immunization with plasmid DNA encoding an influenza virus protein. This study demonstrated that DNA-based immunization stimulates both components of the immune system and helped to establish that DNA immunization is capable of inducing a protective response against infection.

There are two basic ways to inoculate with plasmid-based vaccines. The first involves direct inoculation into muscle tissue, with the plasmid DNA suspended in a saline (salt) solution ("naked" DNA). The DNA is eventually taken up into nearby cells and processed to express the encoded antigen. The other method uses a high-pressure device, a so-called gene gun, to propel DNA-coated gold particles into cells in the skin. This method is sometimes referred to as **biolistic** particle inoculation. Both methods are widely used, and newer methods for the delivery of plasmid DNA vaccines are currently in development.

As of December 2001 there are several clinical studies in progress to evaluate the effectiveness of DNA vaccination. Most of these studies were targeted against viral infectious agents, such as HIV, hepatitis B, and influenza virus. However, there are also studies in progress to develop DNA vaccines against malaria and tuberculosis. There are even several efforts to develop DNA vaccines against various forms of cancer, an approach which seems to offer significant hope for the future. SEE ALSO IMMUNE SYSTEM GENETICS; PLASMID.

Darrell R. Galloway

antigen a foreign substance that provokes an immune response

plasmid a small ring of DNA found in many bacteria

biolistic firing a microscopic pellet into a biological sample (from biological/ballistic)

Bibliography

Tang, D. C., M. Devit, and S. A. Johnston. "Genetic Immunization Is a Simple Method for Eliciting an Immune Response." *Nature* 356 (1992): 152–154.

Ulmer, Jeffrey B., John J. Donnelly, and Margaret A. Liu. "DNA Vaccines Promising: A New Approach to Inducing Protective Immunity." *ASM News* 62 (1996): 476–479.

Weiner, David B., and Ronald C. Kennedy. "Genetic Vaccines." *Scientific American* 281 (1999): 50–57.

Wolff, J. A., et al. "Direct Gene Transfer into Mouse Muscle in vivo." *Science* 247 (1990): 1465–1468.

trisomy presence of three, instead of two, copies of a particular chromosome

Dominance *See Inheritance Patterns*

Down Syndrome

Down syndrome, also called **trisomy** 21, is the single most common genetic cause of moderate mental retardation. It occurs in about one of every eight hundred live births. It is caused by the inheritance of an extra copy of chromosome 21. The condition was named after an English physician, J. Langdon Down, who in 1866 published the first report describing patients with similar facial features and mental retardation. The chromosomal basis of Down syndrome was not determined until nearly a century later.

Clinical Features

Down syndrome is associated with a characteristic physical appearance, mental retardation, and specific birth defects or health conditions. The facial features, in addition to low muscle tone (called hypotonia), are often the first signs that alert a physician to a potential diagnosis of Down syndrome. These features include an up-slant of the outer corners of the eyes, small skin folds over the inner corners of the eyes, a small nose with a flat nasal bridge, a flat profile, and a large, grooved tongue that often protrudes from the mouth. Other physical characteristics can include a short neck, excess skin on the back of the neck, short hands with a single palmar crease, a wide gap between the first and second toes, and short stature. There are many individuals without Down syndrome who may have one or more of these features. It is only when the features occur together and the appropriate genetic test (chromosome studies) confirms clinical suspicion that a diagnosis of Down syndrome is made.

All individuals with Down syndrome have mental retardation, usually mild to moderate. The degree of learning disability may not be immediately apparent, since social ability generally exceeds scholastic ability. Early intervention programs are important for giving people with Down syndrome the best chance to maximize their learning potential.

Certain birth defects and health conditions are more common in individuals with Down syndrome. The most common birth defect is a congenital heart defect, affecting 40 percent to 50 percent of newborns with the condition. Although many can be repaired with surgery, congenital heart defects remain the major cause of early death among affected persons. Individuals with Down syndrome have an increased chance of experiencing hearing loss, vision problems, and thyroid disease, as well as an increased susceptibility to infections. Because of such concerns, specific guidelines for the health care of individuals with Down syndrome have been developed.

Chromosomal Basis of Down Syndrome

In 1959 French geneticist Jerome Lejeune recognized that individuals with Down syndrome have forty-seven chromosomes instead of the usual forty-six. Later, it was determined that it is an extra copy of chromosome 21 that causes the condition. It is not yet clear how the extra chromosome causes the clinical features, although it is believed that an "extra dose" of one or more of the genes on the chromosome is responsible.

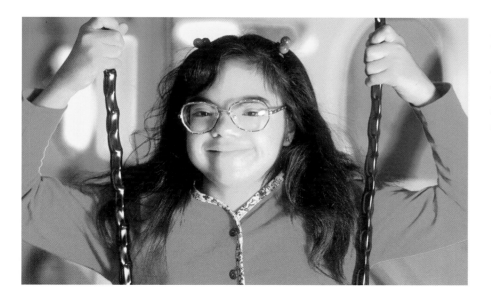

Down syndrome is a chromosome disorder that causes mental retardation. Characteristic features of Down syndrome include a flattened midface, small ears and mouth, and short, broad hands.

There are three types of Down syndrome: trisomy 21, mosaic Down syndrome, and translocation Down syndrome. In 94 percent of cases, the extra copy of chromosome 21 stands alone (is not attached to any other chromosomes) and is present in every cell of the body. This is called trisomy 21, trisomy meaning three.

Trisomy 21 occurs due to a chromosome packaging error. Usually when the body makes its sex cells (egg or sperm cells) during **meiosis**, it packages up one chromosome from each pair. However, sometimes an error (nondisjunction) occurs, causing both chromosomes from a pair to get packaged together. If the sex cell with the extra chromosome is fertilized by a sex cell with the usual chromosome number, the resulting embryo will have a trisomy. If the extra chromosome is chromosome 21, the embryo will have Down syndrome. About 75 percent of embryos with trisomy 21 abort spontaneously before birth. Nondisjunction occurs by chance in the making of both egg and sperm cells, but it happens more often in egg cells as women get older. Thus, the chance of having a baby with Down syndrome increases with increasing maternal age.

meiosis cell division that forms eggs or sperm

Translocation Down syndrome, which accounts for 3 percent to 4 percent of cases, occurs when the extra copy of chromosome 21 is attached to another chromosome. In about one-fourth of the cases where a person has translocation Down syndrome, he or she inherited the translocation from a parent. Therefore it is important to test the parents' chromosomes in these cases, for purposes of future family planning.

The third type of Down syndrome is the mosaic type, which occurs in 2 percent to 3 percent of cases. In mosaic Down syndrome, a person has some cells with an extra copy of chromosome 21 and some cells with the usual two copies. People with mosaic Down syndrome may or may not have milder symptoms than people with "full" trisomy 21.

Testing for Down Syndrome

Cytogenetic analysis looks at the number and structure of a person's chromosomes. This test, which can be performed on a blood sample, is the test used to definitively determine if an individual has Down syndrome.

Risk of Down syndrome and other chromosomal abnormalities increases with maternal age. Adapted from Evans, 1989.

Maternal Age	Risk of Down Syndrome	Total Risk for all Chromosomal Abnormalities
20	1/1667	1/526
21	1/1667	1/526
22	1/1429	1/500
23	1/1429	1/500
24	1/1250	1/476
25	1/1250	1/476
26	1/1176	1/476
27	1/1111	1/455
28	1/1053	1/435
29	1/1000	1/417
30	1/952	1/385
31	1/909	1/385
32	1/769	1/322
33	1/602	1/286
34	1/485	1/238
35	1/378	1/192
36	1/289	1/156
37	1/224	1/127
38	1/173	1/102
39	1/136	1/83
40	1/106	1/66
41	1/82	1/53
42	1/63	1/42
43	1/49	1/33
44	1/38	1/26
45	1/38	1/21
46	1/23	1/16
47	1/18	1/13
48	1/14	1/10
49	1/11	1/8

amniocentesis removal of fluid from the amniotic sac surrounding a fetus, for diagnosis

Prenatal diagnosis for Down syndrome (testing for the condition during pregnancy) is possible. Chromosome studies can be performed on fetal cells collected via chorionic villus sampling (CVS) at ten to twelve weeks of pregnancy or by **amniocentesis** at fifteen to twenty weeks of pregnancy. Because of the link between the mother's age and the chance of Down syndrome, prenatal diagnosis for Down syndrome and other chromosome conditions is routinely offered to women thirty-five and older. Whether to pursue prenatal diagnosis is a personal decision that can only be made by the parents. SEE ALSO BIRTH DEFECTS; CHROMOSOMAL ABERRATIONS; MEIOSIS; MOSAICISM; NONDISJUNCTION; PRENATAL DIAGNOSIS.

Angela Trepanier and Gerald L. Feldman

Bibliography

Evans, Mark I., et al. *Fetal Diagnosis and Therapy: Science, Ethics, and the Law.* Philadelphia, PA: JB Lippincott Co., 1989.

Gardner, R., J. McKinlay, and Grant R. Sutherland. *Chromosome Abnormalities and Genetic Counseling*, 2nd ed. New York: Oxford University Press, 1996.

Nussbaum, Robert L., Roderick R. McInnes, and Huntington F. Willard, eds. *Thompson & Thompson Genetics in Medicine*, 6th ed. Philadelphia, PA: W. B. Saunders, 2001.

Pueschel, Siegfried M., ed. *A Parent's Guide to Down Syndrome: Toward a Brighter Future*, 2nd ed. Baltimore, MD: Paul H. Brooks Publishing, 2001.

Internet Resource

Cohen, William I., ed. "Health Care Guidelines for Individuals with Down Syndrome: 1999 Revision." *Down Syndrome Quarterly* 4, no. 3 (1999): 1–15. <http://www.denison.edu/dsq/health99.shtml>.

Photo Credits

Unless noted below or within its caption, the illustrations and tables featured in *Genetics* were developed by Richard Robinson, and rendered by GGS Information Services. The photographs appearing in the text were reproduced by permission of the following sources:

Volume 1

Accelerated Aging: Progeria (p. 2), Photo courtesy of The Progeria Research Foundation, Inc. and the Barnett Family; *Aging and Life Span* (p. 8), Fisher, Leonard Everett, Mr.; *Agricultural Biotechnology* (p. 10), © Keren Su/Corbis; *Alzheimer's Disease* (p. 15), AP/Wide World Photos; *Antibiotic Resistance* (p. 27), © Hank Morgan/Science Photo Library, Photo Researchers, Inc.; *Apoptosis* (p. 32), © Microworks/Phototake; *Arabidopsis thaliana* (p. 34), © Steinmark/ Custom Medical Stock Photo; *Archaea* (p. 38), © Eurelios/Phototake; *Behavior* (p. 47), © Norbert Schafer/Corbis; *Bioinformatics* (p. 53), © T. Bannor/Custom Medical Stock Photo; *Bioremediation* (p. 60), Merjenburgh/ Greenpeace; *Bioremediation* (p. 61), AP/Wide World Photos; *Biotechnology and Genetic Engineering, History of* (p. 71), © Gianni Dagl Orti/Corbis; *Biotechnology: Ethical Issues* (p. 67), © AFP/Corbis; *Birth Defects* (p. 78), AP/Wide World Photos; *Birth Defects* (p. 80), © Siebert/ Custom Medical Stock Photo; *Blotting* (p. 88), © Custom Medical Stock Photo; *Breast Cancer* (p. 90), © Custom Medical Stock Photo; *Carcinogens* (p. 98), © Custom Medical Stock Photo; *Cardiovascular Disease* (p. 102), © B&B Photos/Custom Medical Stock Photo; *Cell, Eukaryotic* (p. 111), © Dennis Kunkel/ Phototake; *Chromosomal Aberrations* (p. 122), © Pergement, Ph.D./Custom Medical Stock

Photo; *Chromosomal Banding* (p. 126), Courtesy of the Cytogenetics Laboratory, Indiana University School of Medicine; *Chromosomal Banding* (p. 127), Courtesy of the Cytogenetics Laboratory, Indiana University School of Medicine; *Chromosomal Banding* (p.128), Courtesy of the Cytogenetics Laboratory, Indiana University School of Medicine; *Chromosome, Eukaryotic* (p. 137), Photo Researchers, Inc.; *Chromosome, Eukaryotic* (p. 136), © Becker/Custom Medical Stock Photo; *Chromosome, Eukaryotic* (p. 133), Courtesy of Dr. Jeffrey Nickerson/University of Massachusetts Medical School; *Chromosome, Prokaryotic* (p. 141), © Mike Fisher/Custom Medical Stock Photo; *Chromosomes, Artificial* (p. 145), Courtesy of Dr. Huntington F. Williard/University Hospitals of Cleveland; *Cloning Organisms* (p. 163), © Dr.Yorgos Nikas/Phototake; *Cloning: Ethical Issues* (p. 159), AP/Wide World Photos; *College Professor* (p. 166), © Bob Krist/Corbis; *Colon Cancer* (p. 169), © Albert Tousson/Phototake; *Colon Cancer* (p. 167), © G-I Associates/Custom Medical Stock Photo; *Conjugation* (p. 183), © Dennis Kunkel/Phototake; *Conservation Geneticist* (p. 191), © Annie Griffiths Belt/ Corbis; *Delbrück, Max* (p. 204), Library of Congress; *Development, Genetic Control of* (p. 208), © JL Carson/Custom Medical Stock Photo; *DNA Microarrays* (p. 226), Courtesy of James Lund and Stuart Kim, Standford University; *DNA Profiling* (p. 234), AP/Wide World Photos; *DNA Vaccines* (p. 254), Penny Tweedie/Corbis-Bettmann; *Down Syndrome* (p. 257), © Custom Medical Stock Photo.

Volume 2

Embryonic Stem Cells (p. 4), Courtesy of Dr. Douglas Strathdee/University of Edinburgh, Department of Neuroscience; *Embryonic Stem Cells* (p. 5), Courtesy of Dr. Douglas Strathdee/University of Edinburgh, Department of Neuroscience; *Escherichia coli* (*E. coli* bacterium) (p. 10), © Custom Medical Stock Photo; *Eubacteria* (p. 14), © Scimat/Photo Researchers; *Eubacteria* (p. 12), © Dennis Kunkel/Phototake; *Eugenics* (p. 19), American Philosophical Society; *Evolution, Molecular* (p. 22), OAR/National Undersea Research Program (NURP)/National Oceanic and Atmospheric Administration; *Fertilization* (p. 34), © David M. Phillips/Photo Researchers, Inc.; *Founder Effect* (p. 37), © Michael S. Yamashita/Corbis; *Fragile X Syndrome* (p. 41), © Siebert/Custom Medical Stock Photo; *Fruit Fly:* Drosophila (p. 44), © David M. Phillips, Science Source/Photo Researchers, Inc.; *Gel Electrophoresis* (p. 46), © Custom Medical Stock Photo; *Gene Therapy* (p. 75), AP/Wide World; *Genetic Counseling* (p. 88), © Amethyst/Custom Medical Stock Photo; *Genetic Testing* (p. 98), © Department of Clinical Cytogenetics, Addenbrookes Hospital/Science Photo Library/Photo Researchers, Inc.; *Genetically Modified Foods* (p. 109), AP/Wide World; *Genome* (p. 113), Raphael Gaillarde/Getty Images; *Genomic Medicine* (p. 119), © AFP/Corbis; *Growth Disorders* (p. 131), Courtesy Dr. Richard Pauli/U. of Wisconsin, Madison, Clinical Genetics Center; *Hemoglobinopathies* (p. 137), © Roseman/Custom Medical Stock Photo; *Heterozygote Advantage* (p. 147), © Tania Midgley/Corbis; *HPLC: High-Performance Liquid Chromatography* (p. 167), © T. Bannor/Custom Medical Stock Photo; *Human Genome Project* (p. 175), © AFP/Corbis; *Human Genome Project* (p. 176), AP/Wide World; *Individual Genetic Variation* (p. 192), © A. Wilson/Custom Medical Stock Photo; *Individual Genetic Variation* (p. 191), © A. Lowrey/Custom Medical Stock Photo; *Inheritance, Extranuclear* (p. 196), © ISM/Phototake; *Inheritance Patterns* (p. 206), photograph by Norman Lightfoot/National Audubon Society Collection/Photo Researchers, Inc.; *Intelligence* (p. 208), AP/Wide World Photos.

Volume 3

Laboratory Technician (p. 2), Mark Tade/Getty Images; *Maize* (p. 9), Courtesy of Agricultural Research Service/USDA; *Marker Systems* (p. 16), Custom Medical Stock Photo; *Mass Spectrometry* (p. 19), Ian Hodgson/© Rueters New Media; *McClintock, Barbara* (p. 21), AP/Wide World Photos; *McKusick, Victor* (p. 23), The Alan Mason Chesney Medical Archives of The Johns Hopkins Medical Institutions; *Mendel, Gregor* (p. 30), Archive Photos, Inc.; *Metabolic Disease* (p. 38), AP/Wide World Photos; *Mitochondrial Diseases* (p. 53), Courtesy of Dr. Richard Haas/University of California, San Diego, Department of Neurosciences; *Mitosis* (p. 58), J. L. Carson/Custom Medical Stock Photo; *Model Organisms* (p. 61), © Frank Lane Picture Agency/Corbis; *Molecular Anthropology* (p. 66), © John Reader, Science Photo Library/PhotoResearchers, Inc.; *Molecular Biologist* (p. 71), AP/Wide World Photos; *Morgan, Thomas Hunt* (p. 73), © Bettmann/Corbis; *Mosaicism* (p. 78), Courtesy of Carolyn Brown/Department of Medical Genetics of University of British Columbia; *Muller, Hermann* (p. 80), Library of Congress; *Muscular Dystrophy* (p. 85), © Siebert/Custom Medical Stock Photo; *Muscular Dystrophy* (p. 84), © Custom Medical Stock Photo; *Nature of the Gene, History* (p. 103), Library of Congress; *Nature of the Gene, History* (p. 102), Archive Photos, Inc.; *Nomenclature* (p. 108), Courtesy of Center for Human Genetics/Duke University Medical Center; *Nondisjunction* (p. 110), © Gale Group; *Nucleotide* (p. 115), © Lagowski/Custom Medical Stock Photo; *Nucleus* (p. 120), © John T. Hansen, Ph.D./Phototake; *Oncogenes* (p. 129), Courtesy of National Cancer Institute; *Pharmacogenetics and Pharmacogenomics* (p. 145), AP/Wide World Photos; *Plant Genetic Engineer* (p. 150), © Lowell Georgia/Corbis; *Pleiotropy* (p. 154), © Custom Medical Stock Photo; *Polyploidy* (p. 165), AP/Wide World Photos; *Population Genetics* (p. 173), AP/Wide World Photos; *Population Genetics* (p. 172), © JLM Visuals; *Prenatal Diagnosis* (p. 184), © Richard T. Nowitz/Corbis; *Prenatal Diagnosis* (p. 186), © Brigham Narins; *Prenatal Diagnosis* (p. 183), © Dr. Yorgos Nikas/Phototake; *Prion* (p. 188), AP/Wide World Photos; *Prion* (p. 189), AP/

Wide World Photos; *Privacy* (p. 191), © K. Beebe/Custom Medical Stock Photo; *Proteins* (p. 199), NASA/Marshall Space Flight Center; *Proteomics* (p. 207), © Geoff Tompkinson/ Science Photo Library, National Aubodon Society Collection/Photo Researchers, Inc.; *Proteomics* (p. 206), Lagowski/Custom Medical Stock Photo.

Volume 4

Quantitative Traits (p. 2), Photo by Peter Morenus/Courtesy of University of Connecticut; *Reproductive Technology* (p. 20), © R. Rawlins/Custom Medical Stock Photo; *Reproductive Technology: Ethical Issues* (p. 27), AP/Wide World Photos; *Restriction Enzymes* (p. 32), © Gabridge/Custom Medical Stock Photo; *Ribosome* (p. 43), © Dennis Kunkel/ Phototake; *Rodent Models* (p. 61), © Reuters NewMedia Inc./Corbis; *Roundworm:*

Caenorhabditis elegans (p. 63), © J. L. Carson/Custom Medical Stock Photo; *Sanger, Fred* (p. 65), © Bettmann/Corbis; *Sequencing DNA* (p. 74), © T. Bannor/Custom Medical Stock Photo; *Severe Combined Immune Deficiency* (p. 76), © Bettmann/Corbis; *Tay-Sachs Disease* (p. 100), © Dr. Charles J. Ball/ Corbis; *Transgenic Animals* (p. 125), Courtesy Cindy McKinney, Ph.D./Penn State University's Transgenic Mouse Facility; *Transgenic Organisms: Ethical Issues* (p. 131), © Daymon Hartley/Greenpeace; *Transgenic Plants* (p. 133), © Eurelios/Phototake; *Transplantation* (p. 140), © Reuters New Media/Corbis; *Twins* (p. 156), © Dennis Degnan/Corbis; *X Chromosome* (p. 175), © Gale Group; *Yeast* (p. 180), © Dennis Kunkel; *Zebrafish* (p. 182), Courtesy of Dr. Jordan Shin, Cardiovascular Research Center, Massachusetts General Hospital.

Glossary

α the Greek letter alpha

β the Greek letter beta

γ the Greek letter gamma

λ the Greek letter lambda

σ the Greek letter sigma

E. coli the bacterium *Escherichia coli*

"-ase" suffix indicating an enzyme

acidic having the properties of an acid; the opposite of basic

acrosomal cap tip of sperm cell that contains digestive enzymes for penetrating the egg

adenoma a tumor (cell mass) of gland cells

aerobic with oxygen, or requiring it

agar gel derived from algae

agglutinate clump together

aggregate stick together

algorithm procedure or set of steps

allele a particular form of a gene

allelic variation presence of different gene forms (alleles) in a population

allergen substance that triggers an allergic reaction

allolactose "other lactose"; a modified form of lactose

amino acid a building block of protein

amino termini the ends of a protein chain with a free NH_2 group

amniocentesis removal of fluid from the amniotic sac surrounding a fetus, for diagnosis

amplify produce many copies of, multiply

anabolic steroids hormones used to build muscle mass

anaerobic without oxygen or not requiring oxygen

androgen testosterone or other masculinizing hormone

anemia lack of oxygen-carrying capacity in the blood

aneuploidy abnormal chromosome numbers

angiogenesis growth of new blood vessels

anion negatively charged ion

anneal join together

anode positive pole

anterior front

antibody immune-system protein that binds to foreign molecules

antidiuretic a substance that prevents water loss

antigen a foreign substance that provokes an immune response

antigenicity ability to provoke an immune response

apoptosis programmed cell death

Archaea one of three domains of life, a type of cell without a nucleus

archaeans members of one of three domains of life, have types of cells without a nucleus

aspirated removed with a needle and syringe

aspiration inhalation of fluid or solids into the lungs

association analysis estimation of the relationship between alleles or genotypes and disease

asymptomatic without symptoms

ATP adenosine triphosphate, a high-energy compound used to power cell processes

ATPase an enzyme that breaks down ATP, releasing energy

attenuation weaken or dilute

atypical irregular

autoimmune reaction of the immune system to the body's own tissues

autoimmunity immune reaction to the body's own tissues

autosomal describes a chromosome other than the X and Y sex-determining chromosomes

autosome a chromosome that is not sex-determining (not X or Y)

axon the long extension of a nerve cell down which information flows

bacteriophage virus that infects bacteria

basal lowest level

base pair two nucleotides (either DNA or RNA) linked by weak bonds

basic having the properties of a base; opposite of acidic

benign type of tumor that does not invade surrounding tissue

binding protein protein that binds to another molecule, usually either DNA or protein

biodiversity degree of variety of life

bioinformatics use of information technology to analyze biological data

biolistic firing a microscopic pellet into a biological sample (from biological/ballistic)

biopolymers biological molecules formed from similar smaller molecules, such as DNA or protein

biopsy removal of tissue sample for diagnosis

biotechnology production of useful products

bipolar disorder psychiatric disease characterized by alternating mania and depression

blastocyst early stage of embryonic development

brackish a mix of salt water and fresh water

breeding analysis analysis of the offspring ratios in breeding experiments

buffers substances that counteract rapid or wide pH changes in a solution

Cajal Ramon y Cajal, Spanish neuroanatomist

carcinogens substances that cause cancer

carrier a person with one copy of a gene for a recessive trait, who therefore does not express the trait

catalyst substance that speeds a reaction without being consumed (e.g., enzyme)

catalytic describes a substance that speeds a reaction without being consumed

catalyze aid in the reaction of

cathode negative pole

cDNA complementary DNA

cell cycle sequence of growth, replication and division that produces new cells

centenarian person who lives to age 100

centromere the region of the chromosome linking chromatids

cerebrovascular related to the blood vessels in the brain

cerebrovascular disease stroke, aneurysm, or other circulatory disorder affecting the brain

charge density ratio of net charge on the protein to its molecular mass

chemotaxis movement of a cell stimulated by a chemical attractant or repellent

chemotherapeutic use of chemicals to kill cancer cells

chloroplast the photosynthetic organelle of plants and algae

chondrocyte a cell that forms cartilage

chromatid a replicated chromosome before separation from its copy

chromatin complex of DNA, histones, and other proteins, making up chromosomes

ciliated protozoa single-celled organism possessing cilia, short hair-like extensions of the cell membrane

circadian relating to day or day length

cleavage hydrolysis

cleave split

clinical trials tests performed on human subjects

codon a sequence of three mRNA nucleotides coding for one amino acid

Cold War prolonged U.S.-Soviet rivalry following World War II

colectomy colon removal

colon crypts part of the large intestine

complementary matching opposite, like hand and glove

conformation three-dimensional shape

congenital from birth

conjugation a type of DNA exchange between bacteria

cryo-electron microscope electron microscope that integrates multiple images to form a three-dimensional model of the sample

cryopreservation use of very cold temperatures to preserve a sample

cultivars plant varieties resulting from selective breeding

cytochemist chemist specializing in cellular chemistry

cytochemistry cellular chemistry

cytogenetics study of chromosome structure and behavior

cytologist a scientist who studies cells

cytokine immune system signaling molecule

cytokinesis division of the cell's cytoplasm

cytology the study of cells

cytoplasm the material in a cell, excluding the nucleus

cytosol fluid portion of a cell, not including the organelles

de novo entirely new

deleterious harmful

dementia neurological illness characterized by impaired thought or awareness

demography aspects of population structure, including size, age distribution, growth, and other factors

denature destroy the structure of

deoxynucleotide building block of DNA

dimerize linkage of two subunits

dimorphism two forms

diploid possessing pairs of chromosomes, one member of each pair derived from each parent

disaccharide two sugar molecules linked together

dizygotic fraternal or nonidentical

DNA deoxyribonucleic acid

domains regions

dominant controlling the phenotype when one allele is present

dopamine brain signaling chemical

dosage compensation equalizing of expression level of X-chromosome genes between males and females, by silencing one X chromosome in females or amplifying expression in males

ecosystem an ecological community and its environment

ectopic expression expression of a gene in the wrong cells or tissues

electrical gradient chemiosmotic gradient

electrophoresis technique for separation of molecules based on size and charge

eluting exiting

embryogenesis development of the embryo from a fertilized egg

endangered in danger of extinction throughout all or a significant portion of a species' range

endogenous derived from inside the organism

endometriosis disorder of the endometrium, the lining of the uterus

endometrium uterine lining

endonuclease enzyme that cuts DNA or RNA within the chain

endoplasmic reticulum network of membranes within the cell

endoscope tool used to see within the body

endoscopic describes procedure wherein a tool is used to see within the body

endosymbiosis symbiosis in which one partner lives within the other

enzyme a protein that controls a reaction in a cell

epidemiologic the spread of diseases in a population

epidemiologists people who study the incidence and spread of diseases in a population

epidemiology study of incidence and spread of diseases in a population

epididymis tube above the testes for storage and maturation of sperm

epigenetic not involving DNA sequence change

epistasis suppression of a characteristic of one gene by the action of another gene

epithelial cells one of four tissue types found in the body, characterized by thin sheets and usually serving a protective or secretory function

Escherichia coli common bacterium of the human gut, used in research as a model organism

estrogen female horomone

et al. "and others"

ethicists a person who writes and speaks about ethical issues

etiology causation of disease, or the study of causation

eubacteria one of three domains of life, comprising most groups previously classified as bacteria

eugenics movement to "improve" the gene pool by selective breeding

eukaryote organism with cells possessing a nucleus

eukaryotic describing an organism that has cells containing nuclei

ex vivo outside a living organism

excise remove; cut out

excision removal

exogenous from outside

exon coding region of genes

exonuclease enzyme that cuts DNA or RNA at the end of a strand

expression analysis whole-cell analysis of gene expression (use of a gene to create its RNA or protein product)

fallopian tubes tubes through which eggs pass to the uterus

fermentation biochemical process of sugar breakdown without oxygen

fibroblast undifferentiated cell normally giving rise to connective tissue cells

fluorophore fluorescent molecule

forensic related to legal proceedings

founder population

fractionated purified by separation based on chemical or physical properties

fraternal twins dizygotic twins who share 50 percent of their genetic material

frontal lobe one part of the forward section of the brain, responsible for planning, abstraction, and aspects of personality

gamete reproductive cell, such as sperm or egg

gastrulation embryonic stage at which primitive gut is formed

gel electrophoresis technique for separation of molecules based on size and charge

gene expression use of a gene to create the corresponding protein

genetic code the relationship between RNA nucleotide triplets and the amino acids they cause to be added to a growing protein chain

genetic drift evolutionary mechanism, involving random change in gene frequencies

genetic predisposition increased risk of developing diseases

genome the total genetic material in a cell or organism

genomics the study of gene sequences

genotype set of genes present

geothermal related to heat sources within Earth

germ cell cell creating eggs or sperm

germ-line cells giving rise to eggs or sperm

gigabase one billion bases (of DNA)

glucose sugar

glycolipid molecule composed of sugar and fatty acid

glycolysis the breakdown of the six-carbon carbohydrates glucose and fructose

glycoprotein protein to which sugars are attached

Golgi network system in the cell for modifying, sorting, and delivering proteins

gonads testes or ovaries

gradient a difference in concentration between two regions

Gram negative bacteria bacteria that do not take up Gram stain, due to membrane structure

Gram positive able to take up Gram stain, used to classify bacteria

gynecomastia excessive breast development in males

haploid possessing only one copy of each chromosome

haplotype set of alleles or markers on a short chromosome segment

hematopoiesis formation of the blood

hematopoietic blood-forming

heme iron-containing nitrogenous compound found in hemoglobin

hemolysis breakdown of the blood cells

hemolytic anemia blood disorder characterized by destruction of red blood cells

hemophiliacs a person with hemophilia, a disorder of blood clotting

herbivore plant eater

heritability proportion of variability due to genes; ability to be inherited

heritability estimates how much of what is observed is due to genetic factors

heritable genetic

heterochromatin condensed portion of chromosomes

heterozygote an individual whose genetic information contains two different forms (alleles) of a particular gene

heterozygous characterized by possession of two different forms (alleles) of a particular gene

high-throughput rapid, with the capacity to analyze many samples in a short time

histological related to tissues

histology study of tissues

histone protein around which DNA winds in the chromosome

homeostasis maintenance of steady state within a living organism

homologous carrying similar genes

homologues chromosomes with corresponding genes that pair and exchange segments in meiosis

homozygote an individual whose genetic information contains two identical copies of a particular gene

homozygous containing two identical copies of a particular gene

hormones molecules released by one cell to influence another

hybrid combination of two different types

hybridization (molecular) base-pairing among DNAs or RNAs of different origins

hybridize to combine two different species

hydrogen bond weak bond between the H of one molecule or group and a nitrogen or oxygen of another

hydrolysis splitting with water

hydrophilic "water-loving"

hydrophobic "water hating," such as oils

hydrophobic interaction attraction between portions of a molecule (especially a protein) based on mutual repulsion of water

hydroxyl group chemical group consisting of -OH

hyperplastic cell cell that is growing at an increased rate compared to normal cells, but is not yet cancerous

hypogonadism underdeveloped testes or ovaries

hypothalamus brain region that coordinates hormone and nervous systems

hypothesis testable statement

identical twins monozygotic twins who share 100 percent of their genetic material

immunogenicity likelihood of triggering an immune system defense

immunosuppression suppression of immune system function

immunosuppressive describes an agent able to suppress immune system function

in vitro "in glass"; in lab apparatus, rather than within a living organism

in vivo "in life"; in a living organism, rather than in a laboratory apparatus

incubating heating to optimal temperature for growth

informed consent knowledge of risks involved

insecticide substance that kills insects

interphase the time period between cell divisions

intra-strand within a strand

intravenous into a vein

intron untranslated portion of a gene that interrupts coding regions

karyotype the set of chromosomes in a cell, or a standard picture of the chromosomes

kilobases units of measure of the length of a nucleicacid chain; one kilobase is equal to 1,000 base pairs

kilodalton a unit of molecular weight, equal to the weight of 1000 hydrogen atoms

kinase an enzyme that adds a phosphate group to another molecule, usually a protein

knocking out deleting of a gene or obstructing gene expression

laparoscope surgical instrument that is inserted through a very small incision, usually guided by some type of imaging technique

latent present or potential, but not apparent

lesion damage

ligand a molecule that binds to a receptor or other molecule

ligase enzyme that repairs breaks in DNA

ligate join together

linkage analysis examination of co-inheritance of disease and DNA markers, used to locate disease genes

lipid fat or wax-like molecule, insoluble in water

loci/locus site(s) on a chromosome

longitudinally lengthwise

lumen the space within the tubes of the endoplasmic reticulum

lymphocytes white blood cells

lyse break apart

lysis breakage

macromolecular describes a large molecule, one composed of many similar parts

macromolecule large molecule such as a protein, a carbohydrate, or a nucleic acid

macrophage immune system cell that consumes foreign material and cellular debris

malignancy cancerous tissue

malignant cancerous; invasive tumor

media (bacteria) nutrient source

meiosis cell division that forms eggs or sperm

melanocytes pigmented cells

meta-analysis analysis of combined results from multiple clinical trials

metabolism chemical reactions within a cell

metabolite molecule involved in a metabolic pathway

metaphase stage in mitosis at which chromosomes are aligned along the cell equator

metastasis breaking away of cancerous cells from the initial tumor

metastatic cancerous cells broken away from the initial tumor

methylate add a methyl group to

methylated a methyl group, CH_3, added

methylation addition of a methyl group, CH_3

microcephaly reduced head size

microliters one thousandth of a milliliter

micrometer 1/1000 meter

microsatellites small repetitive DNA elements dispersed throughout the genome

microtubule protein strands within the cell, part of the cytoskeleton

miscegenation racial mixing

mitochondria energy-producing cell organelle

mitogen a substance that stimulates mitosis

mitosis separation of replicated chromosomes

molecular hybridization base-pairing among DNAs or RNAs of different origins

molecular systematics the analysis of DNA and other molecules to determine evolutionary relationships

monoclonal antibodies immune system proteins derived from a single B cell

monomer "single part"; monomers are joined to form a polymer

monosomy gamete that is missing a chromosome

monozygotic genetically identical

morphologically related to shape and form

morphology related to shape and form

mRNA messenger RNA

mucoid having the properties of mucous

mucosa outer covering designed to secrete mucus, often found lining cavities and internal surfaces

mucous membranes nasal passages, gut lining, and other moist surfaces lining the body

multimer composed of many similar parts

multinucleate having many nuclei within a single cell membrane

mutagen any substance or agent capable of causing a change in the structure of DNA

mutagenesis creation of mutations

mutation change in DNA sequence

nanometer 10^{-9}(exp) meters; one billionth of a meter

nascent early-stage

necrosis cell death from injury or disease

nematode worm of the Nematoda phylum, many of which are parasitic

neonatal newborn

neoplasms new growths

neuroimaging techniques for making images of the brain

neurological related to brain function or disease

neuron nerve cell

neurotransmitter molecule released by one neuron to stimulate or inhibit a neuron or other cell

non-polar without charge separation; not soluble in water

normal distribution distribution of data that graphs as a bell-shaped curve

Northern blot a technique for separating RNA molecules by electrophoresis and then identifying a target fragment with a DNA probe

Northern blotting separating RNA molecules by electrophoresis and then identifying a target fragment with a DNA probe

nuclear DNA DNA contained in the cell nucleus on one of the 46 human chromosomes; distinct from DNA in the mitochondria

nuclear membrane membrane surrounding the nucleus

nuclease enzyme that cuts DNA or RNA

nucleic acid DNA or RNA

nucleoid region of the bacterial cell in which DNA is located

nucleolus portion of the nucleus in which ribosomes are made

nucleoplasm material in the nucleus

nucleoside building block of DNA or RNA, composed of a base and a sugar

nucleoside triphosphate building block of DNA or RNA, composed of a base and a sugar linked to three phosphates

nucleosome chromosome structural unit, consisting of DNA wrapped around histone proteins

nucleotide a building block of RNA or DNA

ocular related to the eye

oncogene gene that causes cancer

oncogenesis the formation of cancerous tumors

oocyte egg cell

open reading frame DNA sequence that can be translated into mRNA; from start sequence to stop sequence

opiate opium, morphine, and related compounds

organelle membrane-bound cell compartment

organic composed of carbon, or derived from living organisms; also, a type of agriculture stressing soil fertility and avoidance of synthetic pesticides and fertilizers

osmotic related to differences in concentrations of dissolved substances across a permeable membrane

ossification bone formation

osteoarthritis a degenerative disease causing inflammation of the joints

osteoporosis thinning of the bone structure

outcrossing fertilizing between two different plants

oviduct a tube that carries the eggs

ovulation release of eggs from the ovaries

ovules eggs

ovum egg

oxidation chemical process involving reaction with oxygen, or loss of electrons

oxidized reacted with oxygen

pandemic disease spread throughout an entire population

parasites organisms that live in, with, or on another organism

pathogen disease-causing organism

pathogenesis pathway leading to disease

pathogenic disease-causing

pathogenicity ability to cause disease

pathological altered or changed by disease

pathology disease process

pathophysiology disease process

patient advocate a person who safeguards patient rights or advances patient interests

PCR polymerase chain reaction, used to amplify DNA

pedigrees sets of related individuals, or the graphic representation of their relationships

peptide amino acid chain

peptide bond bond between two amino acids

percutaneous through the skin

phagocytic cell-eating

phenotype observable characteristics of an organism

phenotypic related to the observable characteristics of an organism

pheromone molecule released by one organism to influence another organism's behavior

phosphate group PO_4 group, whose presence or absence often regulates protein action

phosphodiester bond the link between two nucleotides in DNA or RNA

phosphorylating addition of phosphate group (PO_4)

phosphorylation addition of the phosphate group PO_4^{3-}

phylogenetic related to the evolutionary development of a species

phylogeneticists scientists who study the evolutionary development of a species

phylogeny the evolutionary development of a species

plasma membrane outer membrane of the cell

plasmid a small ring of DNA found in many bacteria

plastid plant cell organelle, including the chloroplast

pleiotropy genetic phenomenon in which alteration of one gene leads to many phenotypic effects

point mutation gain, loss, or change of one to several nucleotides in DNA

polar partially charged, and usually soluble in water

pollen male plant sexual organ

polymer molecule composed of many similar parts

polymerase enzyme complex that synthesizes DNA or RNA from individual nucleotides

polymerization linking together of similar parts to form a polymer

polymerize to link together similar parts to form a polymer

polymers molecules composed of many similar parts

polymorphic occurring in several forms

polymorphism DNA sequence variant

polypeptide chain of amino acids

polyploidy presence of multiple copies of the normal chromosome set

population studies collection and analysis of data from large numbers of people in a population, possibly including related individuals

positional cloning the use of polymorphic genetic markers ever closer to the unknown gene to track its inheritance in CF families

posterior rear

prebiotic before the origin of life

precursor a substance from which another is made

prevalence frequency of a disease or condition in a population

primary sequence the sequence of amino acids in a protein; also called primary structure

primate the animal order including humans, apes, and monkeys

primer short nucleotide sequence that helps begin DNA replication

primordial soup hypothesized prebiotic environment rich in life's building blocks

probe molecule used to locate another molecule

procarcinogen substance that can be converted into a carcinogen, or cancer-causing substance

procreation reproduction

progeny offspring

prokaryote a single-celled organism without a nucleus

promoter DNA sequence to which RNA polymerase binds to begin transcription

promutagen substance that, when altered, can cause mutations

pronuclei egg and sperm nuclei before they fuse during fertilization

proprietary exclusively owned; private

proteomic derived from the study of the full range of proteins expressed by a living cell

proteomics the study of the full range of proteins expressed by a living cell

protists single-celled organisms with cell nuclei

protocol laboratory procedure

protonated possessing excess H^+ ions; acidic

pyrophosphate free phosphate group in solution

quiescent non-dividing

radiation high energy particles or waves capable of damaging DNA, including X rays and gamma rays

recessive requiring the presence of two alleles to control the phenotype

recombinant DNA DNA formed by combining segments of DNA, usually from different types of organisms

recombining exchanging genetic material

replication duplication of DNA

restriction enzyme an enzyme that cuts DNA at a particular sequence

retina light-sensitive layer at the rear of the eye

retroviruses RNA-containing viruses whose genomes are copied into DNA by the enzyme reverse transcriptase

reverse transcriptase enzyme that copies RNA into DNA

ribonuclease enzyme that cuts RNA

ribosome protein-RNA complex at which protein synthesis occurs

ribozyme RNA-based catalyst

RNA ribonucleic acid

RNA polymerase enzyme complex that creates RNA from DNA template

RNA triplets sets of three nucleotides

salinity of, or relating to, salt

sarcoma a type of malignant (cancerous) tumor

scanning electron microscope microscope that produces images with depth by bouncing electrons off the surface of the sample

sclerae the "whites" of the eye

scrapie prion disease of sheep and goats

segregation analysis statistical test to determine pattern of inheritance for a trait

senescence a state in a cell in which it will not divide again, even in the presence of growth factors

senile plaques disease

serum (pl. sera) fluid portion of the blood

sexual orientation attraction to one sex or the other

somatic nonreproductive; not an egg or sperm

Southern blot a technique for separating DNA fragments by electrophoresis and then identifying a target fragment with a DNA probe

Southern blotting separating DNA fragments by electrophoresis and then identifying a target fragment with a DNA probe

speciation the creation of new species

spindle football-shaped structure that separates chromosomes in mitosis

spindle fiber protein chains that separate chromosomes during mitosis

spliceosome RNA-protein complex that removes introns from RNA transcripts

spontaneous non-inherited

sporadic caused by new mutations

stem cell cell capable of differentiating into multiple other cell types

stigma female plant sexual organ

stop codon RNA triplet that halts protein synthesis

striatum part of the midbrain

subcutaneous under the skin

sugar glucose

supercoiling coiling of the helix

symbiont organism that has a close relationship (symbiosis) with another

symbiosis a close relationship between two species in which at least one benefits

symbiotic describes a close relationship between two species in which at least one benefits

synthesis creation

taxon/taxa level(s) of classification, such as kingdom or phylum

taxonomical derived from the science that identifies and classifies plants and animals

taxonomist a scientist who identifies and classifies organisms

telomere chromosome tip

template a master copy

tenets generally accepted beliefs

terabyte a trillion bytes of data

teratogenic causing birth defects

teratogens substances that cause birth defects

thermodynamics process of energy transfers during reactions, or the study of these processes

threatened likely to become an endangered species

topological describes spatial relations, or the study of these relations

topology spatial relations, or the study of these relations

toxicological related to poisons and their effects

transcript RNA copy of a gene

transcription messenger RNA formation from a DNA sequence

transcription factor protein that increases the rate of transcription of a gene

transduction conversion of a signal of one type into another type

transgene gene introduced into an organism

transgenics transfer of genes from one organism into another

translation synthesis of protein using mRNA code

translocation movement of chromosome segment from one chromosome to another

transposable genetic element DNA sequence that can be copied and moved in the genome

transposon genetic element that moves within the genome

trilaminar three-layer

triploid possessing three sets of chromosomes

trisomics mutants with one extra chromosome

trisomy presence of three, instead of two, copies of a particular chromosome

tumor mass of undifferentiated cells; may become cancerous

tumor suppressor genes cell growths

tumors masses of undifferentiated cells; may become cancerous

vaccine protective antibodies

vacuole cell structure used for storage or related functions

van der Waal's forces weak attraction between two different molecules

vector carrier

vesicle membrane-bound sac

virion virus particle

wet lab laboratory devoted to experiments using solutions, cell cultures, and other "wet" substances

wild-type most common form of a trait in a population

Wilm's tumor a cancerous cell mass of the kidney

X ray crystallography use of X rays to determine the structure of a molecule

xenobiotic foreign biological molecule, especially a harmful one

zygote fertilized egg

Topic Outline

Genetically Modified Foods
HPLC: High-Performance Liquid Chromatography
Pharmaceutical Scientist
Plant Genetic Engineer
Polymerase Chain Reaction
Recombinant DNA
Restriction Enzymes
Reverse Transcriptase
Transgenic Animals
Transgenic Microorganisms
Transgenic Organisms: Ethical Issues
Transgenic Plants

CAREERS

Attorney
Bioinformatics
Clinical Geneticist
College Professor
Computational Biologist
Conservation Geneticist
Educator
Epidemiologist
Genetic Counselor
Geneticist
Genomics Industry
Information Systems Manager
Laboratory Technician
Microbiologist
Molecular Biologist
Pharmaceutical Scientist
Physician Scientist
Plant Genetic Engineer
Science Writer
Statistical Geneticist
Technical Writer

CELL CYCLE

Apoptosis
Balanced Polymorphism
Cell Cycle
Cell, Eukaryotic
Centromere
Chromosome, Eukaryotic
Chromosome, Prokaryotic
Crossing Over
DNA Polymerases
DNA Repair
Embryonic Stem Cells
Eubacteria
Inheritance, Extranuclear

Linkage and Recombination
Meiosis
Mitosis
Oncogenes
Operon
Polyploidy
Replication
Signal Transduction
Telomere
Tumor Suppressor Genes

CLONED OR TRANSGENIC ORGANISMS

Agricultural Biotechnology
Biopesticides
Biotechnology
Biotechnology: Ethical Issues
Cloning Organisms
Cloning: Ethical Issues
Gene Targeting
Model Organisms
Patenting Genes
Reproductive Technology
Reproductive Technology: Ethical Issues
Rodent Models
Transgenic Animals
Transgenic Microorganisms
Transgenic Organisms: Ethical Issues
Transgenic Plants

DEVELOPMENT, LIFE CYCLE, AND NORMAL HUMAN VARIATION

Aging and Life Span
Behavior
Blood Type
Color Vision
Development, Genetic Control of
Eye Color
Fertilization
Genotype and Phenotype
Hormonal Regulation
Immune System Genetics
Individual Genetic Variation
Intelligence
Mosaicism
Sex Determination
Sexual Orientation
Twins
X Chromosome
Y Chromosome

DNA, GENE AND CHROMOSOME STRUCTURE

Antisense Nucleotides
Centromere
Chromosomal Banding
Chromosome, Eukaryotic
Chromosome, Prokaryotic
Chromosomes, Artificial
DNA
DNA Repair
DNA Structure and Function, History
Evolution of Genes
Gene
Genome
Homology
Methylation
Multiple Alleles
Mutation
Nature of the Gene, History
Nomenclature
Nucleotide
Overlapping Genes
Plasmid
Polymorphisms
Pseudogenes
Repetitive DNA Elements
Telomere
Transposable Genetic Elements
X Chromosome
Y Chromosome

DNA TECHNOLOGY

In situ Hybridization
Antisense Nucleotides
Automated Sequencer
Blotting
Chromosomal Banding
Chromosomes, Artificial
Cloning Genes
Cycle Sequencing
DNA Footprinting
DNA Libraries
DNA Microarrays
DNA Profiling
Gel Electrophoresis
Gene Targeting
HPLC: High-Performance Liquid Chromatography
Marker Systems
Mass Spectrometry
Mutagenesis

Nucleases
Polymerase Chain Reaction
Protein Sequencing
Purification of DNA
Restriction Enzymes
Ribozyme
Sequencing DNA

ETHICAL, LEGAL, AND SOCIAL ISSUES

Attorney
Biotechnology and Genetic Engineering, History
Biotechnology: Ethical Issues
Cloning: Ethical Issues
DNA Profiling
Eugenics
Gene Therapy: Ethical Issues
Genetic Discrimination
Genetic Testing: Ethical Issues
Legal Issues
Patenting Genes
Privacy
Reproductive Technology: Ethical Issues
Transgenic Organisms: Ethical Issues

GENE DISCOVERY

Ames Test
Bioinformatics
Complex Traits
Gene and Environment
Gene Discovery
Gene Families
Genomics
Human Disease Genes, Identification of
Human Genome Project
Mapping

GENE EXPRESSION AND REGULATION

Alternative Splicing
Antisense Nucleotides
Chaperones
DNA Footprinting
Gene
Gene Expression: Overview of Control
Genetic Code
Hormonal Regulation
Imprinting
Methylation
Mosaicism
Nucleus

Operon
Post-translational Control
Proteins
Reading Frame
RNA
RNA Interference
RNA Polymerases
RNA Processing
Signal Transduction
Transcription
Transcription Factors
Translation

GENETIC DISORDERS

Accelerated Aging: Progeria
Addicition
Alzheimer's Disease
Androgen Insensitivity Syndrome
Attention Deficit Hyperactivity Disorder
Birth Defects
Breast Cancer
Cancer
Carcinogens
Cardiovascular Disease
Chromosomal Aberrations
Colon Cancer
Cystic Fibrosis
Diabetes
Disease, Genetics of
Down Syndrome
Fragile X Syndrome
Growth Disorders
Hemoglobinopathies
Hemophilia
Human Disease Genes, Identification of
Metabolic Disease
Mitochondrial Diseases
Muscular Dystrophy
Mutagen
Nondisjunction
Oncogenes
Psychiatric Disorders
Severe Combined Immune Deficiency
Tay-Sachs Disease
Triplet Repeat Disease
Tumor Suppressor Genes

GENETIC MEDICINE: DIAGNOSIS, TESTING, AND TREATMENT

Clinical Geneticist
DNA Vaccines

Embryonic Stem Cells
Epidemiologist
Gene Discovery
Gene Therapy
Gene Therapy: Ethical Issues
Genetic Counseling
Genetic Counselor
Genetic Testing
Genetic Testing: Ethical Issues
Geneticist
Genomic Medicine
Human Disease Genes, Identification of
Pharmacogenetics and Pharmacogenomics
Population Screening
Prenatal Diagnosis
Public Health, Genetic Techniques in
Reproductive Technology
Reproductive Technology: Ethical Issues
RNA Interference
Statistical Geneticist
Statistics
Transplantation

GENOMES

Chromosome, Eukaryotic
Chromosome, Prokaryotic
Evolution of Genes
Genome
Genomic Medicine
Genomics
Genomics Industry
Human Genome Project
Mitochondrial Genome
Mutation Rate
Nucleus
Polymorphisms
Repetitive DNA Elements
Transposable Genetic Elements
X Chromosome
Y Chromosome

GENOMICS, PROTEOMICS, AND BIOINFORMATICS

Bioinformatics
Combinatorial Chemistry
Computational Biologist
DNA Libraries
DNA Microarrays
Gene Families
Genome
Genomic Medicine

Archaea
Cell, Eukaryotic
Eubacteria
Evolution, Molecular
Fruit Fly: *Drosophila*
HIV
Maize
Model Organisms
Nucleus
Prion
Retrovirus
Rodent Models
Roundworm: *Caenorhabditis elegans*
Signal Transduction
Viroids and Virusoids
Virus
Yeast
Zebrafish

POPULATION GENETICS AND EVOLUTION

Antibiotic Resistance
Balanced Polymorphism
Conservation Biologist
Conservation Biology: Genetic Approaches
Evolution of Genes
Evolution, Molecular
Founder Effect
Gene Flow
Genetic Drift
Hardy-Weinberg Equilibrium

Heterozygote Advantage
Inbreeding
Individual Genetic Variation
Molecular Anthropology
Population Bottleneck
Population Genetics
Population Screening
Selection
Speciation

RNA

Antisense Nucleotides
Blotting
DNA Libraries
Genetic Code
HIV
Nucleases
Nucleotide
Reading Frame
Retrovirus
Reverse Transcriptase
Ribosome
Ribozyme
RNA
RNA Interference
RNA Polymerases
RNA Processing
Transcription
Translation

Volume 1 Index

Boyer, Herbert, 71, 72, 152

Brain development
    anencephaly, 78
    apoptosis and, 32, 33

Brain disorders. *See* specific disorders and syndromes

BRCA1 and *BRCA2* genes, mutations, 91, 92, 95, 100

Breast disorders
    gynecomastia, 79
    See also Cancer, breast

Brenner, Sydney, 193, 205

Bridges, Calvin Blackman, 130–131

Brown, Louise Joy, 69

BT. *See* Bacillus thuringiensis

Bubble boy (David Vetter), 77

Buchnera sp., chromosome size and gene number, 142

Butterflies and moths, impact of *Bacillus thuringiensis*, 58

C

Cadmium, 60

Caenorhabditis elegans. See Roundworm

Cairns, John, 139

CAIS (complete androgen insensitivity syndrome), 25

Calgene company, genetic engineering, 29, 73

Calpain 10 protease gene, 210

Campylobacter jejuni, chromosome size and gene number, 142

Cancer, **92–97**
    aging and, 7, 8, 92
    angiogenesis, 94
    antisense nucleotide tools, 29, 30
    apoptosis and, 29, 32, 33, 96, 155, 168
    bioinformatics tools, 56
    blood group and, 84
    cell cycle deregulation and, 107–108, 167–169
    cell senescence and, 96
    cell type classification, 94
    complex traits, 213
    cytogenetics analysis, 119
    damaged DNA role, 239, 242
    diagnosis and treatment, 97
    DNA microarray tools, 227–229
    DNA vaccines, 255
    environmental factors and, 92, 179–180
    genetic predisposition to, 90, 91–92, 97–98, 99
    inheritance patterns, 97
    lifestyle and, 6

malignant and benign tumors, 92–93
    metastasis, 94
    organ of origin, 94
    retinoblastomas, 96–97
    tissue-related tumor types, 93
    and treatments, 69, 97, 153
    See also Carcinogens; Oncogenes; Tumors

Cancer, bile duct, 169

Cancer, bladder, 93, 169

Cancer, breast, **89–92**
    alcohol consumption and, 91
    atypical cells, 91
    cell cycle regulation and, 107
    diet and, 91
    environmental factors, 90
    exercise and, 91
    genetic factors, 90, 91–92, 100
    geographic differences, 90
    hormone replacement therapy and, 90
    lifestyle and, 90
    in males, 91
    mammograms, *90*
    menstrual cycle and, 90
    multiplicative gene effects, 180
    pregnancies and, 89–90
    racial and ethnic differences, 90, 91
    radiation and, 91
    statistics, 89, 93

Cancer, bone marrow, 93

Cancer, colon, **166–170**
    adenoma-carcinoma sequence, 167
    colectomies, 168
    diet and, 170
    dividing cancer cell, *169*
    DNA microarray tools, 229
    familial adenomatous polyposis, 167–169
    genetic predisposition to, 99, 167–170
    hereditary nonpolyposis cancer, 167, 169
    inflammatory bowel disease and, 170
    polyps, 167–169, *167*
    prevention, 170
    statistics, 93, 166

Cancer, esophagus, 93

Cancer Genome and Anatomy Project, 229

Cancer, intestines, 169

Cancer, kidney, 93

Cancer, liver, 93

Cancer, lung
    and smoking, *98*, 180
    statistics, 93

Cancer, ovarian, 93, 169

Cancer, pancreas, 93

Cancer, prostate gland, 93

Cancer, renal pelvis, 169

Cancer, skin
    statistics, 93
    susceptibility to, 99

Cancer, stomach
    and colon cancer, 169
    statistics, 93

Cancer, testicular
    and AIS, 25
    statistics, 93

Cancer, ureters, 169

Cancer, uterus, 93

Candidate-gene studies
    behavior genetics, 48
    cardiovascular disease, 101–103

Captive breeding programs, 187, *191*

Carcinogens, **97–100**
    Ames test, **19–21**, *20*, 100
    cigarette smoke, 92, *98*, 100, 244–245
    defined, 245
    in diet, 91, 92
    industrial toxins, 92, 97
    procarcinogens, 99
    radiation, ionizing, 91, 92, 97, 99, 119
    radiation, ultraviolet, 97, 99, 100
    reducing exposure, 100
    repair gene mutations, 95–96
    soot, 98–99, 244
    and spontaneous mutations, 97
    tumor suppressor gene mutations, 99–100
    viruses, 92, 97, 99–100
    See also Mutagens

Carcinomas
    adenoma-carcinoma sequence, 167
    defined, 93

Cardiovascular disease, **101–103**
    aging and, 7, 8
    birth defects, 75
    candidate-gene studies, 101, 102–103
    causative genes, 101
    complex traits, 101–103, 178, 213
    diabetes and, 212
    disease-susceptibility genes, 101

Cytoskeleton, structure and function, 113
Cytosol
defined, 109, 112
glucose breakdown, 111

D

D4 and D5 dopamine receptor genes (*DRD4* and *DRD5*), 41
DAPI dyes, 127
Darwin, Charles, 129, 131
DAT (dopamine transporter) gene, 41
Data Deficient population status, 186
Databases
analysis, 55–56
clinical trials, 102–103
data acquisition, 54
data integration, 56
development, 54
disease registry, 102
gene and protein sequences, 52–54
relational, 55
DBH mice, 5
De novo mutations, 120
De Vries, Hugo, 130
Dead Sea, Archaea adaptations, 37
Deafness, aging and, 7
Deamination errors, DNA damage, 240–242, *240*
Death, as biological process, 31
Dehydroepiandrosterone sulfate, and breast cancer, 90
Deinococcus radiodurans, genetic studies, 139–140
Delbrück, Max, 192, **203–204**, *204*
Dementia
aging and, 7
defined, 1, 7, 15
non-Alzheimer causes, 15
progeria and, 1
See also Alzheimer's disease; Brain disorders; Psychiatric disorders
Demography, defined, 191
Denaturation
in cycle sequencing, 198
DNA unwinding, 218
Deoxynucleotides, defined, 230
Deoxyribonucleic acid. *See* DNA
Deoxyribose sugars
discovery, 249
and DNA structure, 215–220
Dependence. *See* Addiction
Depurination, 240

Depyrimidination, 240
DeSalle, Rob, 190
Desipramine, for ADHD, 39
Detergents, genetically engineered proteases, 63–64, 73
Deuteranopia/anomaly, 172
Development, defined, 204
Development, genetic control of, **204–209**
apoptosis role, 208–209
axes determinations in fruit flies, 206–207
European and American models, 205–206
homeobox sequence, 208
homeotic genes and segmentation, 207–208
morphogen gradients, 206
transcription factor role, 205
See also Sexual development
Diabetes, **209–212**
and birth defects, 77, 82
and cardiovascular disease, 101, 209
complex traits, 178, 210, 213
congenital generalized lipodystrophic (CGL), 211
diet and, 210
excess blood glucose, impact, 209
familial partial lipodystrophic (FPLD), 211
genetic susceptibility to complications, 212
gestational (GDM), 209, 212
insulin and, 63, 66, 72, 153–154, 209, 211
maternally inherited diabetes and deafness (MIDD), 211
maturity onset (MODY), 211
nongenic, 210
physical activity and, 210
racial and ethnic differences, 210
statistics, 209
thiamine-responsive megaloblastic anemia syndrome (TRMA), 212
transient neonatal (TNDM), 212
type 1 (T1DM), 209–210
type 2 (T2DM), 209, 210, 212
wolfram syndrome, 212
Diagnostic testing kits, genetically engineered products, 64
Diamond v. *Chakrabarty*, 72
Dichromat color vision defects, 171, 172
Dideoxynucleotides, 43, 198, 199
Diet. *See* Nutrition

Diethylstilbestrol, and breast cancer risk, 90
Dimethylsulfate, as mutagen, 242
Disaster victims, DNA profiling to identify, *234*, 238
Disease concordance, defined, 214
Disease registry databases, 102
Disease-resistant crops
genetically engineered, 63
nematode resistance, 149
statistics, 57–58
Diseases
as bioterrorism tools, 69–70
DNA vaccines, *253–255*
genetically engineered testing tools, 64
See also specific diseases
Diseases, genetics of, **213–215**
antisense nucleotide tools, 29, 30–31
chaperone role, 118
clinical geneticist role, 149–151
complex traits, **177–181**
damaged DNA role, 239
DNA microarray analysis, 227–229
early studies, 250
vs. environmental factors, 213, 214
familial aggregation studies, 213–214
Human Genome Project contributions, 214–215
impact of mutagens, 19
and longevity, 6–8
Mendelian and complex disorders, 213
molecular chaperones, 118
prokaryote chromosome mapping, 139–140, 142
relative risk ratios, 214
twin and adoption studies, 214
unequal crossovers and, 197
Diseases, genetics of specific
addiction, 4–6
AIS, **21–26**
Alzheimer's disease, 16–18
birth defects, **74–82**
blood group association, 84
cancer, 97–100
cancer, breast, 90, 91–92, 99, 100
cancer, colon, 167–170
cardiovascular disease, 101–103
cystic fibrosis, 200–203
diabetes, **208–212**
Down syndrome, **256–258**
Kennedy disease, 26

major and minor groove sides, 217, *217*, *218*

phosphodiester bonds, 216, *216*, 244

quadruplex, 219, *219*

right-handed helix, 217

slipped-strand, 218, *219*

unwound, 218, *219*

van der Waal's forces, 217

DNA structure and function, history, 62, 215, **248–253**

discovery of DNA, 249–250

DNA as transforming factor, 250–251, *250*

DNA replication studies, 251–252

gene coding studies, 252–253

Watson and Crick's model, 62, 192–193, 251

DNA synthesis

cell cycle phase, 104–105

following conjugation, 183

polymerases and, **230–233**

DNA, transformed (T-DNA), 35

DNA triplets, as universal, 61, 152

DNA vaccines, **253–255**

advantages, 254

clinical studies, 255

testing on cattle, *254*

uses, 72, 255

vaccination techniques, 254–255

DnaK chaperones, energy requirements, 117

DNase

cystic fibrosis treatment, 201

DNA footprinting role, 220

Dolly the sheep, 158, 160, *162*, 163–164

Dominance, 147–148

See also Autosomal dominant disorders; Inheritance patterns

Donahue's syndrome, progeroid aspects, 2

Dopamine

ADHD and, 39–41, *41*

neurotransmitter systems, 5

Dopamine transporter (*DAT*) gene, 41

Down, J. Langdon, 256

Down syndrome, **256–258**

average IQ, 79

chromosomal basis, 79, 256–257

clinical features, 3, 256, *257*

cytogenetic analysis, 257

history, 119, 256

mosaic, 257

prenatal diagnosis, 258

progeroid aspects, 2–3

senescence, 3

statistics, 75

translocation, 257

trisomy 21, 257

DRD2 mice, 5

Drosophila (fruit flies)

addiction studies, 5–6

aging studies, 7

developmental processes, 204–208, *207*, *208*

gene identification studies, 155

nondisjunction studies, 131

number of genes, 11–12

polytene chromosomes, 126, 136–137, *136*

restriction maps, 155–156, *156*

X-linked inheritance studies, 130–131

Drug dependence

and birth defects, 77, 81

defined, 4

See also Addiction

Drugs (medications)

and Alzheimer's disease, 181

and birth defects, 80, 206

for cancer, 97, 153

and colon cancer, 170

combinatorial chemistry to synthesize, 175, 176–177

estrogen replacement, 90

See also Antibiotic resistance; Pharmacogenetics and pharmacogenomics

DSM-IV clinical criteria, 4

DsRNA. *See* RNA, double-stranded

Dupont company, Genesis 3000 sequencer, 44

Dwarfism, achondroplasia, 75–76

Dyskeratosis congenita, progeroid aspects, 2

E

E. coli bacterium. *See* Escherichia coli

E2F family transcription factors, *106*, 107

Ecology. *See* Conservation biology, genetic approaches

Ecosystem, defined, 59

EGFR gene (Epidermal Growth Factor Receptor), 95

Eggs

chromosomal aberrations, 121

enucleated, cloned organisms, 158, 162–164

human, size, 108

meiosis and, 115

Electron transport chain, described, 111

Electrophoresis. *See* Gel electrophoresis

EMBL database, 52

Embryonic development. *See* Development, genetic control of

Embryonic stem cells. *See* Stem cells, embryonic

Embryos, human

cloning, 69, 158–160, 161

eight-cell stage, *163*

Employment discrimination, genetic testing and, 18, 68

EMS (ethyl methanesulfonate), 19

Endangered species

cloning, 158, 161, 164

endangered population status, 186, 190

evolutionary relationship determinations, 188, 189

extinction vortex, 188

genetic diversity, 187

illegal trading in, 186, 189–190

population bottlenecks, 188–189

relocation and reintroduction, 186–187, 190

threatened population categories, 186

viability analyses, 187

Endogenous, defined, 242

Endometrium, defined, 169

Endonucleases

apurinic/apyrimidinic, 244

restriction, 70–71, 155

Endoplasmic reticulum (ER), *105*

defined, 117

structure and function, 112–113

Endoscopic, defined, 168

Environmental Protection Agency (EPA)

biotechnology regulations, 11

oil spill cleanups, 59

Enzymes

to activate procarcinogens, 99

combinatorial chemistry to synthesize, 175, *175*

defined, 45, 63, 82, 99, 137, 175, 220

to detoxify poisons, 45–46

extremozymes, 38, 199

industrial, genetically engineered, 63–64, 154

See also Proteins; *specific enzymes*

EPA. *See* Environmental Protection Agency

Epidemiologic, defined, 101

Genetic testing and screening, specific diseases (continued)
diabetes, 210
Down syndrome, 256, 257–258
fragile X syndrome, 76
Genetic testing, ethical issues, 68–69
clinical geneticist role, 149–150
genetic discrimination, 18, 68
Genetically modified (GM) foods
benefits, 9–11, 73
biopesticides, 57–58
ethical issues, 11, 67–68, 67, 73
labeling, 67, 68
Geneticists
clinical, 74, **149–152**, 213–215
conservation, **190–192**
Genetics, origin of term, 130
See also Classical hybrid genetics; Mendelian genetics; Population genetics
Genetics professors. *See* College professors
Genital and urinary tract disorders, birth defects, 75
Genitals
AIS individuals, 25
normal human, 21, 22, 23
Genomes, defined, 9
Genomic clones, *vs.* cDNA clones, 154
Genotypes
of AIS individuals, 21, 23
blood groups, 83
defined, 21, 147
dominant, 147–148
recessive, 147–148
GenPharm, transgenic cows, 73
Gestational diabetes (GDM), 209, 212
Giemsa stains, 127, 135
Gilbert, Walter, 72
Glucokinase, maturity onset diabetes and, 211
Glucose
breakdown in mitochondria, 111–112
defined, 110
diabetes and, 209, 212
Glycoproteins
blood group substances, 83
defined, 82
structure and function, 113
Glycosidic bonds, 216, *216*
GM foods. *See* Genetically modified foods
GNG3 (seipin) gene, and diabetes, 211

Goats, genetically engineered, 10
Gold, hyperaccumulators of, 61
Golgi apparatus, *105*, 113
Gonads
cancer, 169
defined, 21
development, AIS, 23–26
development, normal human, 21, 22, 23
hypogonadism, 79
ovaries, 21, 22, 90, 91, 93, 169
testis, 21–23, 25, 93
Gonorrhea, antibiotic resistance, 26, 27, 28
Gram negative bacteria, defined, 27
Granulocytes, cancerous tumors, 94
Great Salt Lake, Archaea adaptations, 37
Green cone pigment (*GCP*) genes, 171
Greenhouse gases, methane, 37
GroEl chaperones, 117
GTG banding of chromosomes, 127
Guanine, *216*
depurination, 240
and DNA structure, 215–220, 250–251
origin of term, 249
pronated, 218
reactions with aflatoxins, 245
See also Base pairs
Gurdon, John, 158, 161–162
Gynecomastia, defined, 79

H

H substance, 83
Habitat loss, implications for wild populations, 186–187, 188, 190
Haemophilus influenzae, chromosome size and gene number, 142, 143
Halobacterium, 37
HATs (histone acetyltransferases), 137–138
Health care and biotechnology, ethics, 68–69
Hearing loss
congenital, 76
and diabetes, 211
impact of aging, 7
Hearing, role of alternative splicing, 12, 13–14
Heart disease. *See* Cardiovascular disease
Heat-shock protein chaperone systems (Hsp70 and Hsp60), 116, 117, *118*
Helicobacter pylori, chromosome size and gene number, 142

Hematopoietic cells, replication, 103
Hematopoietic, defined, 103
Hemochromatosis, and diabetes, 212
Hemoglobinopathies
balanced polymorphism, 46
blood cancers, 93, 94
coagulation factor IX, 163–164
cytogenetics analysis, 119
transgenic microorganisms to combat, 63
Hemolysis, defined, 85
Hemolytic anemia, defined, 85
Hepatitis
clinical testing tools, 64
DNA vaccines, 72, 255
Hepatocyte nuclear factor (*HNF*) genes, and diabetes, 211
Herbicide-resistant crops, 57, 58
Herbivores, defined, 57
Hereditary non polyposis cancer (HNPCC), 167, 169
Hertwig, Oskar, 249
Hessian fly, wheat resistant to, 58
Heterochromatin
banding techniques for, 128, 135
constitutive *vs.* facultative, 134–135
defined, 114
pericentric, 114
repetitive DNA sequences, 235
Heterozygote advantage, 46, 131
Heterozygotes, defined, 147
Heterozygous, defined, 35, 131
High-throughput screening, combinatorial chemistry role, 175, *176*
His-bacteria, 19–20
Hispanics, disease prevalence
cystic fibrosis, 202
diabetes, 210
Histidine, *his*- bacteria, 19–20
Histocompatibility leukocyte antigen (HLA) genes, and diabetes, 209–210
Histone acetyltransferases (HATs), 137–138
Histone deacetylases, 138
Histone proteins
gene clusters, 132
lacking in bacteria, 141
in nucleosome complexes, 126, 133–134
octamers, 133
Hitachi Corporation, automated sequencers, 45
HIV/AIDS
antisense nucleotide tools, 30, 31
clinical testing tools, 64

congenital, statistics, 75

DNA vaccines, 255

genetic components, 213

HLA (histocompatibility leukocyte antigene) genes, and diabetes, 209–210

HLA (human leukocyte antigen), in centenarians, 8

HNF (hepatocyte nuclear factor) genes, and diabetes, 211

HNPCC (hereditary non polyposis cancer), 167, 169

Homeobox sequence, 208

Homeodomains, 208

Homeotic genes

in fruit flies, 207–208

Hox genes, 208

Homologous, defined, 45, 123

Homologues

crossing over, 194–196

defined, 130

Homozygous

in classical hybrid genetics, 147

defined, 35, 131

Hood, Leroy, 44

Hoogsteen base pairs, 218, *218*

Horizontal evolution, 27

Hormonal regulation, AIS, **21–26**

Hormone therapy

estrogen replacement, 90

to treat Turner's syndrome, 79

Hormones

defined, 110, 242

human growth, 63

reproductive, and breast cancer, 89–91

signal transduction role, 110

Hox genes, 208

Hsp70 and Hsp60 (heat-shock protein chaperone systems), 116, 117, *118*

Hudson River, PCB cleanup, 59

Human genetic diversity, alternative splicing and, 11–12

Human Genome Project

artificial chromosome tools, 145–146

BAC library tools, 222

behavior genetics, 48

bioinformatics, 52, 54

clinical geneticist role, 149–151

computational biologist role, 181

data acquisition, 54

disease genetics contributions, 214–215

drug discovery applications, 54

E. coli model, 142

as gene cloning tool, 157

history, 73–74

Human growth hormones

antibodies against, 255

from cloned genes, 153

from genetically engineered bacteria, 63

Human leukocyte antigen (HLA), 8

Human papillomavirus, and cancer, 92

Hunkapiller, Michael, 44

Huntingtin gene, 179

Huntington's disease

apoptosis and, 33

as simple traits, 177, 179

Hutchinson-Gilford syndrome, *2*

diagnosis, 1, 3

telomere shortening, 3

Hybridization (molecular). *See* In situ *hybridization*

Hybridization (plant and animal breeding)

Arabidopsis thaliana, 33–35

classical hybrid genetics, **146–149**

Hybrids/hybridize, defined, 11, 46, 70, 146

Hydrogen bonds

defined, 215

and DNA structure, 215–217, *217,* 251

Hydrolysis

defined, 117, 240

errors, DNA damage, 240, *240*

Hydrophilic, defined, 109

Hydrophobic, defined, 109, 245

Hydrophobic interactions

in base-pair stacking, 217

defined, 217

Hyper- and hypoacetylation of histones, 137–138

Hyperaccumulator plants, 60, 62

Hyperactivity. *See* Attention deficit hyperactivity disorder

Hyperplastic cells, defined, 95

Hypertension

and cardiovascular disease, 101

defined, 101

Hypogonadism, defined, 79

I

ICD10 clinical criteria, 4

Immune system genetics

of cheetah populations, 189

combinatorial chemistry tools, 174

diabetes type 1, 209–210

Down syndrome, 3

human leukocyte antigens, 8

Rh blood incompatibility, 85–86

Immune systems

antigens, 82–86, *85,* 253–255

vaccination principles, 253–254

See also Antibodies

In situ hybridization (molecular)

in blotting procedures, 87–89

defined, 87, 225

described, 136–137

in DNA microarrays, 225–226

fluorescence *(FISH),* 128–129, 145

In vitro fertilization

cystic fibrosis and, 201

ethical issues, 69

In vivo, defined, 116

Inbreeding, defined, 188

Inbreeding depression, 188–189

Incubating, defined, 20

Industrial toxins

and birth defects, 77

and cancer, 92

and chromosomal aberrations, 119, 123

and diabetes, 210

and DNA damage, 239, 242

role in disease, 214

Inflammatory bowel disease, and colon cancer, 170

Influenza, DNA vaccines, 255

Inheritance, extranuclear

chloroplast genomes, 112

mitochondrial genomes, 112, 160, 164, 232

Inheritance patterns

Arabidopsis research, 34–36

behavior genetics, **46–51**

blood groups, 83–84

chromosomal aberrations, 120–121, 125

chromosomal theory of inheritance, **129–132**

classical hybrid genetics, **146–149**

complex traits, **177–181**

de novo mutations, 120

genetic discrimination, 68

Mendelian and complex disorders, 213–214

microsatellites, 189

multifactorial and polygenic, 77–78

Punnett squares, *148*

single-gene mutations, 75

Radioactive materials, hyperaccumu-
lators of, 61
Ras and ras genes and proteins,
mutations, 99–100
R-banding of chromosomes, 128
RCP (red cone pigment) genes, 171
Reading frames, open (ORFs), 154
Recessive alleles
familial adenomatous polyposis,
168–169
inheritance patterns, 147–148
Recombinant DNA. *See* DNA,
recombinant
Recombination. *See* Crossing over
Recombination nodules, 195
Red cone pigment (*RCP*) genes, 171
Red-green color blindness, 171–172
Relative risk ratios, 214
Renal pelvis, cancer, 169
Repetitive DNA sequences. *See*
DNA repetitive sequences
Replication
cell cycle, **103–108**
defined, 19, 104
impact of mutagens, 19
rolling, *184*
by viruses, 204
See also Chromosomes, eukaryotic; Chromosomes, prokaryotic; DNA replication; Meiosis;
Mitosis
Reproduction, asexual, as cloning,
161
Reproductive technology, cloning
organisms, **161–165**
Reproductive technology, ethical
issues
cloning, **158–161**
in vitro fertilization, 69, 201
Resistance to antibiotics. *See* Antibiotic resistance
Resistance to pesticides, agricultural
biotechnology and, 11
Resistance to pests and pathogens
Arabidopsis research, 35
breeding for, 149
disease-resistant crops, 57–58,
63, 149
insect-resistant crops, 54, 57, 58,
63, 64, 68
Respiratory tract disorders, birth
defects, 75
Restriction endonucleases, 70–71,
155
Restriction enzymes
in blotting procedures, 88
defined, 70, 88, 152
in gene cloning, 152, 154, 155,
157

in VNTR analysis, 236
Restriction fragment length polymorphisms (RFLPs), 236
Restriction maps (genes), 155–156,
156
Restriction point, in cell cycle,
104–105
Retinas, function, 170
Retinoblastoma gene (*Rb*), mutations, 96–97, *96*
Retinoblastoma protein (pRb), phosphorylation, 106–107, *106*
Retinoblastomas, 96–97, *96*
Retinoic acid, risk during pregnancy,
206
Reverse transcriptase
as gene cloning tool, *153*, 154
role in DNA libraries, *223*, 224
Rh blood group
discovery, 84
genetics of, 85–86
as identification, 233
incompatibility, 75, 84–86
Rhodopsin (*RHO*) genes, 171
Ribose, discovery, 249
Ribosomes, *105*
defined, 29
impact of antibiotics on, 28
stop codons, 13
structure and function, 112
See also RNA, ribosomal (rRNA)
Rice
genetically engineered, 10, *10*
genome size, 34
Rickettsia prowazekli, chromosome
size and gene number, 142
Risch, Neil, 214
RNA
antisense nucleotides, 29, *30*
defined, 224
oligonucleotides, 176–177
RNA, double-stranded (dsRNA), 31
RNA interference, 31
RNA, messenger (mRNA)
alternative splicing, 11–13
antisense nucleotides, 29, *30*, 97
of cloned organisms, 164–165
Crick's contributions, 193, 252
defined, 12, 97, 206
introns lacking in, 154, 157, 224
molecular hybridization mapping
of, 225–226
synthesis, 112
RNA polymerases, function, 112,
205
RNA processing, alternative splicing, 11–13

RNA, ribosomal (rRNA)
banding techniques, 128
function, 112
gene clusters, 132
phylogenetic relationships, 36
RNA sequencing, blotting procedures, **86–89**, 152
RNA, transfer (tRNA)
Crick's contributions, 193, 252
function, 112
RNA triplets, defined, 61
RNase, as gene cloning tool, *153*,
154
Rodent models
addiction studies, 5
aging studies, 7
behavior genetics, 48–49
cloned mice, 160, 164
DAT gene knockouts, 41
developmental processes, 205
Hox gene knockouts, 208
Rods, role in color vision, 171, 172
Rothmund's syndrome, progeroid
aspects, 2
Roundworm (*Caenorhabditis elegans*)
aging studies, 7
apoptosis studies, 32–33
developmental processes, 204,
205, 206, 208
restriction maps, 155–156
Rous, Peyton, 99–100
Rous sarcoma virus, and cancer, 100
Roux, Wilhelm, 132
Rubella (German measles), and birth
defects, 75, 81

S

Saccharomyces sp. *See* Yeasts
S-adenosylmethionine (SAM), 242
SAGE (serial analysis of gene
expression) analysis, 228–229
Saline conditions, Archaea adapta-
tions, 37
Salmonella Newport, antibiotic resis-
tance, 28
Salmonella typhimurium, Ames test,
19–20, *20*
SAM (S-adenosylmethionine), 242
Sanger, Fred, 43, 72, 198
Sanger method. *See* Chain termina-
tion method
Sarcomas, defined, 93
Schrıdinger, Erwin, 192, 203
SCID (severe combined immune
deficiency), 77
Scurvy, cause, 63
Se and *se* genes, 83–84